Introduction to Physics and Chemistry of Combustion

Michael Liberman

Introduction to Physics and Chemistry of Combustion

Explosion, Flame, Detonation

Springer

Michael Liberman
Uppsala University
Dept. Physics
SE-751 21 Uppsala
Sweden
michael.liberman@fysik.uu.se

ISBN: 978-3-540-78758-7 e-ISBN: 978-3-540-78759-4
DOI: 10.1007/978-3-540-78759-4

Library of Congress Control Number: 2008926558

Printed on acid-free paper

9 8 7 6 5 4 3 2 1

springer.com

Contents

1 Basic Concepts of Thermodynamics 1
- 1.1 The Entropy 1
- 1.2 Work and Quantity of Heat: First Law of Thermodynamics 4
- 1.3 Temperature 7
- 1.4 Pressure 9
- 1.5 The Free Energy and the Thermodynamic Potentials 12
- 1.6 The Enthalpy 13
- 1.7 The Nernst's Theorem 14
- 1.8 Carnot's Cycle and Carnot's Theorem 16
- 1.9 Le Chatelier Principle 18
- 1.10 Dependence of the Thermodynamic Quantities on the Number of Particles 18
- 1.11 Ideal Gases 21
- 1.12 Ideal Gases with Constant Specific Heat: Equation of Poisson Adiabatic 22
- Problems 25

2 Chemical Thermodynamics 27
- 2.1 Introduction and Definitions 27
- 2.2 Properties of Substances 30
- 2.3 Heats of Reactions and Formation 32
- 2.4 Origin of the Combustion Heat; Molecular Bonds 37
- 2.5 Adiabatic Flame Temperature 40
- 2.6 The Equilibrium Constant 43
- 2.7 Chemical Equilibrium and Adiabatic Flame Temperature 46
- Problems 50

3 Combustion Chemistry 53
- 3.1 Chemical Reactions 53
- 3.2 Non-branching Chain Reaction: The Hydrogen Chlorine 59

3.3 Oxidation of Nitrogen in Combustion 63
3.4 Chain-Branching Reactions: Explosions 65
3.5 Hydrogen-Oxygen Reactions: Explosion Limits 66
Problems 71

4 Self-Accelerating Reactions, Explosions 73
4.1 Self-Accelerating Reactions 73
4.2 Thermal Self-Ignition 76
4.3 The Frank-Kamenetskii Transformation 79
4.4 Semenov's Theory of Thermal Explosions 81
4.5 Critical Conditions for a Thermal Explosion 84
4.6 Spark Ignition and Minimum Ignition Energy 85
Problems 87

5 Velocity and Temperature of Laminar Flames 89
5.1 Reaction Waves Propagation Through a Combustible Mixture 89
5.2 Velocity and Thickness of Laminar Flames 91
5.3 Temperature and Concentration Distributions in Flames 97
5.4 Normal Velocity of Flame Propagation. Zel'dovich – Frank-Kamentskii Theory 101
5.5 Consequences of the Formula for Normal Flame Velocity 106
Problems 107

6 Introduction to Hydrodynamics of Ideal Fluids 109
6.1 The Fluid Dynamics 109
6.2 The Equation of Continuity 111
6.3 The Euler Equation 113
6.4 Conservation of Energy 116
6.5 The Equation of State 120
6.6 Hydrostatics 123
6.7 A Stationary Flow. The Bernoulli Equation 126
6.8 The Conservation of Velocity Circulation 129
6.9 Potential Flow 131
6.10 Linear Waves and Instabilities 134
6.11 The Gravity Waves 137
6.12 The Rayleigh-Taylor Instability 141
6.13 Sound Waves 143
6.14 One-Dimensional Traveling Waves 148
6.15 Flow in a Pipe Ahead of the Moving Piston 152
Problems 154

7 Energy Dissipation in Gases and Liquids 157
7.1 Viscous Fluids 157
7.2 Energy Dissipation in Viscous Fluids 159

7.3 Thermal Conduction . . . 161
7.3.1 The Equation of Thermal Conduction . . . 161
7.3.2 Thermal Conduction in an Incompressible Fluid . . . 163
7.3.3 Heat Propagation . . . 165
7.4 Stationary Flow of Incompressible Viscous Fluid . . . 168
7.4.1 Flow of Viscous Incompressible Fluid in a Duct . . . 168
7.4.2 The Poiseuille Flow . . . 170
7.4.3 A Stationary Viscous Flow in a Cylindrical Tube . . . 173
7.5 Dimensional Analysis: The Law of Similarity . . . 175
7.6 Flow with Small Reynolds Numbers . . . 178
7.7 Turbulence: Stability of Steady Viscous Flow . . . 179
7.7.1 Instability of Steady Flow at Large Reynolds Numbers . . . 181
7.7.2 Fully Developed Turbulence: Kolmogorov Theory . . . 184
7.7.3 The Velocity Correlation Function . . . 187
7.8 Boundary Layer . . . 190
Problems . . . 196

8 Detonation and Shock Waves . . . 197
8.1 Surfaces of Discontinuity . . . 197
8.2 The Shock Adiabatic . . . 199
8.3 Detonation Adiabatic . . . 202
8.4 Velocity of Detonation Wave . . . 206
8.5 The Chapman-Jouguet Detonation . . . 207
8.6 The Propagation of a Detonation Wave . . . 209
8.7 Strong Explosion in Homogenous Atmosphere . . . 210
Problems . . . 213

9 Hydrodynamics of Propagating Flame . . . 215
9.1 Complete System of Equations for Propagating Flame . . . 215
9.2 Isobaric Approximation . . . 216
9.3 Theory of Planar Flames . . . 219
9.4 The Model of a Discontinuous Flame Front . . . 222
9.5 The Darrieus-Landau Instability of a Thin Flame Front . . . 224
9.6 The Linear Theory of Instability of the Thin Flame Front . . . 228
9.7 Thermal Stabilization of the DL Instability . . . 231
9.8 Flame Instability in a Gravitational Field . . . 237
9.9 Thermal-Diffusive Instability . . . 241
9.10 Nonlinear Stabilization of the Flame Instability . . . 243
9.11 Curved Stationary Flames in Tube . . . 244
9.12 Nonlinear Equation for Stationary Flames . . . 245
9.13 Stability Limits of a Curved Stationary Flame in Tubes . . . 250
9.14 Spherically Expanding Flames . . . 254
9.15 Flame in a Gravitational Field: Theory of Rising Bubbles . . . 258
9.16 Flame in Horizontal and Vertical Tubes . . . 261

9.17 Flame in a Closed Burning Chamber 265
9.18 Flame Quenching and Flammability Limits 267
Problems 269

10 Regimes of Premixed Flames 271
10.1 Combustion Waves in Laminar Flow 271
10.2 Regimes of Premixed Turbulent Combustion 275
10.3 Velocity of Turbulent Flame 279
10.4 Spontaneous Reaction in Nonuniform Combustible Mixture. 281
10.5 The Deflagration to Detonation Transition 283
10.6 Mechanism of Deflagration to Detonation Transition 293
10.7 Formation a Preheat Zone and Deflagration to Detonation Transition 302
10.8 Thermonuclear Burning of the Stars: White Dwarfs 309
10.9 Burning Ignition and Transition to Detonation in White Dwarfs 312
Problems 317

11 Internal Combustion Engines 319
11.1 Spark Ignition Engine (Otto-Engine) 320
11.2 Engine Operating Cycles 323
11.3 Diesel Cycle 327
11.4 Knock in SI-Engines 330

12 Combustion and Environmental Concerns 335
12.1 Formation of Hydrocarbons and Soot 335
12.2 Processes of Particulate and Soot Formation 337
12.3 NO_x Formation and Reduction 339

Appendix A: Conversion Formulas and Constants 343
A.1 Fundamental Constants 343
A.2 Density, Thermal Conductivity, Viscosity 344

Appendix B: Useful Formulas of Vector Analysis 345
B.1 Cylindrical Coordinates 346
B.2 Spherical Coordinates 346

Appendix C: Equations of Fluid Mechanics in Curvilinear Coordinate Systems 349
C.1 Cylindrical Coordinates 349
C.2 Spherical Coordinates 350

Index 351

Preface

Combustion is a truly interdisciplinary subject, and as such it requires merging of different areas of science: hydrodynamics, chemical kinetic, thermodynamics, statistical physics, kinetic theory, and quantum theory. Due to such a complexity the in-depth scientific investigation of combustion is a recent phenomenon, even though it has always had a great impact on all types of human activities. The elements of combustion, such as flame and explosion are, of course, known for a long time but somewhat surprisingly their first analytical description has been obtained not such a long time ago and for only a limited range of conditions. A more coherent picture has begun to emerge in recent times, as a result of a number of experimental and theoretical studies, sponsored by energy production and industrial needs.

Combustion has a wide variety of uses. Chemical combustion is used for energy's production in power plants, gas turbines and engines. Similar process of thermonuclear combustion is a heat source for the Sun and stars. Recently, astronomers were using exploding stars known as Type Ia supernovas as a cosmic standard light markers to analyze the fate of the universe. They found that the universe expansion is speeding up rather than slowing down, whereas previously it was expected that gravity would be slowing down the expansion. Combustion is involved in explosions for both industrial and military purposes. The humanity cannot live without combustion processes but they also have harmful effects, such as unwanted fire and explosions, pollutants and greenhouse gases, which cause global warming.

Combustion is a process of heat release in exothermal reactions, which is accompanied by mass and heat transfer. The chemical combustion involves a chemical transformation between a substance or substances called fuels and other chemicals called oxidizers. In the process, nuclei of the entering substances are not altered but bonds, involving the electrons of the molecules and atoms, are changed. In many cases, it leads to heat being released, while other cases, heat is required to form the bonds. Heat release in the process of bonding change is the most interesting since this energy can be captured and usefully exploited. Combustion may involve all phases of matter – solid, liquid and gas: solid rocket propellants, liquid droplets burning in diesels, and gaseous combustion in SI-engines. The principal difficulties in understanding combustion systems are the wide range of time and space scales involved, chemical complexity and multidimensional nature of the flow configuration. In turbulent combustion, the strong non-linear coupling of the turbulence and the chemistry further compounds difficulties. These turbulence–chemistry interactions arise from the fact that in most combustion systems, mixing processes are not fast compared with rates of chemical reaction, and there are large spatial and temporal variations in species' composition and temperatures. Chemical reaction's rates are strongly coupled to molecular diffusion at the smallest scales of the turbulence. Furthermore, the heat release associated with combustion affects the turbulent flow, both from variations in the mean density field and from the effects of local dilatation. In the face of such difficulties for a direct analytical approach, engineers have traditionally resorted to empirical methods to develop a combustor. While in the last century the empirical approach sufficed, today there is a much stricter control of pollutant emission and a need for a much more effective burning of fuel, which makes empirical approach no longer viable.

Most of the material covered in this book will deal with the gas phase and with premixed gas combustion. Premixed gas combustion is combustion of gaseous reactants, perfectly premixed prior to ignition. In that case, one has only to ignite the mixture in order to initiate a reaction. The most distinctive feature of premixed combustion is its ability to form a self-sustained reaction wave propagating with a well defined speed, which is either larger (detonation wave) or much less (deflagration wave, "flame") than sound velocity. Thus, we can say that these two regimes of reaction wave, deflagration and detonation, appear to be stable attractors each being linked to its own base of initial data. Premixed gas combustion is obviously of utmost practical importance in engines, modern gas turbines and explosions. There the fuel and air are premixed, and combustion occurs by the propagation of a front separating unburned mixture from fully burned mixture. The emphasis in the present course will be placed on regimes of premixed combustion due to its key importance for practical applications.

While there are several outstanding combustion text-books, they are too advanced and may be difficult for introducing the subject at the undergraduate level. These texts are primarily aimed at the audience at a more advanced level. The present book is primarily meant for 2 nd and 3rd year university students, and for PhD students working in the field of physics, chemistry, mechanical engineering, computer science, mathematics and astrophysics. However, many researchers who already work in the combustion will find some useful background material as well as an overview of recent major developments in this field. This book has been developed through modification of my lecture notes of the combustion course that I have taught for the last several years. The book is focused mainly on theoretical modeling and fundamentals of physics and chemistry of combustion processes and on physics mechanisms for various combustion and combustion-related phenomena in gaseous combustible mixture. The combustion of a gas mixture (flame, explosion, detonation) is necessarily accompanied by motion of the gas. The process of combustion is therefore not only a chemical phenomenon but involves the study of gas dynamics. Therefore, we have included elements of fluid dynamics, which are usually missing from university courses. While assuming that first year physics, chemistry, calculus and thermodynamics have been taken, the necessary concepts in thermodynamics and fluid mechanics are presented during the course.

A vast number of combustion research has been published in multi-disciplinary scientific journals, such as *Physical Review* as well as in specialized combustion journals, such as*Combustion and Flame*, *Combustion Theory and Modeling*, *Combustion Science and Technology*, *Journal of Fluid Mechanics*, etc., as well as many volumes of Proceedings of Combustion Institute (International), so that a list of references would consist of thousands of studies. Instead of giving a list of enormous number of references, I offer a short list of books for further study for an interested reader.

1. Lewis, B. & Von Elbe, G., *Combustion, Flames and Explosion of Gases*, Academic Press, 1987.
2. Zel'dovich, Ya. B., Barenblatt, G. I., Librovich, V. B., & Makhviladze, G. M., *The Mathematical Theory of Combustion and Explosion*, Plenum, New York, 1985.
3. Kuo, K. K., *Principles of Combustion*, John Wiley, 2005.
4. Poinsot, T. & Veynanete, D., *Theoretical and Numerical combustion*, Edwards, 2001.
5. Heywood, J. B., *Internal Combustion Engine Fundamentals,* McGraw-Hill, New York, 1988.
6. Ramos, I. I., *Internal Combustion Engine Modeling*, Hemisphere Publishing, New York, 1989.

I want to thank many of my colleagues and PhD students whose interest in and devotion to the combustion research have enormously helped in forming the newer concepts of combustion wave and assisted with the illustrations used in the book. I am gratefully acknowledge help and discussions with A. Ivanov and M. Kuznetsov of the problem of deflagration-to-detonation transition, and M. Kuznetsov and I. Matsukov providing me with their valuable experimental data of deflagration-to-detonation transition.

Uppsala
January 2008

Michael Liberman

Sources and Use of Energy

Today's world energy consumption is about 1 Q/year with annual growth of about 5%. The unit Q, which is used for such large amounts of energy, roughly equals to $1.05 \cdot 10^{21}$ J. This is a huge amount of energy, which is enough to vaporize approximately 4 trillion of cubic meters of water, the amount of a very large lake. Solar radiation annually absorbed by the Earth is approximately 15,000 Q. To avoid drastic change of climate the energy input produced additionally by human activity must be limited to about 0.1% of this value. This means that upper limit for energy annually produced by human activity should not exceed approximately 15 Q. With the present increase of energy consumption we shall have reached this limit in about 30–40 years. Yet a more serious problem is pollutant emission caused by combustion. Today world energy production is dominated by carbon-based fuels, such as oil, natural gas and coal and hence combustion provides over 85% of the total energy. As such, it is also the principal contributor to air pollution.

It is unlikely that by the end of the twenty-first century other energy sources will have become serious competitors to hydrocarbon combustion. Nuclear energy has a limited appeal because of prejudice and, in some cases, political reasons, and so it will probably not become a significant source of energy. Improved breeder nuclear reactors would be able to provide a temporary relief for the energy problem but this would also mean an easier unauthorized access to the weapon-grade nuclear materials. The reserves of nuclear fuel are limited, even compared to coal. Perspectives for thermonuclear energy production remain uncertain. Up-to-date progress in controlled fusion research has turned out to be much more modest than was expected earlier, in the 1960s. One does not presently expect that the controlled fusion becomes an industrial energy source before the end of the century. The solar energy's use requires huge investments, in particular, for producing new materials allowing a substantial reduction in solar energy costs. Additionally, solar energy production is a viable option in a relatively limited geographic area, mainly outside of the developed countries, and hence relying on it would make developed countries dependent on the political stability of those potential solar energy providers. Even then, countries, which are enjoying a lot of sunshine, still do not have particular encouraging prospects for solar energy.

It has been anticipated that fuel cell systems will play an increasing role in power generation, reducing green house gases' emissions and slowing the global warming. In general, vehicles and power plants, which would use stored hydrogen as fuel, could ultimately reduce local petroleum consumption and local air pollution. However, to achieve this goal one requires a practical, economical hydrogen source that does not generate carbon dioxide. The development of such a hydrogen source, as well as hydrogen's storage and distribution, present a major challenge for fuel-cell technology. At present, producing hydrogen by electrolysis would require to double the total generation of electrical energy to power cars by fuel-cells. Much more energy is needed to produce a quantity of hydrogen compared to the energy than can be obtained from it by combustion or by reactions in a fuel cell. Indeed, energy required is the difference between the heat of combustion of the resulting hydrogen and the heat of combustion of the reformed feedback. This difference sets the lower limit on the energy required to produce an alternative fuel. In practice, the overall efficiency of the process – that is, the energy content of the hydrogen produced divided by the total energy consumed by the reforming process is less than 60%. This means that to produce an amount of hydrogen with the energy content 1 MJ, we must spend more than 1.6 MJ of energy. But, only 0.167 MJ must be expanded to produce a quantity of gasoline with energy content of 1 MJ. Thus, use of hydrogen produced by reformation does not free us from dependence on hydrocarbons. Since nuclear power and renewable energy sources, such as hydropower, solar and wind, are not expected to expand enough to support the electrolysis of seawater, the only realistic source for hydrogen fuel is through reforming of petroleum or natural gas. The process of extracting hydrogen from fossil hydrocarbons using very hot steam, for example, will produce as much carbon dioxide as if the fuel had been burned conventionally. If that CO_2 is not sequestered by some means, preferably near the hydrogen plant, its release into the atmosphere will cause as much global warming as if it had come from conventional car or thermal power plant.

Recent climate catastrophes such as hurricanes Katrina and Rita, which hit New Orleans, Texas and Florida, as well as storms and floods in Asia and Europe, all present us with a serious warning of the climate change, caused by the global warming due to release of pollutants into the atmosphere. The main danger for societies and their economies is not a shortage of oil and gas and resulting rise in prices of hydrocarbon fuels but rather the global warming. It is generally accepted that burning of the hydrocarbon fuel, which is vitally necessary for industry and transportation, is the main source of the pollutants, which are causing global warming. Combustion is responsible for nearly all of the emission of NO_x, CO, CO_2, aerosols, and many other chemical species that are harmful to human health and the environment. According to the environmentalists, the first and most reliable signs of an impeding climate catastrophe are the upsurge in the most violent storms. There is a sustained increase in the number and intensity of catastrophic storms over the last 30 years. Carbon dioxide and other greenhouse gases produced in combustion elevate the atmospheric temperatures causing global warming. As a result, when ocean temperature rises, the amount of water vapor in atmosphere will rise as well. A moister atmosphere helps fuel storms as they have more to "spit out" in form of rain and also by helping drive the convection that gives them lethal spin.

Typical combustion processes are inherently multi-scale, involving complex spatio-temporal phenomena, associated with chemical reactions, molecular transport, and turbulence. The considerable disparity between various scales poses formidable difficulties both for theoretical analysis and numerical simulations, and their effective resolution is one of the main issues of the combustion research. Nowadays science and engineering have reached the point where the ability to simulate processes at very small scales of space and time is essential for furthering our understanding of the whole processes. The ability to simulate processes involving a wide range of spatio-temporal scales and adequate chemical kinetics is essential for furthering understanding of burning processes and for developing the tools required for development of new energy technologies. This understanding is crucial for realizing the long-term goal of creating an environmentally and economically sustainable energy source. The anticipated advances in computational power together with theoretical combustion models offer an opportunity to revolutionize the design and performance of combustion systems, which will considerably lower emissions and increase thermodynamic efficiency of new combustion technologies.

In conclusion, we will give some representative numbers of the world reserves of hydrocarbon fuels and their distribution. The world reserves of coals is 144 Q; of oil, 7–8 Q; of natural gas, 1–2 Q. With the present level of consumption, the reserves of oil and gas will have been depleted in 50 years. All reserves of hydrocarbon fuels may be exhausted in about 100 years, although neither the exact energy needs nor the precise reserves are known with the accuracy which is needed to make definitive predictions.

Chapter 1
Basic Concepts of Thermodynamics

Combustion is one of the most complex subjects that involve primarily such disciplines as physics, chemistry, thermodynamics and fluid mechanics. Thermodynamics enables us to calculate the energies of system changes in composition. As such it enables us to determine, for example, the temperature and pressure changes when a system undergoes a chemical transformation. It will be seen that thermodynamics can also be used to tell us what the composition change will be when a system undergoes a reaction. It is not used however to determine rates of chemical transition. That is the subject of chemical kinetics. The subject of thermodynamics is only concerned with beginning and end thermodynamic states for a system, with no concern for the process path between them. Therefore it will be important to recall the necessary basic concepts of thermodynamics in this course. The purpose of this chapter is to make students life easier, by providing review of the main basic concepts of thermodynamics and necessary definitions. Similarly, in several chapters below the necessary concepts of fluid mechanics will be given, some of them usually missed in standard courses of fluid mechanics.

1.1 The Entropy

Thermodynamic physical quantities are those, which describe macroscopic states of system. They include some, which have both a thermodynamic and a purely mechanical meaning, such as energy, volume, density, etc. There are also, however, quantities of another kind, which appear as a result of purely statistical laws and have no meaning when applied to non-macroscopic systems, for example, entropy. In what follows we shall define a number of relations between thermodynamic quantities. When thermodynamic quantities are discussed, the negligibly small fluctuations to which they are subject are usually of no interest, so that, we shall entirely ignore such fluctuations, regarding the thermodynamic quantities as varying only with the macroscopic state of the system.

M.A. Liberman, *Introduction to Physics and Chemistry of Combustion*,
DOI: 10.1007/978-3-540-78759-4_1,

Entropy is one of the main thermodynamic functions. To introduce meaning of the entropy let us consider a closed system for a period of time long compared to its relaxation time, assuming that the system is in complete statistical equilibrium. Let us divide the system into a large number of macroscopic parts, which we call subsystems. Let $w = w(E_n)$ be the distribution function for the subsystem. The probability that the subsystem has energy between within the interval between E and E + dE is the product of $w = w(E_n)$ on the number of quantum states with energies in this interval. Let $\Gamma(E)$ be the number of quantum states with energies less than or equal to E. Then the required number of states with energy between E and E + dE can be written as $\frac{d\Gamma(E)}{dE}dE$, and the energy probability distribution is

$$W(E) = \frac{d\Gamma(E)}{dE} w(E), \tag{1.1.1}$$

which must satisfy the normalization condition

$$\int W(E)dE = 1. \tag{1.1.2}$$

The function W(E) has a very sharp maximum at the average value of the energy, $E = \bar{E}$, being different from zero only in the immediate neighborhood of the point $E = \bar{E}$. We may define the "width" E of the curve $W = W(E)$ as the width of a rectangle whose height is equal to the value of the function W(E) at the maximum and whose area is unity: $W(\bar{E})\Delta E = 1$. Using the expression (1.1.1), we can write this definition as

$$w(\bar{E})\Delta\Gamma = 1, \tag{1.1.3}$$

where

$$\Delta\Gamma = \frac{d\Gamma(\bar{E})}{dE}\Delta E \tag{1.1.4}$$

is the number of quantum states in the interval ΔE of energy. The quantity $\Delta\Gamma$ represents the "degree of broadening" of the macroscopic state of the subsystem with respect to its microscopic states, with ΔE being in order of magnitude equal to the mean fluctuation of energy of the subsystem. Taking into account a correspondence between the volume of a region of phase space and the "corresponding" number of quantum states we can say that a "cell" of volume $(2\pi\hbar)^s$ (where s is the number of degrees of freedom of the system) "corresponds" in phase space to each quantum state. It is clear that in the quasi-classical case the number of states $\Delta\Gamma$ may be written as

$$\Delta\Gamma = \frac{\Delta p \Delta q}{(2\pi\hbar)^s}, \tag{1.1.5}$$

where p and q are classical momentum and coordinate ($p = p_1, \ldots p_s; q = q_1, \ldots q_s$).

The quantity $\Delta\Gamma$ is called the statistical weight of the macroscopic state of the subsystem, and its logarithm

$$S = \ln(\Delta\Gamma) = \ln\left(\frac{\Delta p \Delta q}{(2\pi\hbar)^s}\right) \tag{1.1.6}$$

is called the *entropy* of the subsystem. The entropy thus defined is dimensionless, like the statistical weight itself. Since the number of states $\Delta\Gamma$ is always larger than unity, the entropy cannot be negative. The concept of entropy is one of the most important in statistical physics and thermodynamics.

It is worth to mention that in classical statistics the entropy is defined only to within an additive constant, which depends on the choice of units, and only differences of entropy, i.e. changes of entropy in a given process, are definite quantities independent of the choice of units. This accounts for the appearance of the Max Plank quantum constant $\hbar$ in the definition of the entropy for classical statistics. Only the concept of the number of discrete quantum states enables us to define a dimensionless statistical weight and therefore to give an unambiguous definition of the entropy.

From the general principles of quantum mechanics, using the Liouville theorem in a way analogous to the classical mechanics, it can be shown that logarithm of the energy distribution for subsystem is a linear function of energy

$$\ln w(E_n) = \alpha + \beta E_n, \tag{1.1.7}$$

which can therefore be rewritten as

$$\ln w(\bar{E}) = \alpha + \beta \bar{E} \tag{1.1.8}$$

and also can be viewed as an average value of $\ln w(E_n)$, i.e., $\langle \ln w(E_n) \rangle$. This means that according to our definition of the entropy (1.1.6) and taking into account (1.1.3), we can write expression for entropy as

$$S = \ln(\Delta\Gamma) = -\ln w(\bar{E}) = -\langle \ln w(E_n) \rangle, \tag{1.1.9}$$

For a closed system the statistical weight $\Delta\Gamma$ is a product of the statistical weights of all the subsystems $\Delta\Gamma = \prod \Delta\Gamma_i$, and we conclude that entropy of the closed system is an additive quantity; entropy of the closed system is sum of the entropies of its subsystems

$$S = \sum_i S_i, \tag{1.1.10}$$

If a closed system is not in an equilibrium state, its macroscopic state will vary in time, until ultimately the system reaches a state of complete equilibrium. If each macroscopic state of the system is described by the distribution of energy between the various subsystems, we can say that the sequence of states successively traversed by the system corresponds to more and more probable distributions of energy. The increase in probability is exponential, with the probability given by exp (S), the exponent being the entropy of the system. We can therefore say that the process approaching the equilibrium in a non-equilibrium closed system going in such a way that the system continually passes from states of lower entropy to those of higher entropy until finally the entropy reaches the maximum possible value, corresponding to complete statistical equilibrium. Thus, if a closed system is at some instant in a non-equilibrium macroscopic state, the most probable consequence at later instants is a steady increase in the entropy of the system. This is *the law of increase of entropy* also known as *second law of thermodynamics* (R. Clausius, 1857). The statistical explanation of the law of increase of entropy was given by L. Boltzmann, 1876. The law of increase of entropy can be formulated also as follows: if at some instant the entropy of a closed system does not have its maximum value, then at subsequent instants the entropy will not decrease; it will either increase or at least remain constant.

1.2 Work and Quantity of Heat: First Law of Thermodynamics

The first law of thermodynamics is the principle of conservation of the total energy of a closed, isolated system. This means that the change of energy of the system during a process must be equal to the energy, which system gains or losses during this process. The energy must be understood in general as the total energy of the system, including the kinetic energy of its macroscopic motion and the internal energy of the system, corresponding to the microscopic motion of atoms and molecules, i.e. the quantity of heat contained in the system.

The external forces applied to a body can do work on it, which is determined, according to the general rules of mechanics, as the products of these forces and the displacements, which they cause. This work may serve to bring the body into a state of macroscopic motion or in general to change its kinetic energy, or to move the body in an external field, or to change volume and shape of the body.

We shall everywhere regard as positive an amount of work R done on a given body by external forces. Negative work, $R < 0$, will correspondingly mean that the body itself does work (equal to R) on some external objects. Bearing in mind that the force per unit area of the surface of the body is the pressure, and that the product of the area of a surface element and its displacement is the volume swept out by it, $dsdl = dV$, we find that the work done on the body per unit time when its volume changes is

$$\frac{dR}{dt} = -P\frac{dV}{dt} \tag{1.2.1}$$

(in case of compression, $dV/dt < 0$, so that $dR/dt > 0$). This formula is applicable to both reversible and irreversible processes. Only one condition need be satisfied, namely that throughout the process the body must be in a state of mechanical equilibrium, i.e. at each instant the pressure must be constant throughout the body.

If the body is thermally isolated, the whole change in its energy is due to the work done on it. In the general case of a body not thermally isolated, in addition to the work done, the body gains or loses energy by direct transfer the heat from or to other bodies in contact with it. This part of the change in energy is the quantity of heat Q gained or lost by the body. Thus the change in the energy of the body per unit time may be written

$$\frac{dE}{dt} = \frac{dR}{dt} + \frac{dQ}{dt}. \tag{1.2.2}$$

Like the work, the heat will be regarded as positive if gained by the body from external sources. The energy E in (1.2.2) must, in general, be understood as the total energy of the body, including the kinetic energy of its macroscopic motion and its internal energy. We shall, however, usually consider the work corresponding to the change in volume of a body at rest, in which case the energy reduces to the internal energy of the body, i.e. the external forces compress the body but leave it at rest as a whole.

Taking into account that the work is defined by formula (1.2.1), we have from (1.2.2) for the quantity of heat

$$\frac{dQ}{dt} = \frac{dE}{dt} + P\frac{dV}{dt}. \tag{1.2.3}$$

For infinitely small quantities we can write (1.2.2) and (1.2.3) as

$$dE = dR + dQ, \quad dQ = dE + PdV. \tag{1.2.4}$$

A body, e.g. gas or liquid, can be characterized by two from three quantities: volume, V, pressure, P, and temperature, T, which can be measured macroscopically. Any other function characterizing the state of the system, for example, energy will be function of these two variables. If we shall take T and V as independent variables, then E will be function of these variables, and we can write for the energy differential

$$dE = \left(\frac{\partial E}{\partial T}\right)_V dT + \left(\frac{\partial E}{\partial V}\right)_T dV. \qquad (1.2.5)$$

The expression for heat differential as function of variables T, V will be

$$dQ = \left(\frac{\partial E}{\partial T}\right)_V dT + \left[\left(\frac{\partial E}{\partial V}\right)_T + P\right] dV. \qquad (1.2.6)$$

In a similar way we may take T and P as independent variables

$$dQ = \left[\left(\frac{\partial E}{\partial T}\right)_P + P\left(\frac{\partial V}{\partial T}\right)_P\right] dT + \left[\left(\frac{\partial E}{\partial P}\right)_T + P\left(\frac{\partial V}{\partial P}\right)_T\right] dP. \qquad (1.2.7)$$

Finally, taking V and P as independent variables we obtain

$$dQ = \left(\frac{\partial E}{\partial P}\right)_V dP + \left[\left(\frac{\partial E}{\partial V}\right)_P + P\right] dV. \qquad (1.2.8)$$

The quantity of heat, which must be gained in order to raise the temperature of the body by one unit is called its specific heat. This clearly depends on the conditions under which the heating takes place. Or we can say that specific heat is the ratio of infinitely small amount of heat to the small temperature rise caused by the heat. A distinction is usually made between the specific heat at constant volume C_V and that at constant pressure C_P. According to their definition we have (note that these formulas are for one mol of a gas)

$$C_V = \left(\frac{\partial Q}{\partial T}\right)_V = \left(\frac{\partial E}{\partial T}\right)_V. \qquad (1.2.9)$$

$$C_P = \left(\frac{\partial Q}{\partial T}\right)_P = \left(\frac{\partial E}{\partial T}\right)_P + P\left(\frac{\partial V}{\partial T}\right)_P. \qquad (1.2.10)$$

The second term in (1.2.10) is contribution of the work done by the system to the specific heat. For example, for the ideal gas, the equation of state for one mol of a gas is $PV = RT$, where $R = 1.986\,\text{cal/mol}\cdot\text{K} = 8.314\cdot 10^7\,\text{erg/mol}\cdot\text{K} = 8.314\,\text{J/mol}\cdot\text{K}$ is the universal gas constant, then we obtain

$$C_P - C_V = R. \qquad (1.2.11)$$

From kinetic theory follows that for ideal one-atomic gas $C_V = \frac{3}{2}R$, and for two-atomic gases, $C_V = \frac{5}{2}R$. Correspondingly, for ideal one-atomic gas

$$C_V = \frac{3}{2}R, \ C_P = \frac{5}{2}R, \tag{1.2.12}$$

and for two-atomic gas $C_V = \frac{5}{2}R, \ C_P = \frac{7}{2}R.$

1.3 Temperature

As we already said, there are thermodynamic physical quantities, which appear as a result of purely statistical laws and have no meaning when applied to non-macroscopic systems, entropy introduced at the beginning of this chapter is such quantity, temperature is another example of such quantity.

Let us consider two bodies in thermal equilibrium with each other, forming a closed system. Then the entropy S of this system has its maximum value for a given energy E of the system. The energy E is the sum $E = E_1 + E_2$ of the energies E_1 and E_2 of the two bodies. The same is true for the entropy S of the system, and the entropy of each body is a function of its energy: $S = S_1(E_1) + S_2(E_2)$. Since $E_2 = E - E_1$ being a constant, S is really a function of one independent variable, and the necessary condition for a maximum can be written as

$$\frac{dS}{dE_1} = \frac{dS_1}{dE_1} + \frac{dS_2}{dE_2}\frac{dE_2}{dE_1} = \frac{dS_1}{dE_1} - \frac{dS_2}{dE_2} = 0, \tag{1.3.1}$$

thus, we came to conclusion that for closed body in thermal equilibrium

$$\frac{dS_1}{dE_1} = \frac{dS_2}{dE_2}. \tag{1.3.2}$$

This conclusion can easily be generalized to any number of bodies in equilibrium with one another. Thus, if a system is in a state of thermodynamic equilibrium, the derivative of the entropy with respect to the energy is the same for every part of the system, i.e. it is constant throughout the system. A quantity, which is the reciprocal of the derivative of the entropy S of a body with respect to its energy, is called the absolute temperature (or simply the temperature) of the body, T:

$$\frac{1}{T} = \frac{dS}{dE}. \tag{1.3.3}$$

The temperatures of bodies in equilibrium with one another are therefore equal: $T_1 = T_2$. Like the entropy, the temperature is a purely statistical quantity, which has meaning only for macroscopic bodies.

Let us next consider two bodies forming a closed system but not in equilibrium with each other with their temperatures T_1 and T_2 being different. In the course of time, equilibrium will be established between the bodies, and their temperatures will gradually become equal. During this process, their total entropy $S = S_1 + S_2$ must increase, i.e. its time derivative is positive:

$$\frac{dS}{dt} = \frac{dS_1}{dt} + \frac{dS_2}{dt} = \frac{dS_1}{dE_1}\frac{dE_1}{dt} + \frac{dS_2}{dE_2}\frac{dE_2}{dt} > 0.$$

Since the total energy is conserved, $\frac{dE_1}{dt} + \frac{dE_2}{dt} = 0$, we obtain

$$\frac{dS}{dt} = \left(\frac{dS_1}{dE_1} - \frac{dS_2}{dE_2}\right)\frac{dE_1}{dt} = \left(\frac{1}{T_1} - \frac{1}{T_2}\right)\frac{dE_1}{dt} > 0.$$

Let the temperature of the second body be greater than that of the first one, i.e. $T_2 > T_1$. Then $dE_1/dt > 0$, and $dE_2/dt < 0$, which means that the energy of the second body with greater temperature decreases and that of the first increases. This property of the temperature may be formulated as follows: energy passes from bodies at higher temperature to bodies at lower temperature.

The entropy is a dimensionless quantity. The definition (1.3.3) therefore shows that the temperature has the dimensions of energy, and thus can be measured in energy units, for example ergs, joules etc. In ordinary circumstances, however, the erg is too large a quantity, and in practice the temperature is customarily measured in its own units, called degrees Kelvin or simply degrees. The conversion factor between ergs and degrees, i. e. the number of ergs per degree, is called Boltzmann's constant and is usually denoted by $k: \ k = R/N_0 = 1.38 \cdot 10^{-16}\,\mathrm{erg/deg} = 1.38 \cdot 10^{-23}\,\mathrm{J/K}$, $N_0 = 6.023 \cdot 10^{23}$ is Avogadro's number. Also in practice the amount of heat is customarily measured in its own units, called calories or in joules. The conversion factors are:

$$1\,\mathrm{cal} = 4.185 \cdot 10^7\,\mathrm{erg}, \ 1\,\mathrm{cal} = 4.185\,\mathrm{J}, \ 1\mathrm{eV} = 1.602 \cdot 10^{-12}\,\mathrm{erg} = 1.602 \cdot 10^{-19}\,\mathrm{J}.$$

Sometime it is convenient to use energy units. The use of the factor k, whose only purpose is to indicate the conventional units of temperature measurement, would merely complicate the formulae. To convert to the temperature measured in degrees, in numerical calculations, we need only replace T by kT. If the temperature is in degrees, the factor k is usually included in the definition of entropy: $S = k \ln \Delta\Gamma$ (Note that $0°C = 273K$).

If the body is subject to no interactions other than changes in external conditions, it is said to be thermally isolated. It must be emphasized that, although a thermally isolated body does not interact directly with any other bodies, it is not in general a closed system, and its energy may vary with time. In a purely mechanical way, a thermally isolated body differs from a closed system only in that its energy depends explicitly on the time because of the variable external field.

Let us suppose that a body is thermally isolated, and is subject to external conditions, which vary sufficiently slowly. Such a process is said to be *adiabatic*.

It can be shown that, in an adiabatic process, the entropy of the body remains unchanged, i.e. the process is *reversible*.

1.4 Pressure

The energy of a body, as a thermodynamic quantity, has the property of being additive: the energy of the body is equal to the sum of the energies of its individual (macroscopic) parts. Another fundamental thermodynamic quantity, the entropy, also has this property. The additivity of the energy and the entropy leads to the following important result. If a body is in thermal equilibrium, we can say that, for a given energy, the entropy depends only on the volume of the body, and not on its shape. Indeed, a change in the shape of the body can be regarded as a rearrangement of its individual parts, and since the entropy and energy, are additive, they will remain unchanged. Thus the macroscopic state of a body at rest in equilibrium is entirely determined by only two quantities, for example the volume and the energy. All other thermodynamic quantities can be expressed as functions of these two. Of course, because of this mutual dependence of the various thermodynamic quantities, any other pair could be regarded as the independent variables.

Let us now calculate the force exerted by a body on the surface bounding its volume. According to the formulae of mechanics, the force acting on a surface element $d\mathbf{s}$ is

$$\mathbf{F} = -\frac{\partial E(q, p; \mathbf{r})}{\partial \mathbf{r}},$$

where $E(q, p; \mathbf{r})$ is the energy of the body as a function of the coordinates and momenta of its particles and of the radius vector $\mathbf{r}$ of the surface element $d\mathbf{s}$, which here acts as an external parameter. Averaging this equation we obtain

$$\mathbf{F} = \overline{\frac{\partial E(q, p; \mathbf{r})}{\partial \mathbf{r}}} = -\left(\frac{\partial E}{\partial \mathbf{r}}\right)_S = -\left(\frac{\partial E}{\partial V}\right)_S \frac{\partial V}{\partial \mathbf{r}} = -\left(\frac{\partial E}{\partial V}\right)_S d\mathbf{s}. \quad (1.4.1)$$

We see that the mean force on a surface element is normal to the element and proportional to its area (Pascal's law). The magnitude of the force per unit area

$$P = -\left(\frac{\partial E}{\partial V}\right)_S \quad (1.4.2)$$

is called the pressure.

In defining the temperature by formula (1.3.3) we were essentially considering a body, which is not in direct contact with any other bodies. It is desired to define the temperature in terms of thermodynamic quantities for the given body only, and then its volume must be regarded as constant. In other words, the temperature is defined as the derivative of the energy of the body with respect to its entropy, taken at constant volume:

$$T = \left(\frac{\partial E}{\partial S}\right)_V. \tag{1.4.3}$$

Equations (1.4.2) and (1.4.3) can also be written together as a relation between differentials, which gives the differential of the function E(S, V) via independent variables S and V instead of E(T, V) given by (1.2.5):

$$dE = TdS - PdV. \tag{1.4.4}$$

Let us show that the pressures of bodies in equilibrium with one another are equal. This follows immediately from the fact that thermal equilibrium necessarily presupposes mechanical equilibrium. In other words, the forces exerted on each other by any two of these bodies at their surface of contact must be equal in magnitude and opposite in direction. The equality of pressures in equilibrium can also be derived from the condition of maximum entropy, in the same way as the equality of temperatures was shown Sect. 1.2. Consider two parts, in contact, of a closed system in equilibrium. One necessary condition for the entropy to be a maximum is that it should be a maximum with respect to a change in the volumes V_1 and V_2 of these two parts when the states of the other parts undergo no change, in particular, $V = V_1 + V_2$ remains constant. We can write similar to Eq. (1.3.1)

$$\frac{dS}{dV_1} = \frac{dS_1}{dV_1} + \frac{dS_2}{dV_2}\frac{dV_2}{dV_1} = \frac{dS_1}{dV_1} - \frac{dS_2}{dV_2} = 0.$$

From the relation (1.4.4) written in the form

$$dS = \frac{1}{T}dE + \frac{P}{T}dV$$

follows that $\partial S/\partial V = P/T$, and so $P_1/T_1 = P_2/T_2$. Since the temperatures are the same in equilibrium, we therefore find that the pressures are equal.

It must be noticed that, when thermal equilibrium is established, the equality of pressures (i.e. mechanical equilibrium) is reached much more rapidly than that of temperatures, and the cases are often met in which the pressure is constant throughout a body but the temperature is not. The reason is that the non-constancy of pressure is due to the presence of uncompensated forces; these bring about macroscopic motion so as to equalize the pressure much more rapidly than to equalize the temperature, which does not involve macroscopic motion.

Let us assume that at every instant throughout the process the body may be regarded as being in a state of thermal equilibrium corresponding to its energy and volume at that instant. The relation (1.4.4) gives the differential of the function $E(S, V)$. Taking into account (1.2.4): $dE = dR + dQ$ and $dQ = dE + PdV$ we obtain

$$dQ = TdS. \tag{1.4.5}$$

The work dR and the quantity of heat dQ gained by the body in an infinitesimal change of state are not the total differentials of any quantities. Only the sum $dR + dQ$, i.e. the change in energy dE, is a total differential. We can therefore speak of the energy E in a given state, but not, for example, of the quantity of heat, which a body possesses in a given state. In other words, the energy of the body cannot be divided into thermal and mechanical parts. This is possible only when considering the change in energy. The change in energy when a body goes from one state to another can be divided into the quantity of heat gained or lost by the body and the work done on it or by it. This division is not uniquely determined by the initial and final states of the body, but depends also on the nature of the process itself. That is, the work and the quantity of heat are functions of the process undergone by the body and not only of its initial and final states. This is seen particularly when the body undergoes a cyclic process, starting and finishing in the same state. The change in energy is then zero, but the body may gain or lose a quantity of heat or work. Mathematically this corresponds to the fact that the integral of the total differential dE around a closed circuit is zero, but the integral of dQ or dR, which are not total differentials, is not zero. For any process around a closed circuit

$$\oint \frac{dQ}{T} \leq 0, \tag{1.4.6}$$

and only for reversible processes we have

$$\oint \frac{dQ}{T} = 0. \tag{1.4.7}$$

This is another expression of the second law of thermodynamic: the entropy of the body either increases or remains unchanged for the reversible processes.

From (1.4.4) and (1.4.5) we can write the specific heat at constant volume C_V and that at constant pressure C_P according to their definition given in Sect. 1.2 in the form:

$$C_V = T\left(\frac{\partial S}{\partial T}\right)_V, \tag{1.4.8}$$

$$C_P = T\left(\frac{\partial S}{\partial T}\right)_P. \tag{1.4.9}$$

Using (1.4.5) and (1.2.6) we can write expression for the entropy as

$$dS = \frac{dQ}{T} = \frac{1}{T}\left(\frac{\partial E}{\partial T}\right)_V dT + \frac{1}{T}\left[\left(\frac{\partial E}{\partial V}\right)_T + P\right] dV. \tag{1.4.10}$$

1.5 The Free Energy and the Thermodynamic Potentials

The first law of thermodynamic says that the work done on a body in an infinitesimal change of state is a differential dR = dE – dQ. In the case of isothermal reversible change we can write this as:

$$dR = dE - dQ = dE - TdS = d(E - TS) \tag{1.5.1}$$

or dR = dF, where we defined

$$F = E - TS \tag{1.5.2}$$

as another function of the state of the body, which is called the *Helmholtz free energy* (or *free energy*).

From the inequality (1.4.6) steams that if the system is in a thermal equilibrium with external media, which is at constant temperature (large reservoir), then the work done on the body when its state changes from 1to2 is $R(1 \to 2) \le E(1) - E(2) - T[S(1) - S(2)]$. If the initial and final temperature of the system is the same then in a reversible isothermal process the work done on the system is equal to the change in its free energy. Substituting dE = TdS – PdV and dF = dE – TdS – SdT, we obtain

$$dF = -SdT - PdV, \quad F = F(T, V) \tag{1.5.3}$$

Hence it follows that

$$S = -\left(\frac{\partial F}{\partial T}\right)_V, \quad \text{and } P = -\left(\frac{\partial F}{\partial V}\right)_T. \tag{1.5.4}$$

Using the relation E = F + TS we can express the energy in terms of the free energy.

If system is in a thermal equilibrium with external media or it is in a vessel with given volume and shape (rigid walls) so that work on the system can not be done, then the free energy has minimum, and the system is at the state of stable equilibrium similar minimum of the potential energy being the condition of the stable equilibrium for mechanical systems. Therefore the free energy can be viewed as the thermodynamic potential at constant volume.

1.6 The Enthalpy

If the process occurs at constant pressure, then the quantity of heat, which is then $dQ = dE + PdV$, can be written as the differential

$$dQ = d(E + PV) = dH \tag{1.6.1}$$

of a quantity

$$H = E + PV, \tag{1.6.2}$$

which is called the *enthalpy* of the body. The change in the enthalpy in processes occurring at constant pressure is therefore equal to the quantity of heat gained by the body, so that the enthalpy is called also heat function.

Taking $dE = TdS - PdV$ and $dH = dE + PdV + VdP$, we find an expression for the total differential of the enthalpy, $H = H(S, P)$

$$dH = TdS + VdP. \tag{1.6.3}$$

From this it follows that

$$T = \left(\frac{\partial H}{\partial S}\right)_P, \text{ and } V = \left(\frac{\partial H}{\partial P}\right)_S. \tag{1.6.4}$$

It follows from (1.6.1) that, in processes involving a thermally isolated body ($dQ = 0$) and occurring at constant pressure, the enthalpy is conserved, i.e. $H =$ constant.

We obtained for the specific heat at constant volume the expression (1.2.9)

$$C_V = \left(\frac{\partial E}{\partial T}\right)_V. \tag{1.6.5}$$

Using expression (1.4.9) for the specific heat at constant pressure C_P, and substituting T from (1.6.4), we obtain

$$C_P = T\left(\frac{\partial S}{\partial T}\right)_P = \left(\frac{\partial H}{\partial S}\right)_P\left(\frac{\partial S}{\partial T}\right)_P = \left(\frac{\partial H}{\partial T}\right)_P. \tag{1.6.6}$$

This formula shows that at constant pressure the enthalpy has properties similar to those of the energy at constant volume.

Thus, if we know any of the quantities E, F, H as a function of the corresponding two variables and take its partial derivatives, we can determine all the remaining thermodynamic quantities. For this reason E, H and F are called thermodynamic potentials: the energy E with respect to the variables S, V; the enthalpy H with respect to S, P, and the free energy F with respect to V, T.

We still lack a thermodynamic potential with respect to another pair of variables: P, T. To derive this we substitute in (1.6.2) $PdV = d(PV) - VdP$, take d(PV) to the left-hand side of the equation, and obtain

$$dG = -SdT + VdP \tag{1.6.7}$$

with a new quantity, $G = G(T, P)$

$$G = E - TS + PV = F + PV = H - TS, \tag{1.6.8}$$

called the *Gibbs free energy*.

It is clear that if temperature and pressure of the system are constant, then the state corresponding to minimum of the Gibbs free energy is the stable equilibrium state.

From (1.6.7) we clearly have

$$S = -\left(\frac{\partial G}{\partial T}\right)_P, \text{ and } V = \left(\frac{\partial G}{\partial P}\right)_T. \tag{1.6.9}$$

Thus, we have the following thermodynamic potentials for all pairs of variables: $E = E(S, V)$; $H = H(S, P)$; $F = F(V, T)$; $G = G(P, T)$. In practice the most convenient, and the most widely used pairs of thermodynamic variables are (V, T) and (P, T). It is therefore necessary to transform various derivatives of the thermodynamic quantities with respect to one another to different variables, both dependent and independent. If (V, T) are used as independent variables, the results of the transformation can be conveniently expressed in terms of the pressure P and the specific heat C_V (as functions of V and T). The equation, which relates the pressure, volume and temperature, is called *the equation of state* for a given body. Thus, the purpose of the formulae in this case is to make it possible to calculate various derivatives of thermodynamic quantities from the equation of state and the specific heat C_V. Similarly, when (P, T) are taken as the basic variables the results of the transformation should be expressed in terms of V and C_P as functions of P and T. One should remember that the dependence of C_V on V or of C_P on P (but not on the temperature) can itself be determined from the equation of state.

1.7 The Nernst's Theorem

It can be proved that the specific heats C_V and C_P are always positive. The fact that the specific heat, C_V is positive means that the energy is a monotonically increasing function of the temperature. Conversely, when the temperature decreases the energy decreases monotonically, and therefore, when the temperature has its least possible value, i.e. at absolute zero, a body must be in the state of least possible energy. Regarding the energy of a body as the sum of the energies of its parts, we can say that each of these parts will also be in the state of least energy. It is clear that the minimum value of the sum must correspond to the minimum value of each term. Thus at absolute zero any part of the body must be in a particular quantum state – the ground state. In other words, the statistical weights of these parts are equal to unity, and the same is true for their product, i.e. the statistical weight of the macroscopic state of the body as a whole. The entropy of the body, being the logarithm of its statistical weight, is therefore equal to zero. This result that the entropy of any body vanishes at the absolute zero of temperature is called *Nernst theorem* (W. Nernst, 1906). It should be emphasized that this theorem is a deduction from quantum statistics, in which the concept of discrete quantum states is of essential importance. The theorem cannot be proved in purely classical statistics, where the entropy is determined only to within an arbitrary additive constant.

Nernst's theorem enables us to draw conclusions also concerning the behavior of certain other thermodynamic quantities as $T \to 0$. For instance, for $T = 0$ the specific heats C_P and C_V both vanish:

$$C_P = C_V = 0 \quad \text{for} \quad T = 0 \tag{1.7.1}$$

This follows immediately from the definition of the specific heat in the form

$$C = T\frac{\partial S}{\partial T} = \frac{\partial S}{\partial \ln T}.$$

When $T \to 0$, $\ln T \to -\infty$, and since S tends to a finite limit (zero), it is clear that the derivative tends to zero.

The thermal expansion coefficient also tends to zero:

$$\left(\frac{\partial V}{\partial T}\right)_P = 0 \quad \text{for} \quad T = 0. \tag{1.7.2}$$

This derivative is equal to the derivative $-(\partial S/\partial P)_T$, which vanishes for $T = 0$, since S is zero for $T = 0$ and any pressure. This is easily seen from (1.6.9) which gives

$$\left(\frac{\partial S}{\partial P}\right)_T = -\frac{\partial}{\partial P}\left(\frac{\partial G}{\partial T}\right)_P = -\frac{\partial}{\partial T}\left(\frac{\partial G}{\partial P}\right) = -\frac{\partial V}{\partial T}.$$

If the specific heat of a body is known for all temperatures, the entropy can be calculated by integration, and Nernst's theorem gives the value of the constant of integration. For example, the dependence of the entropy on temperature for a given pressure is determined by

$$S = \int_0^T \frac{C_P}{T} dT. \tag{1.7.3}$$

The corresponding formula for the enthalpy is

$$H = H_0 + \int_0^T C_P dT \tag{1.7.4}$$

where H_0 is the value of the enthalpy for $T = 0$.

1.8 Carnot's Cycle and Carnot's Theorem

Consider processes in a thermally isolated system involving subsystems, which are not in thermal equilibrium with each other. According to the second law of thermodynamic entropy of the system increases and the system approaches equilibrium. During a way as the system passes to the equilibrium the work can be done by the system on surroundings. However, path to the equilibrium as well as the final equilibrium state, its entropy and energy may be different. Correspondingly, work done by the system also depends on the particular path how the system approaches equilibrium. Carnot's theorem establishes the maximum amount of work that can be obtained depending on the difference between the hot and cold temperature reservoirs of the system, or the efficiency of transforming thermal energy into mechanical energy. Obviously, heat can be transformed to work only if there are at least two reservoirs with different temperatures. We also assume that the volume of the system remains unchanged, i.e. we are interested in the work which is produced due to heat is transformed to work when system approached equilibrium.

According to the first thermodynamic law the work done by a thermally isolated system is difference between the initial energy of the system, E_0, and its energy at the equilibrium state, that is function of entropy in this state, E(S)

$$|R| = E_0 - E(S). \tag{1.8.1}$$

Taking derivative on entropy of the final state we obtain

$$\frac{\partial |R|}{\partial S} = -\left(\frac{\partial E}{\partial S}\right)_V = -T, \tag{1.8.2}$$

where (Sect. 1.3) T is temperature of the final state. The derivative in Eq. (1.8.2) is negative and this means that the work |R| done by the system increases with the decrease of entropy. However, entropy of thermally isolated system cannot decrease, which means that maximum value of |R| can be achieved if entropy remains constant during the process. Thus, we come to the conclusion that the system produces maximum work if its entropy is constant during the transition to the equilibrium state, e.g., transition to the equilibrium state is reversible process. In other words, maximum efficiency is achieved if and only if no new entropy is created in the cycle.

Let us assume that the system contains two bodies of different temperatures T_1 and T_2 so that $T_2 > T_1$, and work is done due to heat transforming between two bodies. It must be emphasized that if the heat transfer occurs directly between the bodies, no work can be done. Such process is irreversible and the entropy of the system increases by $\Delta S = \Delta Q(1/T_1 - 1/T_2)$, assuming that volume did not change. In order to accomplish a reversible transfer of the heat and therefore to maximize the work, some additional – working medium must be used, which in fact is executing a reversible cycle process. This process must be fulfilled in such a way that the bodies between which direct transfer of energy occurs be at the same temperature. The working body at temperature T_2 is brought in contact with the heat reservoir at the same temperature and receives form it isothermally a certain amount of energy. Then it is adiabatically cooled to temperature T_1 transfers at this temperature energy to the reservoir at temperature T_1, and finally returned adiabatically to its initial state. The working body does work on surroundings in the expansions involved in the process. The described cycle is called Carnot cycle. It is convenient to present this cycle on the P,V diagram. Let the initial state with temperature T_2 corresponds to the point P_A, V_A. The first phase of the expansion is isotherm (no heat transfer) to the point P_B, V_B. The second phase is adiabatic cooling to temperature $T_1 : P_B, V_B \to P_D, V_D$. Next is adiabatic compression along isotherm at temperature $T_1 : P_D, V_D \to P_C, V_C$, and the final phase is adiabatic heating to $T_2 : P_C, V_C \to P_A, V_A$. According to the second law of thermodynamic the work done during the process is equal to the difference of the heat received to and transferred from the working body. Since the working body returned to its initial state it may be ignored. During phase AB the working body gain amount of heat Q_2 from reservoir at temperature T_2, and it lose amount of heat Q_1 to another reservoir at temperature at T_1 during the phase DC. Thus

$$|R_{\max}| = Q_2 - Q_1. \tag{1.8.3}$$

When transforming thermal energy into mechanical energy, the thermal efficiency of a process is the percentage of energy that is transformed into work. Thermal efficiency is

$$\eta_{\max} = \frac{R}{Q_2} = \frac{Q_2 - Q_1}{Q_2} = 1 - \frac{Q_1}{Q_2}. \tag{1.8.4}$$

Writing $Q_2 = T_2 \Delta S_2$ and $Q_1 = -T_1 \Delta S_1$, and taking into account that since process is reversible the sum of the entropies $\Delta S_1 + \Delta S_2 = 0$ remains constant, we obtain

$$\eta_{\max} = \frac{T_2 - T_1}{T_2} = 1 - \frac{T_1}{T_2}. \tag{1.8.5}$$

Carnot's theorem sets a limit on the maximum amount of efficiency of any possible heat engine, which solely depends on the difference between the hot and cold temperature reservoirs. The maximum efficiency possible by any sort of engine operating between two heat reservoirs has a limit defined by the efficiency (1.8.5). The theoretical maximum efficiency of a heat engine equals the difference in temperature between the hot and cold reservoir divided by the absolute temperature of the hot reservoir.

1.9 Le Chatelier Principle

If system at equilibrium experiences a change in concentration, temperature or total pressure the equilibrium will shift in order to minimize that change. This qualitative law makes it possible to envisage the displacement of equilibrium. The general principle was formulated by Henry Louis Le Chatelier and Karl Ferdinand Braun who discovered it independently. When a given system in an equilibrium state is disturbed by the application of some action, a direct reaction occurs in such a way as to diminish the action. The principle is the physical interpretation of the mathematical conditions for stability of the equilibrium state. These conditions can be derived in the form of inequalities for derivatives of entropy as function of two independent variables, which must be maximum at the equilibrium. Suppose two bodies A and B are in equilibrium at the same temperature, T. If the equilibrium is disturbed by an increase of the temperature of body A, the heat starts to flow from body A to body B, tending to reduce the temperature difference, so that the increase in the temperature of body A, brings the process tending to decrease it. For example, if the body is disturbed from equilibrium by a change in its volume at constant temperature, then, in particular, its pressure is changed. The restoration of equilibrium in the body leads to a decrease in the absolute value of the change in pressure. The decreasing or increasing in the volume of the body brings about processes in it, which tend to lower or to increase the pressure respectively.

1.10 Dependence of the Thermodynamic Quantities on the Number of Particles

As well as the energy and entropy, such thermodynamic quantities as H, F ,G also have the property of additivity. This follows directly from their definitions if we bear in mind that the pressure and temperature are constant throughout a body in equilibrium. From this property we can draw certain conclusions concerning the manner in which each of these quantities depends on the number of particles in the body. Here we shall consider bodies consisting of identical particles (molecules), but all the results can be immediately generalized to mixtures of different particles.

The additivity of a quantity signifies that, when the amount of matter and therefore the number N of particles is changed by a given factor, the quantity is changed by the same factor. In other words, we can say that an additive thermodynamic quantity must be a homogeneous function of the first order with respect to the additive variables.

Let us express the energy, E, of the body as a function of the entropy, volume, and number of particles. Since S and V are themselves additive, this function must be of the form

$$E = Nf\left(\frac{S}{N}, \frac{V}{N}\right), \tag{1.10.1}$$

which is the most general homogeneous function of the first order in N, S and V.

The free energy, F, is a function of N, T and V. Since the temperature is constant throughout the body, and the volume is additive, a similar argument gives

$$F = Nf\left(\frac{V}{N}, T\right). \tag{1.10.2}$$

In the same way we have for the enthalpy, expressed as a function of N, S and the pressure P,

$$H = Nf\left(\frac{S}{N}, P\right). \tag{1.10.3}$$

Finally, the thermodynamic potential as a function of N, P and T is

$$G = Nf(P, T) \tag{1.10.4}$$

We shall now formally consider N as independent variable. Then the expressions for the differentials of the thermodynamic potentials must include terms proportional to dN. For example, the total differential of the energy will be written

$$dE = TdS - PdV + \mu dN, \tag{1.10.5}$$

where μ, denotes the partial derivative

$$\mu = \left(\frac{\partial E}{\partial N}\right)_{S,V}. \tag{1.10.6}$$

The quantity μ is called the *chemical potential* of the body.

Similarly we have

$$dE = TdS - PdV + \mu dN, \tag{1.10.7}$$

$$dH = TdS + VdP + \mu dN, \tag{1.10.8}$$

$$dF = -SdT - PdV + \mu dN, \tag{1.10.9}$$

$$dG = -SdT + VdP + \mu dN, \tag{1.10.10}$$

with the same μ.

These formulae show that

$$\mu = \left(\frac{\partial E}{\partial N}\right)_{S,V} = \left(\frac{\partial H}{\partial N}\right)_{S,P} = \left(\frac{\partial F}{\partial N}\right)_{T,V} = \left(\frac{\partial G}{\partial N}\right)_{P,T}. \tag{1.10.11}$$

The chemical potential can be obtained by differentiating any of the quantities E, H, F, G with respect to the number of particles, but the result is expressed in terms of different variables in each case.

Differentiating G in the form (1.10.4), we find that $\mu = \partial G/\partial N = f(P, T)$, i.e.

$$G = N\mu. \tag{1.10.12}$$

Thus the chemical potential of a body (consisting of identical particles) is just its thermodynamic potential per molecule. When expressed as a function of P and T, the chemical potential is independent of N. Thus we can write down for the differential of the chemical potential

$$d\mu = -sdT + vdP, \tag{1.10.13}$$

where s and v are the entropy and volume per molecule.

If we consider a definite amount of matter, the number of particles in it is a given constant, while the volume is variable. Let us now take a certain volume within the body, and consider the matter enclosed therein. The number of particles N will now be variable, and the volume V constant. Then, for example, Eq. (1.10.9) reduces to

$$dF = -SdT + \mu dN. \tag{1.10.14}$$

In this case the independent variables are T and N.

1.11 Ideal Gases

In majority cases, in particular, dealing with combustible gaseous mixture, gases can be treated as *ideal*. The meaning of an ideal (or perfect) is a gas in which the interaction between its molecules is negligibly weak. Physically, this approximation is eligible either because the interaction between molecules is small compared to its kinetic energy (temperature is relatively high) or because the gas is sufficiently rarefied. In practice all ordinary molecular or atomic gases can be considered as being ideal. The condition of ideal gas approximation is violated either at very high densities of gases or at very high pressures. Considerable advantage of dealing with ideal gas approximation is that in this case we can calculate all thermodynamic characteristics of the gas and its equation of state in an explicit and simple analytical form. The first immediate feature of an ideal gas is that its internal energy, E, depends only on the temperature, as it will be seen below. The explanation of this is that in an ideal gas the volume occupied by the molecules of the gas and their mutual interactions are negligible, so the mean spacing between molecules cannot matter.

As is known from statistical physics, free energy of an ideal gas, which obeys Boltzmann statistics can be written as (we will use energy units for temperature below)

$$F = -NT\ln\left\{\frac{e}{N}\int e^{-E(p,q)/T}dV\right\}, \tag{1.11.1}$$

where V is volume of the gas with N number of particles, E(p,q) its energies. Though integral in (1.11.1) cannot be calculated in a general form without knowing about energy levels corresponding to internal molecular states etc., however, important fact is that it depends only on temperature and therefore can be written as

$$F = -NT\ln\left(\frac{eV}{N}\right) + Nf(T), \tag{1.11.2}$$

where f(T) some function of temperature.

Using expression (1.11.2), and substituting it into second formula in (1.5.4) we find for pressure

$$P = -\left(\frac{\partial F}{\partial V}\right)_T = NT/V, \tag{1.11.3}$$

which is nothing but well-known equation of state of an ideal gas (Clapeyron equation). For temperature measured in degrees we can write

$$PV = NkT. \tag{1.11.4}$$

Knowing free energy potential F, the other thermodynamic potentials can be found using formulas of Sect. 1.6. In particular, expression (1.11.2) we obtain for Gibbs free energy

$$G = -NT\ln\left(\frac{eV}{N}\right) + Nf(T) + PV. \qquad (1.11.5)$$

Since G must be expressed as a function of P and T, we must substitute here $V = NT/P$ from (1.11.3), and using a new function $\chi(T) = f(T) - T\ln T$, we obtain

$$G = NT\ln P + N(f(T) - T\ln T)) = NT\ln P + N\chi(T). \qquad (1.11.6)$$

In a similar way we obtain

$$S = -\frac{\partial F}{\partial T} = N\ln\left(\frac{eV}{N}\right) - Nf'(T), \qquad (1.11.7)$$

$$S = -\frac{\partial \Phi}{\partial T} = -N\ln P - N\chi'(T) \qquad (1.11.8)$$

$$E = F + TS = Nf(T) - NTf'(T). \qquad (1.11.9)$$

Notice, that the internal energy is a function only of the temperature of the gas and the same is true for the enthalpy $H = E + PV$. Thus,

$$H = E + PV = Nf(T) - NT(f'(T) - 1). \qquad (1.11.10)$$

Since the specific heats $C_V = (\partial E/\partial T)_V$ and $C_P = (\partial H/\partial T)_P$ are functions of the temperature solely, for the specific heats per molecule we obtain, $C_V = Nc_V$ and $C_P = Nc_P$ we find that

$$c_P - c_V = 1, \qquad (1.11.11)$$

or using ordinary units for temperature we can write (1.11.11) as $c_P - c_V = k$.

1.12 Ideal Gases with Constant Specific Heat: Equation of Poisson Adiabatic

On many cases the specific heat of gases is constant, independent of temperature over a wide temperature interval. In this case the equation of state of ideal gas – relation between the volume, temperature and pressure becomes especially simple. Differentiating (1.11.9) we find $c_V = (\partial E/\partial T)_V = -Tf''(T)$. Integrating this for constant specific heat c_V, we obtain

$$f(T) = -c_V T \ln T - \zeta T + \varepsilon_0, \tag{1.12.1}$$

where ζ and ε_0 are the integrating constants. Substituting this expression for f(T) in the expression for free energy (1.11.2) we find

$$F = N\varepsilon_0 - NT \ln\left(\frac{eV}{N}\right) - Nc_V T \ln T - N\zeta T. \tag{1.12.2}$$

Using formulas for E, G, H, we can write also expressions for all other thermodynamic potentials

$$E = N\varepsilon_0 + Nc_V T, \tag{1.12.3}$$

$$G = N\varepsilon_0 + NT \ln P - Nc_P T \ln T - N\zeta T, \tag{1.12.4}$$

$$H = E + PV = N\varepsilon_0 + Nc_P T. \tag{1.12.5}$$

Finally, differentiating (1.12.2) and (1.12.4) with respect to temperature we find according to formulas of Sects. 1.5 and 1.6 the entropy as function of either T, V or T, P, respectively

$$S = N \ln\left(\frac{eV}{N}\right) + Nc_V \ln T + (\zeta + c_V)N, \tag{1.12.6}$$

$$S = -N \ln P + Nc_P \ln T + (\zeta + c_P)N. \tag{1.12.7}$$

These last expressions give the relation between the volume, temperature and pressure for an ideal gas with constant specific heats undergoing the gas adiabatic compression or expansion.

The process is called adiabatic if it is reversible and the system is thermally isolated, so that there is no heat exchanges between the system and surroundings. For adiabatic process the entropy is constant, therefore we have from (1.12.7)

$$-N \ln P + Nc_P \ln T = \text{constant}, \tag{1.12.8}$$

Hence,

$$T^{C_P}/P = \text{constant}. \tag{1.12.9}$$

Using $c_P - c_V = 1$ from the previous section, and introducing $\gamma = c_P/c_V$, which is called adiabatic constant, we obtain equation of *Poisson adiabatic*

$$T^{\gamma}P^{1-\gamma} = \text{constant}. \tag{1.12.10}$$

With the help of equation of state $PV = NRT$ the Eq. (1.12.10) can be presented also in the form useful for applications

$$T\,V^{\gamma-1} = \text{constant}, \tag{1.12.11}$$

$$PV^{\gamma} = \text{constant}. \tag{1.12.12}$$

Since gas in the adiabatic processes is thermally isolated, $dQ = 0$, so its energy decreases during adiabatic expansion (see Eq. (1.2.4)). This also means that according to (1.12.3) its temperature also decreases with the decrease of its energy.

The second law of thermodynamic for ideal gas can be written with help of Eq. (1.12.3) as

$$C_V dT + PdV = dQ. \tag{1.12.13}$$

Using equation of state for ideal gas, $PV = RT$, we obtain from (1.12.13)

$$(C_V + R)dT - VdP = dQ. \tag{1.12.14}$$

For a process at constant pressure $dP = 0$, and we obtain from (1.12.4) relation between specific heats at constant pressure and constant volume (1.2.11)

$$C_P = \left(\frac{dQ}{dT}\right)_P = C_v + R$$

Substituting P from $PV = RT$ into (1.12.13) we find after integrating

$$\ln T + \frac{R}{C_V}\ln V = \text{constant}, \tag{1.12.15}$$

$$TV^{\frac{R}{C_V}} = TV^{\gamma-1} = \text{constant}. \tag{1.12.16}$$

Simple application of the theory of adiabatic gas expansion is reduction of the air temperature high at the atmosphere. Temperature of the air at atmosphere changes mainly due to convection. Warm air is rising up to the level where pressure is lower and cools. Since thermo-conduction of air is very low, we can consider the process as the adiabatic expansion. In equilibrium the difference in pressure corresponding to dh is $dP = -\rho g dh$, where ρ is density of the air and g is the gravity acceleration. Substituting here density from the equation of state in the form $\rho = m/V = MP/RT$, and Eq. (1.12.10), we obtain

$$\frac{dT}{dh} = -\frac{\gamma - 1}{\gamma}\frac{gM}{R}. \tag{1.12.17}$$

Taking the average molecular weight for air $M = 28.9$, $\gamma = 7/5$, $g = 980 cm/s$, $R = 8.314 \cdot 10^7 erg/grad$, we obtain $\frac{dT}{dh} = -9.8$ grad/km, which means that during summer time there may be snow on the top of mountains higher than 3 km.

Problems

1.1. Show that the temperatures of two bodies in thermal equilibrium are equal. Use principle of maximum entropy ($\frac{1}{T} = \frac{dS}{dE}$).

1.2. Show that the pressures of two bodies in thermal equilibrium are equal. Use principle of maximum entropy ($P = -\left(\frac{\partial E}{\partial V}\right)_S$, $dE = TdS - PdV$).

1.3. Two bodies forming a closed system but not in equilibrium with temperatures $T_1 < T_2$ are connected. In the course of time, equilibrium will be established between the bodies, and their temperatures will become equal. Use principle of maximum entropy to show that energy passes from body at higher temperature to body at lower temperature.

1.4. From the condition that the entropy $S = S(T, V)$ is the total differential show that energy of the ideal gas with the equation of state $PV = RT$ is the function of temperature and does not depend on pressure.

1.5. Two identical ideal gases at the same temperature, T, and containing the same number of atoms, N, but different pressures, P_1 and P_2, are in the vessels of volume V_1 and V_2 correspondingly. Find the change in entropy after the vessels have been connected.

1.6. Two identical ideal gases at the same pressure, P, and containing the same number of atoms, N, but different temperatures, T_1 and T_2, are in the vessels of volume V_1 and V_2 correspondingly. Find the change in entropy after the vessels have been connected.

1.7. Calculate work and amount of heat obtained by the gas, which is compressed from V_1 to V_2, in accordance to the process described by equation $PV^n = C$.

1.8. What is the heat obtained by ideal gas during process at constant volume?

1.9. Find heat and work for the process, which occurs at constant pressure when volume was changed from V_1 to V_2.

1.10. Find expression for work during adiabatic compression of ideal gas when volume was changed from V_1 to V_2.

1.11. Find expression for work made by an ideal gas during isothermal change of its volume from V_1 to V_2.

1.12. Find expression for work made by an ideal gas during isothermal change of its pressure from P_1 to P_2.

1.13. How much heat is necessary to heat 48 m^3 of air in the room from 0 C to 20 C at constant volume. Take density of air $\rho = 0.00129 g/cm^3$, specific heat at constant pressure $c_P = 0.238 cal/g\ deg$, $\gamma = 1.4$.

1.14. How much heat is necessary to heat 48 m^3 of air in the room from 0 C to 20 C at constant pressure of 1 atm. Take density of air $\rho = 0.00129 g/cm^3$, specific heat at constant pressure $c_P = 0.238\ cal/gdeg$.

1.15. How much heat is necessary to heat 48 m^3 of air in the room from 0 C to 20 C if during the heating air is leaking through a hole in the window. Outside pressure is 1 atm. Take density of air at 0 C $\rho = 0.00129\ g/cm^3$, specific heat at constant pressure $c_P = 0.238 cal/gdeg$, $\gamma = 1.4$.

1.16. On PV diagram the Otto cycle (1-2-3-4-1) may be presented as adiabatic compression (1-2), isobaric heating (2-3), adiabatic expansion (3-4), pressure drops at constant volume (4-1). Calculate work and efficiency of the process.

1.17. On PV diagram the Diesel cycle (1-2-3-4-1) may be presented as adiabatic compression (1-2), heating and expansion at constant pressure (2-3), adiabatic expansion (3-4), pressure drops at constant volume (4-1). Calculate work and efficiency of the process.

1.18. The Joule cycle (1-2-3-4-1) on PV diagram consists of: adiabatic processes (1-2) and (3-4) and two transitions at constant pressure (2-3) and (4-1). Calculate work and efficiency of the process.

Chapter 2
Chemical Thermodynamics

2.1 Introduction and Definitions

We shall often be concerned about the atomic weight and molecular weight of atoms and molecules. For example, relative to the carbon, the number 12 is the number of particles in the nucleus, consisting in this case of six positively charged protons and six neutral neutrons. Carbon 12, being electrically neutral, has six negatively charged electrons, giving it the atomic number 6. The atomic weight or molecular weight of any other atom or molecule is the number of particles in the nucleus: total sum of protons and neutrons in the nucleus. The proton and neutron have almost, but not exactly, the same mass and the electron has a nearly negligible mass compared to the proton ($m_e/m_p \cong 0.00054463$). The consequence is that almost any atom or molecule will have its atomic or molecular weight very nearly an integer. As an example, the oxygen 16 atom has an atomic weight of 15.995, which is 16 for practical purposes.

A common set of elements occurring in combustion problems is the C H N-O system. These elements have the atomic weights 12, 1, 14 and 16, respectively. The molecular weight of any compound is merely the sum of the weights of the atomic constituents. For example, the molecular weight of water, H_2O, is $1 + 1 + 16 = 18$. Similarly, the molecular weight of carbon dioxide, CO_2 , is $12 + 2 \times 16 = 44$.

The next concept is that of the mole, which is defined as the mass in grams (g) of an element or compound equal to its atomic or molecular weight. For example, 12 g of carbon is one mole of carbon; 44 g of CO_2 is one mole of the carbon dioxide. It is clear, since the mole and the mass are synonymous, one mole of a compound or element consists of a fixed number of particles of the substance. This number is Avogadro's number, given the symbol N_0 and having the value $N_0 = 6.023 \cdot 10^{23}$ particles.

Let the number of molecules of each species be denoted by N_i, where i is the species identifier. The concentration of species i, c_i, is defined as $c_i = N_i/N_0V$, and has the units of moles per unit volume, where V is the volume. The overall concentration is obtained by summing up the concentrations of the individual components. That is,

M.A. Liberman, *Introduction to Physics and Chemistry of Combustion*,
DOI: 10.1007/978-3-540-78759-4_2, © Springer-Verlag Berlin Heidelberg 2008

$$c = \sum_{i=1}^{M} c_i = \frac{N}{N_0 V} = \frac{1}{N_0 V} \sum_{i=1}^{M} N_i, \qquad (2.1.1)$$

where M is the total number of species in the volume V.

In view of the large magnitude of N_0, it should be clear that even very small volume V contains an enormous number of particles. In fact, V can be shrunk so low that it appears to an observer and to the calculus as a mathematical point, while still containing many particles. This is the continuum limit, where we may speak of concentration at a point, and it may vary from point to point in a flow, for example. This concept we will be using in fluid dynamics.

The mole fraction of species i, X_i is defined as $X_i = N_i/N = c_i/c$, where $N = \sum_{i=1}^{M} N_i$. It is obvious that

$$\sum_{i=1}^{M} X_i = \frac{1}{c} \sum_{i=1}^{M} c_i = \frac{c}{c} = 1.$$

Consequently we conclude that only (M – 1) of the mole fractions are independent.

The molecular weight of species *i* will be given by the symbol W_i. From it and the concentration of species i we may construct the density of species i or the mass per unit volume, as $\rho_i = c_i W_i$. The overall density, ρ will be

$$\rho = \sum_{i=1}^{M} \rho_i$$

The overall molecular weight of a mixture, the mass per unit mole, then follows as

$$W = \frac{\text{mass}}{\text{mole}} = \frac{\sum_{i=1}^{M} c_i W}{c} = \sum_{i=1}^{M} c_i W/c = \sum_{i=1}^{M} X_i W_i.$$

The mass fraction of species i is formed in a fashion similar to the mole fraction and is defined as $Y_i = \rho_i/\rho$.The relation between the mole fraction and the mass fraction then becomes

$$Y_i = \frac{c_i W_i}{\sum_{i=1}^{M} c_i W_i} = \frac{c_i W_i/c}{\sum_{i=1}^{M} c_i W_i/c} = \frac{X_i W_i}{W}.$$

Example 2.1 In a vessel of volume 1 m^3 there are 2 g of molecular hydrogen, H_2, 8 g of molecular oxygen, O_2 and 14 g of molecular nitrogen, N_2. What are the mole fractions, mass fractions and concentrations of the hydrogen, oxygen and nitrogen? What are the overall density and mixture molecular weight?

Solution

$$\text{Total mass} = 2 + 8 + 14 = 24\,\text{g},$$

$$\text{Moles } H_2 = 2\,\text{g}/(2\,\text{g/mole}) = 1\,\text{mol},$$

$$\text{Moles } O_2 = 8/(2 \times 16) = 0.25\,\text{mol},$$

$$\text{Moles } N_2 = 14/(2 \times 14) = 0.5\,\text{mol}.$$

$$\text{Total moles} = 1 + 0.25 + 0.5 = 1.75\,\text{mol}.$$

$$X_{H_2} = 1/1.75 = 0.571;\; X_{O_2} = 0.25/1.75 = 0.1428;\; X_{N_2} = 0.285,$$

$$X_{H_2} + X_{O_2} + X_{N_2} = 1.$$

$$Y_{H_2} = 1/1.75 = 0.083;\; Y_{O_2} = 8/24 = 0.333;\; Y_{N_2} = 0.583.$$

$$c_{H_2} = 1\,\text{mol/m}^3;\; c_{O_2} = 0.25\,\text{mol/m}^3;\; c_{N_2} = 0.5\,\text{mol/m}^3,\; \rho = 24\,\text{g/m}^3.$$

The rule of writing down chemical reactions and the law of stoichiometry is shown below using example of carbon and oxygen reaction

$$(1)C + (1)O_2 \rightarrow (1)CO_2$$

This is to be read as "one molecule (or mole) of carbon plus one molecule (or mole) of molecular oxygen goes to one molecule (or 1 mol) of carbon dioxide." The numbers multiplying the species are called stoichiometric coefficients. When the number is unity it is always omitted. The above reaction would actually be written as

$$C + O_2 \rightarrow CO_2$$

Another example is

$$2C + O_2 \rightarrow 2CO.$$

Insofar as the law of stoichiometry is concerned, there is no violation if the above reactions proceed from left to right or inverse reaction – from right to left as well as. If this is indeed occurring we write the reaction as

$$2C + O_2 \leftrightarrow 2CO.$$

If the reaction only proceeds one way (from left to right, for example) we call the species on the left the reactants and those on the right the products. This distinction is lost if the reaction is reversible and proceeds both ways.

Finally, if we consider reactions between multiple phases, we must distinguish the phases in these cases. If solid carbon, for example, reacts with liquid molecular oxygen to form gaseous carbon monoxide, we may write

$$2C(s) + O_2(l) \rightarrow 2CO(g),$$

where (s), (l) and (g) stand for solid, liquid and gas, respectively. If there is no indication of the phase, it is presumed that only gas phase species are present.

2.2 Properties of Substances

We shall deal with solids, liquids and gases. For solids and liquids we will use the actual measured thermodynamic properties of the substances. One of the most important for practical applications is combustion of gaseous mixtures. For gases, we will always assume them to be ideal (or perfect) gases. The term "ideal" means that interaction between particles in the gas is negligible compared with their kinetic energy. Then, the equation of state is

$$PV = nRT, \tag{2.2.1}$$

where V is volume, P is pressure, T is absolute thermodynamic temperature, and $R = 8.314 J/mol \cdot K$ is the universal gas constant. The symbol n is the number of moles in a fixed mass system

$$n = \sum_{i=1}^{M} (N_i/N) = \sum_{i=1}^{M} n_i.$$

Dividing Eq. (2.2.1) by V, we can rewrite it as

$$P = cRT = \rho RT/W,$$

which is now in terms of variables which are independent of system size and may vary from point to point in a system.

The energy units we shall use through the course are erg, calorie, Joule and eV. The conversion between the calorie, Joule, erg and eV are:

$$1\,cal = 4.186\,J;\ \ 1\,J = 10^7\,erg;\ \ 1\,eV = 1.602 \cdot 10^{-19}\,J = 11606\,K.$$

The perfect (ideal) gases have the property that if a species other than a given species occupying a container is added to the container the pressure of the original species acting on the container walls is not altered. In a mixture of several different ideal gases the pressure on the wall due to each species *i* is P_i. Each of these partial pressures may be computed by

$$P_i = c_i RT \tag{2.2.2}$$

and each P_i is independent of the others.

It is obvious that for ideal gas the pressure from all of the species may be computed by

$$P = \sum_{i=1}^{M} P_i = \sum_{i=1}^{M} c_i RT = RT \sum_{i=1}^{M} c_i = cRT,$$

since RT does not depend upon the summation index i and may be taken from under the summation sign. This is known as Dalton's law of partial pressures.

Turning to the variables describing various kinds of energy, the nomenclature to be used here is that small letters will describe variables on a per unit mass basis. For example, the internal energy on a mass basis will be given the symbol e. The enthalpy on a per unit mass basis is h; where, $h = e + P/\rho$. If these variables are given on a molar basis (per unit mole), capital letters will be used, E and $H = E + P/c$. Similarly, depending upon the basis the entropy will be denoted by S or s.

The ideal gases, which obey Eq. (2.2.1), have the property that e and h (E and H) are functions only of temperature, thus

$$e = \int_{T_{ref}}^{T} c_v(T)dT + e_{ref}, \tag{2.2.3}$$

$$h = e + \frac{p}{\rho} = e(T) + \frac{R}{W}T = h(T) = \int_{T_{ref}}^{T} c_p(T)dT + h_{ref}, \tag{2.2.4}$$

where the subscript "ref" denotes some reference state, that will be defined later, and $c_v(T)$ and $c_p(T)$ are the specific heat at constant volume and at constant pressure, respectively. The specific heats on a molar basis are denoted by C_v and C_p , respectively. The conversion between the molar and mass basis is through the molecular weight, that is, $E = eW$ and $H = hW$.

In a mixture of species we will always assume that there is no interaction between the species. This has the consequence that the quantities evaluated for the entire mixture are merely the sums of the quantities for the individual species – they are *additive quantities*

$$E = \sum_{i=1}^{M} X_i E_i, \quad H = \sum_{i=1}^{M} X_i H_i, \quad e = \sum_{i=1}^{M} Y_i e_i, \quad H = \sum_{i=1}^{M} Y_i H_i,$$

where the E_i and H_i are the same as if the species were alone.

2.3 Heats of Reactions and Formation

Consider the first law of thermodynamics applied to a fixed mass system, where the system undergoes a change of state from condition 1 to 2. It is presumed that changes in kinetic and potential energy are negligible. Then

$$E_2 - E_1 = \Delta E = (\textit{heat added}) - (\textit{work done by the system}).$$

Presume further that there is no work done. That is, there is no shaft work and the system remains at constant volume so that the pressure does no mechanical work on the surroundings. In this case

$$\Delta E = Q_v$$

where Q_v is the heat added at constant volume. An example would be an explosion in a constant volume vessel if the walls did not break.

Another example of application of the first law is for a process taking place at constant pressure. For example, it can be change form the initial state to a final state of the system, consisting of a frictionless perfectly sealed piston of weight W, resting upon the contents in the vessel. Here, work is performed on the surroundings if the contents of the vessel expand in going from state 1 to state 2. The first law applied here yields

$$\Delta E = Q_p - p(V_2 - V_1), \text{or } \Delta H = Q_p, \tag{2.3.1}$$

where Q_p denotes the heat added in a constant pressure process. As we have seen in Chap. 1 it is purely a matter of algebraic convenience, which variable is used the energy or the enthalpy. A constant volume chemical conversion may serve as a good model for the combustion in an automobile cylinder, if the combustion takes place in a short enough time that the piston does not move too far. In steady flow rocket or jet engines the combustion takes place, usually at nearly constant pressure, so the constant pressure assumption is nearly valid.

As an example, consider the reaction of two propellants often used for rocket engine, hydrogen (H_2) and fluorine (F_2). Consider the reactants to be half mole hydrogen and half mole fluorine. The product of reaction is hydrogen fluoride (HF), and the equation is

$$\frac{1}{2}H_2 + \frac{1}{2}F_2 \rightarrow HF.$$

We assume that all hydrogen and fluorine are consumed. Let us assume also that the pressure and temperature of the initial and final states are the same. This assumption contradicts intuition, that tells us that this propellant combination is a powerful explosive and energy should be released. However, such process can be realized if the pressure is maintained constant by the piston and the temperature can be manipulated by the extraction of heat. Thus, an overall constant pressure, constant temperature process can be forced, if the proper amount of heat is removed.

Because of its importance as a reference condition, assume that the pressure and temperature for this process are 1 bar (1 bar $= 10^5$ Pa, with 1 Pa $= 1\,N/m^2$) and 298 K, respectively. Under these conditions the substances behave as perfect gases. The enthalpies are therefore only functions of temperature and, for any species i will be denoted as $(H_T)_i$. Applying Eq. (2.3.1) to this process, we have

$$Q_p = (H_{298})_{HF} - \frac{1}{2}(H_{298})_{H_2} - \frac{1}{2}(H_{298})_{F_2}.$$

If this process is actually carried out, the result will be

$$Q_P = -272.5\,kJ = -64.2\,kcal.$$

The negative sign means that heat is evolved rather than added. Reactions of this kind are called *exothermic*; if heat were required to be added to keep the temperature the same, the reaction is called *endothermic*. The quantity Q_p for this reaction is called the heat of reaction at the reference state. The heat of reaction is positive for exothermic reactions and it is negative for endothermic reactions. If the pressure or temperature were not at 1 bar or 298 K the Q_p would be the heat of reaction at the stated pressure and temperature, which must, in fact, be stated. The heat of reaction will differ from the above number, depending upon the thermodynamic conditions.

According to definition, for solids and liquids, the standard state of a substance is its real state at a pressure of 1 bar, for any temperature T. The standard state for a gas is assumed to be ideal gas state at 1 bar at any temperature T. To understand what is meant here, it should be emphasized that all gases have some pressure and temperature conditions under which they cannot be considered as ideal gases. For example, near the liquefaction condition, the perfect gas law does not hold. As an example, water vapor at the boiling point of liquid water is not a perfect gas. But we define the standard state of water vapor at 1 bar and 373 K (the boiling point) as the fictitious ideal gas state at these conditions. As an example, at 1000 K and 1 bar water vapor is, for all practical purposes, ideal and the thermodynamic properties may be extrapolated to the boiling point assuming it to be ideal. We denote the standard state with a superscript “0”. In the above hydrogen-fluorine example the substances, in fact, behave as ideal gases so that we may write at the 1 bar, 298 K condition

$$Q_p = (H^0_{298})_{HF} - \frac{1}{2}(H^0_{298})_{H_2} - \frac{1}{2}(H^0_{298})_{F_2}.$$

The Q_p comes out as a difference between the enthalpies of the product and the elements from which it is formed.

In principle, all compounds can be formed by reactions by their constituent elements. In thermodynamics we never know absolute values of such variables as, e.g. enthalpy, energy, etc. That is not important because only changes in these variables are dealt with in thermodynamics. To facilitate numerical work, we define a reference set of substances and their thermodynamic properties. We define reference chemicals as the elements in the form most abundantly found in nature when they are found alone. So, for example, oxygen in the air is found as the diatomic molecule O_2 in the gas phase. The reference state for oxygen is therefore the gaseous diatomic molecule. Similarly, reference states for fluorine, hydrogen, and nitrogen are gaseous F_2, H_2, and N_2 respectively. By contrast, the reference state for carbon is C(s), or graphite.

We define formation reactions for a compound or atom as the reaction that forms one mole of the substance from the elements or element in their reference state. The heat of formation for a substance is the standard state enthalpy change for a reaction forming the substance from its reference state elements. Moreover, the heat of formation of the elements in their standard states is defined as zero. The heat of formation may be defined at any temperature and for species i, and it is denoted as $(H_f^0)_{T,i}$. In the above example it was a formation reaction at 298 K at the stated conditions. Therefore,

$$(H_f^0)_{298,HF} = -64.2 \text{ kcal}.$$

Other examples of formation reactions are

$$H_2 + \frac{1}{2}O_2 \rightarrow H_2O(l), \ (H_f^0)_{298,H_2O(l)} = -68.32 \text{ kcal} = -2865.9 \text{ kJ},$$

$$C(s) + O_2(g) \rightarrow CO_2(g), \ (H_f^0)_{298,CO_2(g)} = -94.05 \text{ kcal} = -393.7 \text{ kJ}.$$

The numbers for heats of formation are empirically derived; experiments must be performed to obtain them. Some of the numbers must be obtained by indirect means, using principles of thermodynamics. For example, consider the formation of CO:

$$C(s) + \frac{1}{2}O_2(g) \rightarrow CO(g).$$

This formation reaction cannot be carried out in the laboratory, because CO_2 is formed side by side with CO. However, the formation reaction for CO_2 can be carried out with almost a perfect yield for the carbon dioxide. The reaction is given above with a heat of formation of – 94.0 kcal/mol. It is also possible to form and store pure CO and to do the following reaction at the reference conditions:

$$CO + \frac{1}{2}O_2 \rightarrow CO_2.$$

The heat of reaction at 1 bar and 298 K is –67.6 kcal/mol of CO_2 formed. The reverse reaction must yield a heat of reaction of +67.6.0 kcal. The following reactions may be written, violating no law of thermodynamics or stoichiometry:

$$C(s) + \frac{1}{2}O_2 + \frac{1}{2}O_2 \rightarrow CO_2 \rightarrow CO + \frac{1}{2}O_2.$$

The overall process merely carries 1/2 mol O_2 along "for the ride" with no thermodynamic change. Consequently, the overall process is one of formation of one mole of CO from the elements and the heat of formation of CO must be

$$(H_f^0)_{298,CO} = -94.0 + 67.6 = -26.4\,\text{kcal/mol}.$$

A list of heats of formation for some reactions is given in Table 2.1. Notice that the heat of formation of atoms is positive if the reference state of the atom is diatomic molecule. This occurs because strong chemical bonds must be broken to form the atoms; the process is endothermic. Notice also, that the compounds with the highest negative heat of formation in the C-H-O system are CO_2 and H_2O; the formation process for these compounds is exothermic. An extensive set of thermodynamic properties of substances is provided by the Joint Army Navy NASA Air Force Thermochemical Tables (*JANNAF Tables*).

Let us consider how we can calculate heat of formation if atoms are not at the reference state of 1 bar and 298 K. As an example, consider a mixture of 1/2 mol H_2 and 1/2 mol F_2 at 5 bar pressure and 200 K. What happens if the reaction takes place at constant pressure to form one mole HF and the mixture ends up at 2000 K. How much heat must be liberated or absorbed in this process? Since thermodynamics does not care about the process path, the result must be the same for any path as long as the initial and the end states are specified. We will choose the path so that the calculations are simple and convenient. Consider the following paths. Path (1-a): heat the reactants at constant pressure from temperature of 200 K to the reference temperature of 298 K. Path (a-b): at constant temperature drop the pressure to the standard state pressure of 1 bar. Path (b-c): react the reactants to product at constant (reference) pressure and temperature. Path (c-d): compress the product to 5 bar at constant temperature. Path (d-2): heat up the product to 2000 K at constant pressure. Because the overall process is one at constant pressure, the overall enthalpy change is

$$\Delta H(1-2) = H_2 - H_1 = n_{HF}(\Delta H)_{HF} - n_{H_2}(\Delta H)_{H_2} - n_{F_2}(\Delta H)_{F_2},$$

where the overall enthalpy change is the sum of the enthalpy changes for each step along the path for reagents, products plus the enthalpy of a formation reaction, n_A is number of moles of the species: 1/2 mol H_2, 1/2 mol F_2, and 1 mol of HF.

Table 2.1 Heats of formation at 298 K

Symbol	Name	State	ΔH_f^0(kcal/mol)
H	Hydrogen atom	Gas	52.1
H_2	Hydrogen	Gas	0
O	Oxygen atom	Gas	59.56
O_2	Oxygen	Gas	0
N	Nitrogen atom	Gas	112.95
N_2	Nitrogen	Gas	0
Cl	Chlorine atom	Gas	28.9
Cl_2	Chlorine	Gas	0
H_2O	Water	Gas	–57.8
H_2O	Water	Liquid	–68.3
NO	Nitric oxide	Gas	21.6
OH	Hydroxyl radical	Gas	10.06
C	Carbon	Solid	0
C	Carbon	Gas	171.0
C_2	Carbon	Gas	198.0
CO	Carbon monoxide	Gas	–26.42
CO_2	Carbon dioxide	Gas	–94.05
NO	Nitrogen oxide	Gas	21.6
NO_2	Nitrogen dioxide	Gas	8.0
CH_4	Methane	Gas	–17.85
C_3H_8	Propane	Gas	–24.8
HCl	Hydrogen Chlorine	Gas	–22.1
HF	Hydrogen Fluorine	Gas	–64.2
C_2H_6	Ethane	Gas	–20.24
C_4H_{10}	Butane	Gas	–29.8
C_6H_6	Benzene	Gas	19.8
C_2H_4	Ethene	Gas	12.5
C_8H_{18}	Octane	Liquid	–59.7
C_2H_2	Acetylene	Gas	54.19

Path (1-a) is a constant pressure heating process. The reactants are perfect gases, and the JANNAF Tables gives us the enthalpy change in going from 298 K to any other temperature. For hydrogen and fluorine, the reactants,

$$Q_{p,1-a} = \frac{1}{2}(H^0_{298,H_2} - H^0_{200,H_2} + H^0_{298,F_2} - H^0_{200,F_2}) = \frac{1}{2}(2.77 + 2.99).$$

Path (b-c) is a formation reaction (at the reference state) at 1 bar and 298 K, for which the JANNAF Tables gives

$$Q_{p,b-c} = (\Delta H^0_f)_{298,HF} = -272.5\,\text{kJ}.$$

Path (d-2) is a constant pressure heating process for pure HF, which from the JANNAF Tables requires an enthalpy change of

$$Q_{p,d-2} = (H^0_{2000,HF} - H^0_{298,HF}) = +52.83\,kJ.$$

Paths (a-b) and (c-d) have no enthalpy change associated because for ideal gases the enthalpy only depends upon temperature. Consequently, the overall enthalpy change in going from the state 1 to 2 is

$$\Delta H_{1-2} = 52.83 - 272.55 + \frac{1}{2}(5.76) = Q_{p,1-2} = -216.84\,kJ = -51.8\,kcal.$$

Notice that $Q_{P,1-2}$ is negative, so that the overall process is exothermic, even in view of the fact that the temperature has increased. These reactants, if heat is not removed out of the system, would go to an even higher temperature than 2000 K.

Generalizing what we have done if there are M different species in both the reactants and products, the heat evolved in a constant pressure process is

$$Q_p = \sum_{j=1}^{M} n_j \left[H_{T_2} - H^0_{298} + (\Delta H^0_f)_{298}\right]_j - \sum_{i=1}^{M} n_i \left[H_{T_1} - H^0_{298} + (\Delta H^0_f)_{298}\right]_i \quad (2.3.2)$$

where the index j; denotes products and the index i denotes reactants. Similarly, for a constant volume process the heat evolved is

$$Q_p = \sum_{j=1}^{M} n_j \left[E_{T_2} - E^0_{298} + (\Delta E^0_f)_{298}\right]_j - \sum_{i=1}^{M} n_i \left[E_{T_1} - E^0_{298} + (\Delta E^0_f)_{298}\right]_i \quad (2.3.3)$$

where the energy for any species is obtained from the enthalpy by $E_i = H_i - p/c_i$.

Notice, that hydrogen and oxygen can coexist in a mixture at the reference conditions of 298 K and 1 bar. In order to start reaction an ignition source is needed, such as a spark. We say that such a mixture is in a metastable equilibrium state, requiring a "push" to initiate reaction. Fortunately, all thermodynamics laws apply to systems in such a state of metastable equilibrium, and their thermodynamic state may be calculated as if the mixture were in true equilibrium in the metastable state.

2.4 Origin of the Combustion Heat; Molecular Bonds

Majority of combustion processes involve the combination of fuels containing hydrogen and carbon with the oxygen in air. Together with water, these substances will remain, at least until the end of the twenty-first century, the most accessible and widely used by mankind for production of energy and for chemical and technological needs. Prior to examining the physical and mathematical foundations of combustion, let us consider on a molecular level the origin of heat of formation and energy of combustion whose release controls all the other

Table 2.2 Bond energies

Molecule	Energy [kcal/mole]
$H_2 = 2H$	103.3
$O_2 = 2O$	118.0
$N_2 = 2N$	225.9
$C_2 = 2C$	144.6
$Cl_2 = 2Cl$	58.0
$H_2O = H + OH$	119.2
$HCl = H + Cl$	103.1
$HO = H + O$	110.0
$NO = N + O$	150.9
$NO_2 = N + O_2$	105.1
$NO_2 = NO + O$	73.1
$CO = C + O$	257.2

effects such as heating of the gas and the appearance of active chemical centers. What makes up the heats of basic reactions between a carbon and hydrogen compound and oxygen?

The heats of formation for some reactions are given in Table 2.1. Table 2.2 gives bond energies of molecule, which are the energy needed to break up molecule for atoms.

Let us evaluate the heat of reaction $2H_2 + O_2 = 2H_2O$ using magnitudes of the bond energies given in Table 2.2. As a reasonable assumption, we assume that reaction of formation the water vapor form molecular hydrogen and oxygen is going through the following steps. Dissociation of two molecule of H_2 requires 206.6 kcal/mol and dissociation of one molecular O_2 requires 118.0 kcal/mol. From 4 atoms of hydrogen and 2 atoms of oxygen form two molecule of water, which can be viewed as formation of 4 bonds of H-O. Each bond of H-O according to Table 2.2 corresponds to energy release of 110 kcal/mol. Thus, the energy balance is $206.6 + 118.0 - 440 = -115.44$ kcal/mol. According to Table 2.1 the heat of formation of gaseous H_2O in the reaction is –57.8 kcal/mol, and we obtained $-115.4/2 = -57.7$ kcal/mol. In general case the bond energy depends on the configuration of atoms – what bonds they form in the molecule.

The structures of some typical hydrocarbon fuels are shown below. Most common hydrocarbon fuels are n-octane and are given by the formula C_nH_{2n+2}, such as e.g. methane, propane, etc.

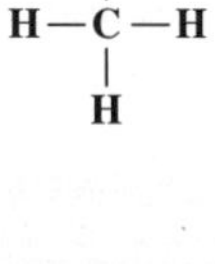

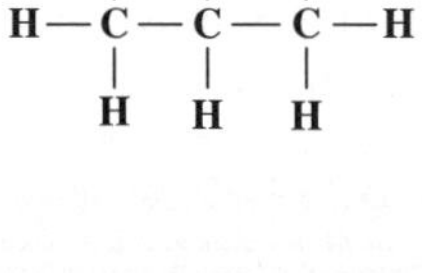

Olefins have open-chain containing one unsaturated double-bond (more hydrogen can be added) and have formula C_nH_{2n}. Acetylenes (alkynes) have open-chain containing one C-C unsaturated triple-bond, $H-C\equiv C-H$; acetylene C_2H_2. For alcohols one hydroxyl (OH) group is substituted for one hydrogen, e.g. methane, CH_4 becomes methyl alcohol, CH_3OH also known as methanol, and ethane becomes ethyl alcohol C_2H_5OH known as ethanol.

Let us consider the energy balance for the consecutive oxidation reactions of solid carbon (graphite):

$$C(s)+\frac{1}{2}O_2=CO+26\,\mathrm{kcal/mol},\ \ CO+\frac{1}{2}O_2=CO_2+68\,\mathrm{kcal/mol}$$

Thus, the complete oxidation of solid carbon yields 94 kcal/mol:

$$C(s)+O_2=CO_2+94\,\mathrm{kcal/mol},$$

which is similar in magnitude to the energy released in reaction hydrogen with oxygen:

$$2H_2+O_2=2H_2O(\mathrm{vapor})+116\,\mathrm{kcal/mol}.$$

According to the Table 2.2 the CO molecule is one of the most stable, with one of the highest bond energy of 257 kcal/mol. The next most stable molecule is N_2, with the bond energy of 226 kcal/mol. In both molecules there are three pairs of bound electrons, or three valence bonds. In the CO molecule one electron initially goes from O to C, after which the O^+ and C^- are similar to nitrogen atoms; this is confirmed by the existence of a dipole moment in the CO molecule. In the CO_2 molecule the second oxygen atom is bound more weakly, so that

$$CO+O=CO_2+127\,\mathrm{kcal/mol}.$$

The bond strength of oxygen in these compounds is comparable to the bond strength of the original oxygen molecule. In the latter case, $O+O=O_2+118\,\mathrm{kcal/mol}$, and only 59 kcal/mol goes to each oxygen atom. The low bond energy of the oxygen molecule is the reason for its chemical activity and the reason that oxidation is used as an energy source.

The bond strength of a carbon atom in the crystalline lattice of graphite (and of diamond and amorphous carbon as well) is very large. The reaction energy in $C(s)+\frac{1}{2}O_2=CO+26\,\mathrm{kcal/mol}$ is the difference of two very large quantities: half of the dissociation energy of O_2 into atoms (59 kcal/mol) and the heat of vaporization of the carbon atom must be subtracted from the binding energy of CO (257.2 kcal/mol). Thus, the energy balance is $257.2-59-171=27.2\,\mathrm{kcal/mol}$, which is close to 26.4 kcal/mol in Table 2.1.

The conversion of solid carbon and gaseous hydrogen into hydrocarbon fuels takes place with a small change in energy. On the other hand, when oxygen is introduced into organic compounds (alcohols, aldehydes, organic acids, carbohydrates, etc.) nearly as much energy is released as during complete combustion (to CO_2 and H_2O), when an equal amount of oxygen is used. Roughly (100–120) kcal is released per mole of consumed oxygen during complete combustion of any organic fuel. The exceptions are certain endothermic energy-rich compounds, such as, for example, acetylene, whose heats of combustion are greater.

Incomplete combustion is energetically uneconomical both per molecule of fuel and per molecule of oxygen that is consumed. Only 52 kcal/mol are released in the reaction $2C(s) + O_2 = 2CO$ instead 94 kcal/mol during burning of CO with the complete oxidation of carbon. The strong bond of the C atom in solid carbon means that carbon leaves the solid state only in combination with oxygen in the form of CO or CO_2. Relative to a free carbon atom the reaction $2CO = CO_2 + C(s) - 129$ kcal/mol has a large energy barrier. The reaction $2CO = CO_2 + C(s) + 41$ kcal/mol is energetically favorable only in reference to solid carbon for incomplete combustion and at low reaction temperatures. That's why soot and carbon black are formed during combustion by the disintegration of organic molecules with a carbon skeleton, rather than from CO.

Consider now reaction with nitrogen, which constituting 78.1% by volume of Earth's atmosphere, and therefore inevitably participate in combustion. Since nitrogen molecule N_2, with dissociation energy of 226 kcal/mol, is very stable, the reaction of N_2 and O_2 to form 2 NO is endothermic and it could occur only at high temperatures. The higher oxides (NO_2, N_2O_3, N_2O_4, N_2O_5) are formed from nitrogen and oxygen with almost no change in energy compared with the bonding energies of N_2 andO_2. Therefore, oxygen packed in compounds with nitrogen ($CH_3 - ONO_2$-nitroether, $CH_3(C_6H_2)(NO_2)_3$ – trinitrotoluene), is energetically almost equivalent to gaseous oxygen. Oxygen that is built into an organic molecule but bonded to nitrogen creates a material that releases a large amount of energy when the molecule is regrouped with formation of N_2 and transformation of the oxygen into CO_2 and H_2O molecules. That is reason why compounds, in which oxygen is bound to nitrogen are used as powders and explosives. The energy released during combustion or explosion of powders differs only slightly from the energy of combustion of organic substances in the same amount of oxygen as is bound to the nitrogen.

2.5 Adiabatic Flame Temperature

One of the most important calculations chemical thermodynamics allows us to make is that of the adiabatic flame temperature. This is the temperature achieved in a constant pressure reaction with no heat transfer. For example, the adiabatic flame temperature is achieved in reaction in a vessel, which is insulated to heat transfer but operated at constant pressure. We must set $Q_p = 0$ in Eq. (2.3.2) and it becomes the equation describing the adiabatic process leading to the adiabatic flame temperature. This is the temperature achieved in jet engine and rocket combustion chambers if the flow velocities are small compared to sound velocity, neglecting by minor heat losses, which may slightly low the flame temperature.

Consider as an example the constant pressure reaction of hydrogen and oxygen to form water adiabatically. Let the initial state is at 200 K

$$H_2 + \frac{1}{2}O_2 \rightarrow H_2O.$$

Equation (2.3.2) becomes

$$0 = (1)\left[H_{T_2} - H^0_{298} + (\Delta H^0_f)_{298}\right]_{H_2O} - \left[H_{200} - H^0_{298}\right]_{H_2} - \frac{1}{2}\left[H_{200} - H^0_{298}\right]_{O_2}.$$

Using numbers from the JANNAF Tables we obtain equation for the adiabatic flame temperature T_2

$$0 = \left[H_{T_2} - H^0_{298}\right]_{H_2O} - 241.83 - \left[-2.77 - \frac{1}{2}2.87\right].$$

Looking the table numbers for water and solving equation for $\left[H_{T_2} - H^0_{298}\right]_{H_2O}$, we obtain $T_2 = 4900$ K, which is much higher than $T_2 = 3080$ K, shown in Table 2.3 and not realistic.

In a more simple way, the JANNAF Tables give the standard state specific heat (on a molar basis) for the compound in question. Then, taking into account that C_p is a monotonically increasing function of temperature, we can write

$$H_{T_i} - H^0_{298,i} = \int_{298}^{T} C_{p,i}(T)dT$$

An upper limit estimate for the adiabatic flame temperature will therefore be attained with a lower limit estimate for C_p and vice versa. Using trial and error method, the adiabatic flame temperature can be found with the help of the JANNAF Tables.

At this point we remind the meaning of a stoichiometric mixture, which is a mixture of reactants that can, without violating the law of stoichiometry, react to only products with the highest negative heats of formation in the chemical system being considered. There should be no excess of fuel or oxidizer on the right side of the reaction. In the hydrocarbon- oxygen mixture, for example, the products with the highest negative heats of formation are CO_2 and H_2O, but not

Table 2.3 Flame temperatures of various stoichiometric mixtures; initial temperature 298 K

Fuel	Oxidizer	Pressure (bar)	T(K)
Methane 6%	94% Air	1	1850
Methane 11.6%	Air	1	2100
Methane	Air	20	2270.
Methane	Oxygen	1	3030
Methane	Oxygen	20	3460
Propane 4.02%	Air	1	2240
Propane 7.4%	Oxygen	1	2500
Propane 19.3%	Oxygen	1	3000
Carbon monoxide	Air	1	2400
Carbon monoxide	Oxygen	1	2977
Acetylen	Air	1	2600
Acetylen	Oxygen	1	3342
Butane	Oxygen	1	3100
Hydrogen 20%	Air	1	1910
Hydrogen 30%	Air	1	2300
Hydrogen 57%	Air	1	1850
Hydrogen	Oxygen	1	3080

CO and OH. In the H-F system, HF has the highest negative heat of formation. Therefore, the reactant mixtures

$$H_2 + \frac{1}{2}O_2, H_2 + F_2, \ CH_4 + 2O_2$$

are stoichiometric mixture. These mixtures react toH_2O, HF, and CO_2, with no fuel or oxidizer left over. Contrary, the following reactant mixtures are not stoichiometric mixtures:

$$H_2 + O_2, \ 2H_2 + F_2, \ CH_4 + 3O_2 + N_2.$$

Here there is either excess fuel or oxidizer even though water, carbon dioxide and hydrogen fluoride could be formed. Stoichiometric mixtures have nearly the highest adiabatic flame temperatures possible for the chemical system under consideration. The reason is that excess fuel or oxidizer acts as a diluent and cannot pass to the high negative heat of formation substances. If f denotes the ratio of mass of fuel to the mass of oxidizer in a reactive mixture, the equivalence ratio is $\phi \equiv f/f_s$, where f_s is the stoichiometric mass ratio. If air is the oxidizer, convention is that the nitrogen is included in the mass of the oxidizer, even though it may no react. A typical plot of adiabatic flame temperature versus equivalence ratio is shown in Fig. 2.1. The maximum temperature occurs near the stoichiometric mixture ratio, though the actual maximum is usually slightly shifted to the fuel rich side of stoichiometric.

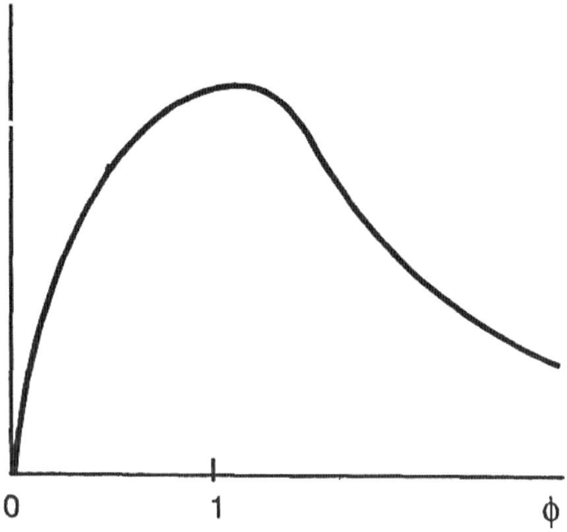

Fig. 2.1 Typical adiabatic flame temperature versus equivalence ratio

2.6 The Equilibrium Constant

In all the reactions considered to this point the products have been specified; they were consistent with the law of stoichiometry and were thermodynamically possible products. However, the extremely high adiabatic flame temperatures calculated in the previous section are unreasonable. The reason is that most combustion systems (exception is reaction in lean mixtures, whose product temperature is below approximately 1250 K) reach rather high temperature, when there is a possibility of dissociation – breaking the molecule bonds of normal stable combustion products such as CO_2, H_2O, O_2 and N_2 . For rich mixtures CO also exists in the products and at high temperatures the molecules dissociate to form H, O, OH, NO. In reality, for the hydrogen-oxygen system, a mixture reacting adiabatically would produce several products besides H_2O, like H_2, H, OH, O_2, O and HO_2 (the hydroperoxl radical). The effects of dissociation are always decrease flame temperatures since dissociation reactions are endothermic and even a few percents dissociation can lower the flame temperature substantially. Notice that actual adiabatic flame temperature of a stoichiometric mixture of hydrogen and oxygen according to Table 2.3 is 3080 K, it is not 4900 K as it was calculated in the previous section. This means that actual calculation of the flame temperature would require more than mass balance but also the equilibrium relations among the combustion products composition.

In order to make more accurate calculations we need to take all these species into account. Thermodynamics also allows to calculate which species are present, and in what abundance. At equilibrium the rate of the forward reaction equals the rate of the backward reaction $H_2 \leftrightarrow 2H$, $O_2 \leftrightarrow 2O$, $H_2 + O_2 \leftrightarrow 2OH$, etc.

To describe the process at chemical equilibrium we consider the Gibbs free energy, and assume that the reaction occurs at constant temperature and pressure. First of all, we shall describe the method for expressing the reaction. If all the terms are taken to one side, than written as symbolic equation, we have

$$\sum_i n_i A_i = 0, \tag{2.6.1}$$

where A_i are the chemical symbols of the reacting substances, and the coefficients n_i are positive or negative integers. For example, the reaction $2H_2 + O_2 = 2H_2O$, we rewrite as $2H_2 + O_2 - 2H_2O = 0$, and the coefficients are $n_{H_2} = 2$, $n_{O_2} = 1$, $n_{H_2O} = -2$.

Let $N_1, N_2, \ldots.$ be the number of particles (atoms, molecules, etc.) of the various species taking part in the reaction. In the process at constant temperature and pressure the Gibbs free energy tends to minimum. Then the necessary condition for G to be a minimum can be written as the vanishing of the total derivative of G for given P and T with respect to one of the N_i for example, N_1:

$$\frac{\partial G}{\partial N_1} + \frac{\partial G}{\partial N_2}\frac{\partial N_2}{\partial N_1} + \frac{\partial G}{\partial N_3}\frac{\partial N_3}{\partial N_1} + \ldots = 0. \tag{2.6.2}$$

The changes in the numbers N_i during the reaction are related by the reaction Eq. (2.6.1). It is therefore clear that if N_1 changes by n_1, each of the other N_i will change by n_i, so that $dN_i = (n_i/n_1)dN_1$, and $dN_i/dN_1 = n_i/n_1$. Therefore, Eq. (2.6.2) can be written as

$$\sum_i \frac{\partial G}{\partial N_i}\frac{n_i}{n_1} = 0. \tag{2.6.3}$$

Since $\dfrac{\partial G}{\partial N_i} = \mu_i$ according to definition of the chemical potential, and multiplying (2.6.3) by n_1, we obtain

$$\sum_i n_i \cdot \mu_i = 0. \tag{2.6.4}$$

This is the required general condition for chemical equilibrium.

Let us apply this general condition of chemical equilibrium to reactions in a gas mixture, assuming that the gas may be regarded as an ideal one. In Chap. 1 we obtained for the Gibbs free energy, the chemical potential and entropy of ideal gas:

$$G = NRT \ln P + N\chi(T) = H - TS,$$
$$\mu = \partial G/\partial N,$$
$$S = -N \ln P + Nc_P \ln T + (\zeta + c_P)N = N\big(S_0(T) - \ln(P/P_{ref})\big).$$

The chemical potential for each gas in the mixture is

$$\mu_i = RT \ln P_i + \chi_i(T), \tag{2.6.5}$$

where P_i is the partial pressure of the ith gas component in the mixture; $P_i = c_i P$, and P is the total pressure of the mixture; $c_i = N_i/N$ is the concentration of the ith gas component in question; $N = \sum_i N_i$ is the total number of molecules in the mixture. Substituting expression (2.6.5) in (2.6.4) and we obtain

$$\sum_i n_i \mu_i = RT \sum_i n_i \ln P_{0i} + \sum_i n_i \chi_i = 0, \tag{2.6.6}$$

where P_{0i} are the partial pressures of the gases in the state of chemical equilibrium.

We can rewrite this last condition as

$$\prod_i P_{0i}^{n_i} = K_p(T), \tag{2.6.7}$$

where

$$K_p(T) = \exp\left[-\sum (n_i \chi_i)/RT\right] \tag{2.6.8}$$

is called *the chemical equilibrium constant.*

Substituting in Eq. (2.6.7) $P_{0i} = c_{0i} P$, where c_{0i} are the concentrations of the gases in chemical equilibrium, we obtain

$$\prod_i c_{0i}^{n_i} = P^{-\sum_i n_i} K_p(T). \tag{2.6.9}$$

Notice that the dependence of the gas reaction equilibrium constant on the pressure is entirely determined by power of the total pressure – this is the factor $P^{-\sum_i n_i}$.

The dependence of the gas reaction equilibrium constant on temperature cannot be obtained as universal law, though it is not differ considerably from the dependence of gases having constant specific heat. In the later case we obtain for the equilibrium constant

$$K_p(T) = \exp\left(\sum (n_i \zeta_i)\right) \cdot T^{\sum (c_{pi} n_i)} \exp\left(-\sum (n_i e_{0i})/T\right), \tag{2.6.10}$$

which is essentially an exponential function of temperature (last exponential term in the right side).

2.7 Chemical Equilibrium and Adiabatic Flame Temperature

Expression (2.6.7) can be presented in more convenient for user form. Since $G = H - TS$, we can write it as

$$G = \sum_{i=1}^{M} n_i H_{T,i} - T\sum_{i=1}^{M} n_i S_i = \sum_{i=1}^{M} n_i \left[H_{T,i} - TS_i^0 + RT\ln P_i\right]$$
$$= \sum_{i=1}^{M} n_i \left[G_i^0 + RT\ln P_i\right], \tag{2.7.1}$$

where we defined the standard state free energy (G at one bar, but at any T of interest)

$$G_i^0(T) \equiv H_{T,i}^0 - TS_i^0.$$

Recall that for ideal gases $H_{T,i} = H_{T,i}^0$, and the entropies for ideal gases are

$$S_i = S_i^0(T) - R\ln\left(\frac{P_i}{P_{ref}}\right) \tag{2.7.2}$$

with the standard state entropy given by

$$S_i^0 = \int_0^T (C_{p,i}^0(T)/T)dT, \tag{2.7.3}$$

where we use expression for the specific heat at constant pressure, $C_P = T(\partial S/\partial T)_P$. The integral in (2.7.3) is given in the JANNAF Tables as well as $H_{T,i}^0$. We take P_{ref} as 1 bar and always work in bars for it is used in the JANNAF Tables. Then Eq. (2.6.7) can be presented in the form

$$\prod_i P_{0i}^{n_i} = K_p(T) = \exp\left[-\frac{\Delta G^0}{RT}\right]. \tag{2.7.4}$$

The change of G^0 for the formation of the reaction product from the reagents is called the *standard state free energy of formation* and is also given in the JANNAF Tables for any compound or atom. One must remember that it is zero for a reference element. For example, for the equilibrium reaction

$$n_A A + n_B B \leftrightarrow n_c C + n_D D$$

the equilibrium is given by

$$K_P(T) = \frac{P_C^{n_C} \cdot P_D^{n_D}}{P_A^{n_A} \cdot P_B^{n_B}} = \exp\left[-\frac{\Delta G^0}{RT}\right]. \tag{2.7.5}$$

In terms of the concentrations of the gases, similar to (2.6.9) we have

$$K_P(T) = \frac{X_C^{n_C} \cdot X_D^{n_D}}{X_A^{n_A} \cdot X_B{}^{n_B}} \left(\frac{P}{P_{ref}}\right)^{n_C+n_D-n_A-n_B}, \tag{2.7.6}$$

where

$$X_A = \frac{n_A}{n_A + n_B + n_C + n_D}, \quad X_B = \frac{n_B}{n_A + n_B + n_C + n_D}, \ldots$$

Notice that the sign of $n_C + n_D - n_A - n_B$ defines how equilibrium composition depends on pressure. With increase of the pressure the total number of molecular compounds in the right hand sizes decreases. With increase of the temperature molecules dissociate more, which means that molecules with positive heat of formation are favored as the temperature rises.

In general the combustion products consist of more than just CO_2, H_2O, and NO_2. For rich mixtures CO also exists in the products and at high temperatures the molecules may dissociate to form H, O, OH, NO via the following reactions

$$H_2 \rightarrow 2H, \quad O_2 \rightarrow 2O, \quad H_2 + O_2 \rightarrow 2OH, \quad O_2 + n_2 \rightarrow 2NO.$$

The opposite direction reactions of recombination are also possible

$$2H \rightarrow H_2, \quad 2O \rightarrow O_2, \quad 2OH \rightarrow H_2 + O_2, \quad 2NO \rightarrow O_2 + N_2.$$

At equilibrium the rate of the forward reaction equals the rate of the backward reaction

$$H_2 \leftrightarrow 2H, \quad O_2 \leftrightarrow 2O, \quad H_2 + O_2 \leftrightarrow 2OH, \quad O_2 + n_2 \leftrightarrow 2NO.$$

At equilibrium the relative proportion of the species mole fraction is fixed and the equilibrium composition for species is defined by the equilibrium constant (2.7.6).

As an example consider distribution of the products in the reaction 1 mol of CO_2 0.5 mol of O_2 and 0.5 mol of N_2 react to form a mixture consisting of CO_2, CO, O_2, N_2 and NO. We want to determine the equilibrium composition of the product mixture assuming that the equilibrium composition is at 1 bar and the final temperature is 3000 K.

$$CO_2 + 0.5O_2 + 0.5N_2 \rightarrow aCO + bNO + cCO_2 + dO_2 + eN_2. \tag{2.7.7}$$

The law of stoichiometry is insufficient to determine coefficients in (2.7.7). From an atomic balance we have 3 equations for 5 unknowns a, b, c, d, e:

(C) $1 = a + c$,
(O) $3 = a + b + 2c + 2d$
(N) $1 = b + 2e$, so that 2 unknowns, say a and b need 2 equilibrium equations to solve the problem. These equations are for equilibrium reactions

$$CO_2 \leftrightarrow CO + 0.5O_2 \;\rightarrow\; K_{P1}(3000K) = 0.3273, \tag{2.7.8}$$

$$0.5O_2 + 0.5n_2 \leftrightarrow NO \;\rightarrow\; K_{P2}(3000K) = 0.1222. \tag{2.7.9}$$

where K_{P1} and K_{P2} are the equilibrium constants for standard reactions (2.7.8–9), which are tabulated in the JANNAF as a function of temperature for different equilibrium reactions.

From the expression for equilibrium constants

$$K_{P1} = \frac{X_{CO} \cdot X_{O_2}^{1/2}}{X_{CO_2}} = 0.3273, \text{ and } K_{P2} = 0.1222 = \frac{X_{NO}}{X_{O_2}^{1/2} \cdot X_{N_2}^{1/2}}$$

The total number of components with the use of equations (C), (O), (N) is $N = a + b + c + d + e = a + b + (1 - a) + 0.5(1 + a - b) + 0.5(1 - b) = 0.5(a + 4)$, so that

$$X_{CO} = \frac{2a}{(4 + a)}, \; X_{O_2} = \frac{(1 + a - b)}{(4 + a)}, \; X_{CO_2} = 2\frac{1 - a}{(4 + a)},$$

and we obtain two equations for a and b:

$$K_{P1} - \frac{X_{CO} \cdot X_{O_2}^{1/2}}{X_{CO_2}} - 0.3273 = \frac{a}{1 - a}\left(\frac{1 + a - b}{4 + a}\right)^{1/2}, \tag{2.7.10}$$

$$K_{P2} = \frac{X_{NO}}{X_{O_2}^{1/2} \cdot X_{N_2}^{1/2}} = 0.1222 = \frac{2b}{[(1 + a - b)(1 - b)]^{1/2}}. \tag{2.7.11}$$

Solving last two equations together with the atom balance equations we obtain $a = 0.3745$, $b = 0.0675$, $c = 0.625$, $d = 0.6535$, $e = 0.4663$. If the products are at high temperature (2000 K or larger) minor species will be present due to the dissociation of the major species CO_2, H_2O, etc.

Let us calculate as an example distribution of the products in the reaction hydrogen-fluorine, assuming that the final pressure is 1 bar and final temperature 3500 K, so that final products include also hydrogen and fluorine atoms

$$H_2 + F_2 \rightarrow n_{HF}HF + n_H H + n_F F. \tag{2.7.12}$$

The law of stoichiometry is insufficient to determine the n_i in (2.7.7). From an atomic balance we have 2 equations

$$n_{HF} + n_H = 2, \; n_{HF} + n_F = 2.$$

We have 3 unknowns n_{HF}, n_H, n_F so need one equilibrium equation to solve the problem.

The equilibrium equation is for the reaction $H + F \leftrightarrow HF$

$$K_p(T) = \left(\frac{P_{HF}}{P_H P_F}\right) = \frac{n_{HF}P}{n}\frac{n}{n_H P}\frac{n}{n_F P} = \frac{n}{P}\frac{n_{HF}}{n_H n_F} \tag{2.7.13}$$

is not listed in the JANNAF, but we can combine the right hand part in the combination of reactions

$$K_p(T) = \left(\frac{P_{HF}}{P_H P_F}\right) = \left(\frac{P_{HF}}{P_{H_2}^{1/2}P_{F_2}^{1/2}}\right)\left(\frac{P_{H_2}^{1/2}}{P_H}\right)\left(\frac{P_{F_2}^{1/2}}{P_F}\right) = \frac{K_{pf,HF}}{K_{pf,H}K_{pf,F}}, \tag{2.7.14}$$

which corresponds to the standard reactions tabulated in the JANNAF

$$\frac{1}{2}H_2 + \frac{1}{2}F_2 \rightarrow HF \; \frac{1}{2}H_2 \rightarrow H, \; \frac{1}{2}F_2 \rightarrow F.$$

For these reactions the JANNAF gives

$$K_{pf,HF} = \left(\frac{P_{HF}}{P_{H_2}^{1/2}P_{F_2}^{1/2}}\right) = 10^{4.221}, \qquad K_{pf,H} = \left(\frac{P_{H_2}^{1/2}}{P_H}\right) = 10^{-0.228},$$

$$K_{pf,F} = \left(\frac{P_{F_2}^{1/2}}{P_F}\right) = 10^{2.121}.$$

Substituting these numbers in (2.7.8) together with

$$n = n_{HF} + n_H + n_F = 2 + n_H, n_F = n_H, n_{HF} = 2 - n_H$$

we come to the equation for n_H:

$$\frac{(2 + n_H)(2 - n_H)}{n_H^2} = 10^{2.238} = 213.$$

From here,

$$n_H = n_F = 0.0187,\ n_{HF} = 1.9813,\ X_H = X_F = 0.0926,\ X_{HF} = 0.9815.$$

It is imperative to realize that all conceivable reactions in the system must be in equilibrium. For this reason, as in the examples above, we have complete liberty in choosing the reactions to consider in making equilibrium calculations.

We now return to calculation of adiabatic flame temperature through use of Eq. (2.3.2), with $Q_p = 0$. Now we know how to calculate the n_j, which in general depend not only upon the pressure of interest, but also the adiabatic flame temperature itself. That is, $n_j = n_j(P, T_2)$ so that at a fixed pressure Eq. (2.3.2) is the equation for T_2. The procedure for adiabatic flame temperature calculations is as following: write down the initial reaction equation with the products that we assume to consider; reduce the number of unknown n_j using condition of an atomic balance; in general there are M-N unknowns for the stoichiometric coefficients on the product side for M values of n_j and N types in the reactants; the equations for these M-N unknowns steam from the equilibrium laws (the formation reactions) for the product species, which are not reference state elements. The equilibrium equations may always be reduced to M-N equations containing the unknowns. The adiabatic flame temperature may be found step-by-step, with the known n_j and guessed T_2, and then repeating all steps of the procedure. Note that dissociation in the products will result in a lower adiabatic flame temperature since dissociation reactions are endothermic. We must check also that partial pressure of excluded products is negligible compared to included products. It is very tiresome calculations to be made by hand. Hand calculations are not practical when many species are considered, one uses a computer program to calculate the product equilibrium composition. There are standard computer programs to calculate the product equilibrium composition. A popular program used for equilibrium calculations is STANJAN or Equilibrium Combustion Solver Applet.

Problems

2.1. In a vessel of volume 1 m^3 there are 2 g of molecular H_2, 6 g of molecular O_2 and 30 g of N_2. What are the mole fractions, mass fraction and concentrations of the hydrogen, oxygen and nitrogen?

2.2. Evaporation of one atom of carbon requires 171 kcal/mol, dissociation of oxygen molecule is 118 kcal/mol. What is the net energy in reaction of solid carbon oxidation $C(s) + \frac{1}{2}O_2 = CO$, if bond energy of CO is 256 kcal/mol? Result: 256-59-171 = 26 kcal.

2.3. Calculate adiabatic flame temperature for the reaction $H_2 + O_2 = n_{H_2O}H_2O + n_{O_2}O_2 + n_OO$. With reagents initially at 298 K and P = 1 bar. Result: $T_f = 3200$ K.

2.4. A steady flow 1 mol/s of CH_4 and 2 mol/s of O_2 enters a reactor and become H_2O and CO_2. Write equation of the reaction. What is the concentration of CO_2 and H_2O vapor density at the exit if the density of gas exiting the reactor is 1 kg/m^3?

2.5. Calculate (1) partial pressures, (2) mole fractions, (3) concentrations, (4) densities of the gaseous components in a vessel, which contains 0.2 kg of H_2, 3 mol of CO, amount of H_2O vapor enough to give partial pressure of 1 bar (10^5 Pa). Pressure and temperature of the perfect gaseous mixture in the vessel: $P = 3$ bar, $T = 500$ K.

2.6. Estimate heat of formation of methane, CH_4 from known heats of formation of CO_2, $\Delta H^0_{f,CO_2} = -94.0$ kcal/mol, and water vapor $\Delta H^0_{f,H_2O} = -57.8$ kcal/mol from the reaction $CH_4 + 2O_2 = CO_2 + 2H_2O$.

2.7. Estimate heat of formation of propane, C_3H_8 from known bond energy of H_2, 104.2 kcal/mol, C – C, 85.0 Kcal/mol, C – H, 98.1 kcal/mol evaporation energy of carbon from crystal to gaseous phase is 171.3 kcal/mol.

2.8. CO is oxidized to CO_2 in an excess of air in afterburner entering at $T_0 = 298$ K. Assuming that air consists of 20% O_2 and 80% N_2 and there is no dissociation find the required air/fuel ratio if the final temperature is 1000 K.

Chapter 3
Combustion Chemistry

3.1 Chemical Reactions

All of the chemical reactions shown as examples in previous chapter almost never take place as written. The true path of products creation is usually consists of many steps at the molecular or atomic level. In a chemical transformation the molecular bonds holding atoms in one kind of molecule are broken and new bonds are formed with other molecules, which appear during the reaction. The number of atoms does not change in a chemical reaction; they are only regrouped in accordance with a specific release of energy. The conservation of the elementary composition can be written down in the form of a balance equation for the number of atoms of each element

$$\sum_{i} n_{ik} N_i = A_k, \tag{3.1.1}$$

where n_{ik} is the number of atoms of the k-th element in the i-th molecule, N_i is the number of molecules of type i in the system, and A_k is the total number of atoms of the given element in the mixture. The number of Eq. (3.1.1) corresponds to the number of elements present in the system. The summation is carried out over all types of molecules. Equation (3.1.1) are satisfied for any state of the reacting mixture: before the onset of the reaction when the excess energy is stored in chemical bonds, after the reaction is finished and complete chemical equilibrium has been achieved, and at any intermediate time during the reaction. The numbers A_k are the same for all states of the reacting mixture.

The rate of the reaction may be expressed in terms of the concentration of any reacting substances or any products of the reaction as the rate of decrease of the concentration of the reactants or as the rate of increase of a product of reaction. If only one reaction occurs in a system, or reaction of arbitrary complexity may be described as a one-step chemical reaction, then Eq. (3.1.1) for the initial and final states can be written in the form of the stoichiometric Eq. (2.6.1) with the stoichiometric coefficients of the reaction:

$$\sum_{i=1}^{M} n_i A_i = \sum_{i=1}^{M'} n_j A_j,$$

M.A. Liberman, *Introduction to Physics and Chemistry of Combustion*,
DOI: 10.1007/978-3-540-78759-4_3, © Springer-Verlag Berlin Heidelberg 2008

where A_i and A_j: are the chemical symbols of the initial reagent i and of the reaction product j and n_i and n_j are the stoichiometric coefficients, which are interrelated by k conditions of the form

$$\sum_{i=1}^{M} n_i n_{ik} = \sum_{i=1}^{M'} n_j n_{jk}.$$

If the initial substances are taken in a proportion such that the chemical transformation can, in principle, convert them fully into the reaction products, then the mixture is referred to as *stoichiometric*. If one of the substances appears in the initial mixture in an amount smaller than that required according to the stoichiometric equation, then this substance is *deficient*.

From the standpoint of thermodynamics, it does not matter what is the true reaction mechanism, because only the initial and final states are of interest. But often we need the true reaction path to study flame phenomena. We also need information on the rate in proceeding from an initial state to a final state. Such rates of chemical reactions are provided by chemical kinetics. The determination of actual reaction paths involves very complex inference from usually incomplete data and theoretical argument. From an engineering viewpoint, complete knowledge of a reaction scheme is often not necessary, and useful approximations can be made. We mostly shall use approximate approach of one-step Arrhenius law for reaction, which yields valuable insight into flame behavior.

The transition from the initial to the final state is characterized by the chemical reaction rates for the various components of the reacting medium. The reaction rate for one of the components is defined as follows:

$$W_i = -\frac{1}{V}\frac{dN_i}{dt}, \quad W'_J = \frac{1}{V}\frac{dN'_j}{dt}, \tag{3.1.2}$$

where N_i and N'_j are the number of molecules (atoms) of the i-th reactant and the j-th product in volume V at time t.

The relationship between the reaction rates for different substances is determined by the conservation of the elementary composition (Eq. (3.1.1)), which implies that

$$\frac{W_i}{n_i} = \frac{W'_j}{n_j} = W. \tag{3.1.3}$$

If the volume is constant in the course of a reaction, then the rates are defined as the time derivatives of the volume concentrations, $c_i = N_i/V$ and $c'_i = N'_j/V$:

$$W_i = -\frac{dc_i}{dt}, \quad W'_j = \frac{1}{V}\frac{dc'_j}{dt}. \tag{3.1.4}$$

The concentration of molecules in reacting gaseous mixtures, however, may also change because of thermal expansion of the gas or gas dynamic processes. In these cases the reaction rate must be calculated according to Eq. (3.1.2).

Atoms or molecules can react only when they collide, and the rate of atomic or molecular collisions is proportional to the number densities (number per unit volume) of the colliding species. If a reaction takes place as a result of the collision of two molecules (bimolecular reaction) of substances A_1 and A_2, then the reaction rate is

$$-\frac{dc_1}{dt} = -\frac{dc_2}{dt} = kc_1c_2.$$

For example, in the reaction

$$O + H \rightarrow OH$$

the rate of reaction for the reactants must be proportional to c_Hc_O. We read the reaction above as "one atom of oxygen collides with one atom of hydrogen to form one molecule of OH." We define the reaction rate, W, through a coefficient, k, called the specific reaction rate constant, by

$$W = kc_Hc_O.$$

For a general reaction, we multiply k by the product of the concentrations of the reacting species. For example, for

$$H + H + H \rightarrow H_2 + H \tag{3.1.5}$$

$W = kc_H^3$. The reaction rate tells how often a reaction occurs and is used to tell the time rate of concentration change of the reacting species. For example, in the first reaction above, the units of k are given by the following equality:

$$\frac{dc_O}{dt} = -(1)kc_Oc_H = -W$$

with the minus sign being chosen because O is being destroyed. In the second example, because H is both being destroyed by the reaction and created on the right side, the concentration change of H is

$$\frac{dc_H}{dt} = -3W + W = -2W = -2kc_H^3. \tag{3.1.5a}$$

The above two cases are examples of two-body and three-body reactions respectively indicating a collision between either two atoms or three atoms must take place before a reaction can occur. Notice, in order for the units to be correct, the units of k depend upon the number of bodies colliding. Two- and three-body reactions are the most common ones encountered. The probability that four or more bodies could get together is too low, especially in ideal gas systems and they never been observed in experiments.

In general the law of mass action states that the rate of disappearance of a chemical species is proportional to the concentration of the reacting chemical species, each concentration raised to a power equal to the corresponding stoichiometric coefficient, i.e.

$$W_i = k\prod_{j=1}^{M}(n_j)^{v_j}, \tag{3.1.6}$$

where the factor k, known as the specific reaction rate constant, is a function of the gas temperature. The stoichiometric coefficient v_j is called the order of the reaction with respect to species j. In practice we use often the overall order of the reaction, which is defined as $\sum_j v_j$. Thus, the rate of change of the concentration of a given species i, in an actual reacting system is given by the equation

$$\frac{dN_i}{dt} = (v_i' - v_i)k\prod_{j=1}^{M}(N_j)^{v}, \tag{3.1.7}$$

which means that v' moles of N_i are formed for every v moles of N_i consumed. For the above example (3.1.5), this will be Eq. (3.1.5a), or $d(H)/dt = -2k(H)^3$. For a single-step reaction $\sum_j v_j$ gives the overall order of the reaction.

The sum of the exponents on the concentrations in the reaction rate is called the molecularity of the reaction, and it indicates the number of particles entering the reaction. In general the molecularity of majority reactions is 2 or 3, however, for complex reactions the concept of molecularity has no sense, and the overall order of the reaction can even be fractional one. For two-body reactions the molecularity is two. Since $c_i = P_i/RT = X_iP/RT$ so that, for fixed composition and temperature, c_i is proportional to pressure. Whenever we discuss the pressure dependence of reaction rates, it is assumed that T and the X_i are fixed. Consequently, two-body reactions have a rate proportional to P^2, and three-body reactions have a rate proportional to P^3. As pressure increases, therefore, three-body reactions increase in rate faster than do the rates of two-body reactions. In general, those reactions involving the combination of active particles – radicals, ions, excited atoms and molecules – which do not require that an activation barrier be overcome and which release substantial amounts of energy cannot take place in bimolecular reactions and require the participation of a third particle. When two active particles collide the complex that is formed contains a large amount of excess energy, which causes it to break down rapidly. In order for the complex to survive as a stable chemical compound, the excess energy must be removed. Radiation does not make a significant contribution here because the probability of a photon emission is rather small (the small parameters which appear in the emission probability are $e^2/\hbar c = 1/137$, and the ratio of the size of the molecule to the wavelength of the emitted light). One possibility remains: the energy is removed by a third particle, which may have a low chemical activity. This kind of process occurs, for example, in hydrogen-oxygen flames in which active radicals (hydrogen atoms) recombine. Reactions involving stable chemical compounds may also occur in three-molecular collisions; however, since the three-body collision rate is substantially below the binary collision rate, even at rather high pressures, the main change will take place in bimolecular reactions.

The temperature dependence of k follows from the fact that very few of the collisions between reactive molecules A_1 and A_2 will lead to the formation of a new molecule, that is, to a chemical reaction. In order for a reaction to occur, the colliding molecules must have a sufficient kinetic energy in order to overcome a potential barrier and to destroy or change the stable chemical bonds and electronic structure of the reactants. This potential barrier is characterized by the activation energy E and the temperature dependence of k owing to the existence of the activation energy is described by the Arrhenius equation

$$k = k_0 \exp(-E/RT). \tag{3.1.8}$$

in which the constant k_0 is called the *pre-exponential factor*, $R = 1.986\,\mathrm{cal/mol} \times \mathrm{deg}$ is the universal gas constant, and T is the temperature in Kelvin. The activation energy, E, may be thought of as the minimum energy required in a collision for the reaction to occur, or as the size of the potential energy barrier that molecules have to overcome to be able to react. The pre-exponential factor may have weak temperature dependence. The law expressed by Eq. (3.1.8) is called the *Arrhenius law*.

The extremely strong temperature dependence of the reaction rate (3.1.8) is a fundamental characteristic of chemical processes under nonisothermal conditions, such as explosions and combustion. In gases the Arrhenius law is a reflection of the fact that, according to a Maxwell-Boltzmann energy distribution, only a very small fraction of the reactive molecules that belong to the high energy "tail" of the distribution can react. In the "tail", for example, molecules with energy of the order E can react. On the other hand, those molecules, which are moving at the same speed in the same direction, will not participate in a chemical reaction even though they may have a large kinetic energy. Thus, even for highly energetic molecules, a chemical reaction is a rare process.

Most chemical reactions are of second order, taking place so that their rates are dominated by collisions of two species. However, there are reactions that are dominated by a decomposition processes, e.g. by loose bond braking step. These reactions as well as isomerization reactions are the first order reactions. The first order reactions occur as a result of a two-step process and therefore their rate is also obey to the Arrhenius law.

For arbitrary second order reaction written as

$$A + B \rightarrow C + D, \text{ for example, } H + O_2 \rightarrow OH + O$$

the rate of reaction for the reactants is proportional to $c_A c_B (c_H c_{O_2})$.

$$W = \frac{dA}{dt} = -kc_A c_B = -\frac{dC}{dt}.$$

The pre-exponential factor k_0 in the Eq. (3.1.8), which is the rate of collisions, may have weak temperature dependence, which can be estimated from simple kinetic theory as

$$[A] \cdot [B] \sigma_{AB} (8\pi k_B T/\mu)^{1/2}, \tag{3.1.9}$$

where σ_{AB} is the cross section for the collisions, $\mu = m_A m_B/(m_A + m_B)$ is the reduced mass, and k_B is the Boltzmann constant. The last term in (3.1.9) is the thermal velocity of the molecules. The pre-exponential factor has a very mild temperature dependency that is generally ignored compared to the exponential dependence.

Most elementary binary reactions exhibit Arrhenius behavior over modest ranges of temperature. However, for the large temperature ranges "non-Arrhenius" behavior is more likely, especially for the processes with a small energy barrier. Finally, it should be noticed that the units for the reaction rate constant k_0 when the reaction order is n will be $\{[(\text{concentration})^{n-1}][\text{time}]\}^{-1}$. For example, for the first order reaction the units of the reaction rate constant are $[\text{sec}^{-1}]$, and for a second order reaction $[\text{cm}^3][\text{moles}]^{-1}[\text{sec}^{-1}]$.

One other type of reaction occurs with highly reactive species in enclosures. The walls of the vessel may absorb some species. In order for this to occur, the species must have diffused to the proximity of the wall and then collided with it. The time to get near a wall is proportional to ρ, which, in turn, is proportional to P. Alternatively, the rate at which molecules appear near the wall is proportional to 1/P. These arguments support the contention that wall reaction rates are dependent upon pressure to no larger than the first power. The consequence is that three reaction types (wall, two-body, and three-body) depend upon pressure in different ways.

The reactions written above indicated only forward reaction – left to right reaction. The reverse reaction can also occur. We denote the rate constants as k_f and k_b, for the forward and reverse reactions, respectively. For example,

$$H + H + H \leftrightarrow^{k_f}_{k_b} H_2 + H$$

with

$$\frac{dc_H}{dt} = -2k_f c_H^3 + 2k_b c_{H_2} c_H \tag{3.1.10}$$

Note the special case of equilibrium where $dc_H/dt = 0$, when from (3.1.10) steams

$$\frac{c_{H_2}}{c_H^2} = \frac{k_f(T)}{k_b(T)} = RT\frac{P_{H_2}}{P_H^2} = RTK_p(T) = \frac{RT}{K_{pf,H}^2}.$$

That is, since the functions of temperature are related through the equilibrium constant and functions cannot change even when out of equilibrium only one of the reaction rate constants is independent. If either the forward or the reverse reaction rate constant is known the other is determined through thermodynamic relation.

Practically all reactions are initiated by bimolecular collisions. However, many of the bimolecular reactions exhibit first-order kinetics. As it was mentioned earlier, this is a result of two-step reaction sequence in which the reacting molecule is activated by collision processes and then the activated species decomposes to products. Another situation when a pseudo-first-order reaction arises is one when one of the reactants (usually the oxidizer in a combustion system – very fuel lean combustion with O_2being in large excess) is in large excess. In this case the concentration of another reagent does not change appreciably, and with this accuracy the reaction appears as overall first order.

3.2 Non-branching Chain Reaction: The Hydrogen Chlorine

Most chemical reactions are complex and include series of elementary steps. A solution requires the integration of an entire system of kinetic equations, which consists of the kinetic equations for each type of molecule or atom that participates in the reaction. Each of these equations must take into account the loss and production of the given type of particle in all the elementary steps of the chemical reaction. Analytic solutions of the system of kinetic equations for complex reactions can only be obtained in the simplest cases. The situation is still more complicated for non-isothermal processes when the system of kinetic equations must be solved along with the energy balance equation. The widely used approximation technique is the method of stationary concentrations. The idea of this method lies in assuming that the concentration of the active intermediate substances (ions, radicals) are constant; that is, their time derivatives in the kinetic equations can be set equal to zero and the corresponding differential equations are replaced by algebraic ones, so that the overall order of the system is reduced. The method of stationary concentrations is based on the high reactivity of the intermediate active substances. They react so rapidly with other substances that over a time in which the concentrations of the principal substances (reagents and reaction products) change by a small amount, the difference in the rates of formation and consumption of the active intermediate particles remains small compared with these rates.

We shall consider the most important class of complex reactions – chain reactions. In chain reactions active particles react with reagents or intermediate substances to form both combustion products and new active particles; that is, active particles are "regenerated" during the reactions. This occurs, for example in the reaction of H_2 with bromine Br_2. Since the Br_2 bond energy is 46 kcal/mol and H_2 bond energy is 104 kcal/mol, the reaction is initiated by *chain initiating step* which is dissociation of the Br_2 with formation of two active radicals of Br

$$M + Br_2 \rightarrow 2Br + M$$

The propagation steps will be reactions

$$Br + H_2 \rightarrow HBr + H,$$

$$H + Br_2 \rightarrow HBr + Br,$$

$$H + HBr \rightarrow H_2 + Br.$$

As an example of a chain reaction we shall consider the reaction of hydrogen with chlorine. In principle, the reaction of hydrogen with chlorine can proceed via simple binary collisions

$$H_2 + Cl_2 = 2HCl. \tag{3.2.1}$$

However a chain process is much faster so that reaction (3.2.1) is of no importance whatever. The end product, hydrogen chloride, is formed as a result of a sequential alternation of reactions involving atomic chlorine and atomic hydrogen as active centers. The first active centers are chlorine atoms formed by dissociation of molecular chlorine by collisions of the type

$$Cl_2 + M = Cl + Cl + M(k_1), \tag{3.2.2}$$

where k_1 is the rate constant and M is some third particle. The formation of atomic hydrogen in the reaction $H_2 + M = H + H + M$ analogous to (3.2.2) can be neglected in view of much greater stability of the H_2 molecular bond (104 kcal/mol compared to 57 kcal/mol for Cl_2). In the theory of chain reactions, processes of the type (3.2.2), in which active centers are formed, are called *chain initiation reactions*. The third particle M, which participates in the initiation reaction, is needed to transfer energy to the Cl_2 molecule to break the Cl – Cl bond.

Reaction (3.2.2) is reversible; that is, two chlorine atoms can recombine into a molecule in a three-body collision:

$$Cl + Cl + M = Cl_2 + M, (k_1') \tag{3.2.3}$$

A reaction of this type, accompanied by the loss of free valence, is called a *chain termination reaction*. In the above example of $H_2 - Br_2$ reaction, the corresponding chain termination reaction is $2Br + M = Br_2 + M$. In the presence of molecular hydrogen, however, the chlorine atoms may react in another way:

$$Cl + H_2 = HCl + H, (k_2) \tag{3.2.4}$$

Although the concentration of Cl atoms is very small and the number of $Cl + H_2$ collisions is much smaller than the number of collisions between chlorine molecules and hydrogen molecules, reaction (3.2.4) takes place much more rapidly than the simple bi-molecular reaction $H_2 + Cl_2$ because of its low activation energy ($E \simeq 5\,kcal/mol$ compared to roughly 40 kcal/mol for the reaction $H_2 + Cl_2$). However, even this fact is insufficient to explain the rapidity of the entire process. Although the chlorine atoms react rapidly, they are formed very slowly. Reaction (3.2.4) shows that formation of the end product, HCl, is accompanied by the appearance of a new active center, a hydrogen atom, which reacts in the process

$$H + Cl_2 = HCl + Cl, (k_3) \tag{3.2.5}$$

even more rapidly than a chlorine atom reacts with hydrogen. In this way any chlorine atom that participates in reaction (3.2.4), which is inevitably followed by reaction (3.2.5), is recovered after the formation of two HCl molecules. This is shown in the overall balance equation for reactions (3.2.4) and (3.2.5)

$$Cl + H_2 + H + Cl_2 = 2HCl + H + Cl. \tag{3.2.6}$$

The regenerated chlorine atom again participates in reaction (3.2.4) and so on. A small amount of chlorine atoms keeps the reaction going constantly and, since these atoms are not in effect consumed, their number is roughly constant. This mechanism ensures that the rate for the whole process is large. Reactions such as (3.2.4) and (3.2.5), which conserve free valency, are known as *chain propagation reactions*.

We shall now determine the concentration of chlorine atoms in the system and derive a formula for the reaction rate using the method of stationary concentrations setting

$$\frac{d[H]}{dt} = k_2[Cl][H_2] - k_3[H][Cl_2] = 0 \tag{3.2.7}$$

$$\frac{d[Cl]}{dt} = 2k_1[Cl_2][M] - 2k_1'[Cl]^2[M] = 0, \tag{3.2.8}$$

where symbol [...] means concentration of species.

The steady-state concentrations of the active centers (hydrogen and chlorine atoms) from (3.2.7) and (3.2.8) are

$$[H] = \frac{k_2}{k_3}[H_2]\left(\frac{k_1}{k_1'}\right)^{1/2}[Cl_2]^{-1/2} \text{ and } [Cl] = \left(\frac{k_1}{k_1'}\right)^{1/2}[Cl_2]^{1/2}. \quad (3.2.9)$$

The overall rate of the process (see (3.2.4) and (3.2.5)) is

$$W = \frac{d[HCl]}{dt} = k_2[Cl][H_2] + k_3[H][Cl_2],$$

which, after substitution of the steady-state concentrations for [H] and [Cl] from (3.2.9), can be written as

$$W = 2k_2\left(\frac{k_1}{k_1'}\right)^{1/2}[Cl_2]^{1/2}[H_2]. \quad (3.2.10)$$

This equation shows that the reaction is of first order in hydrogen and of order 1/2 in chlorine, so that the overall order of the reaction is 1.5. A fractional power of one of the reagent concentrations in a reaction rate formula is an indication that the reaction is complex, usually a chain. In this case this means that the reaction proceeds via a chlorine atom.

To evaluate magnitude of the rate of hydrogen chloride formation we notice that the ratio k_1/k_1' is the equilibrium constant for the reaction $2Cl \leftrightarrow Cl_2$ (JANNAF tables)

$$K_1 = 2.5 \cdot 10^{25} e^{-56800/RT} \text{ cm}^{-3},$$

and $k_2$is the constant for the bimolecular elementary process

$$k_2 = 10^{-10} e^{-6000/RT} \text{ cm}^3/\text{s}.$$

Substituting these values in Eq. (3.3.10), we obtain

$$W = 10^3 e^{-34400/RT}[Cl_2]^{1/2}[H_2].$$

For a temperature $T = 600$ K and the initial concentrations of chlorine and hydrogen $[H_2]_0 = [Cl_2]_0 = 0.7 \cdot 10^{19}$ cm^{-3} (pressure $P = 1$ bar), this equation gives an initial maximum reaction rate of $W_0 \approx 6.5 \cdot 10^{18}$ cm^{-3}s which is about 10^6 times greater than the rate of the simple bimolecular reaction $H_2 + Cl_2$ which is of the order of 1 s. If a hydrogen-chlorine mixture were kept at a constant temperature and there were no atoms present at the initial moment, then the reaction rate would first rise to a value close to W_0 and then fall slowly, approaching zero, as the reagents were consumed. The initial rise in the reaction rate takes place over the time required for production of a sufficient equilibrium quantity of chlorine atoms.

3.3 Oxidation of Nitrogen in Combustion

Another example of a non-branching chain reaction is the oxidation of nitrogen. The oxidation of nitrogen, accompanied by the formation of nitric oxide NO, acquired special attention because of harmful effects of nitric oxide even at very small concentrations. The formation of nitric oxide is unavoidable when air, which contains 78% nitrogen, is used as an oxidant. Moreover, in order to ensure complete combustion of fuels, the fuel-to-air ratio is adjusted so that a slight excess of oxygen remains in the combustion products. At high concentrations NO oxidizes further in the reaction $2NO + O_2 = 2NO_2$. The presence of NO_2 is easily confirmed by the yellowish color it imparts to smoke. Nitrogen dioxide NO_2 is easily absorbed by water or alkaline solutions. However the oxidation reaction of $2NO + O_2$ is a trimolecular reaction, which proceeds very slow at low concentrations of NO. Therefore, the oxidation of nitrogen and an understanding of the conditions under which this reaction is suppressed are of particular importance.

When air is heated to several thousand degrees the chemical reaction

$$\frac{1}{2}N_2 + \frac{1}{2}O_2 + 21.5\ \mathrm{kcal/mol} = NO \tag{3.3.1}$$

takes place. It has been shown by Zel'dovich, Sadovnikov and Frank-Kamenetsky that nitric oxide is produced in a non-branching chain reaction involving free oxygen and nitrogen atoms as active centers:

$$O + N_2 \leftrightarrow NO + N - 75\,\mathrm{kcal/mol}, (k_1, k_1') \tag{3.3.2}$$

and

$$N + O_2 \leftrightarrow NO + O + 32\,\mathrm{kcal/mol}, (k_2, k_2') \tag{3.3.3}$$

Here the heats of reaction have been calculated from the known dissociation energies of N_2 (226 kcal/mol), NO (151 kcal/mol), and O_2 (118 kcal/mol). The rate of production of NO is limited by the first, endothermic stage (3.3.2). The use of oxygen atom in reaction (3.3.2) is compensated by its regeneration in reaction (3.3.3); the N atom formed in the first reaction immediately reacts with molecular oxygen to yield another active oxygen atom. If it would be assumed that the reaction occurs through collisions of N_2 and O_2 molecules, then the calculated reaction rate was a thousand times smaller than the experimentally observed value. The experimental heat of activation is roughly 135 kcal/mol.

The method of stationary concentrations implies that

$$-\frac{d[N]}{dt} = -k_1[O][N_2] + k_1'[NO][N] + k_2[N][O_2] - k_2'[NO][O] = 0 \tag{3.3.4}$$

so that the concentration of N atoms can be expressed in terms of the steady-state concentrations of oxygen atoms as

$$[N] = \frac{k_1[N_2] + k_2'[NO]}{k_1'[NO] + k_2[O_2]}[O] \qquad (3.3.5)$$

Substituting this expression in the kinetic equation for nitric oxide

$$\frac{d[NO]}{dt} = k_1[O][N_2] + k_2[N][O_2] - k_1'[NO][N] - k_2'[NO][O] \qquad (3.3.6)$$

we obtain

$$\frac{d[NO]}{dt} = 2\frac{k_1k_2[N_2][O_2] - k_1'k_2'[NO]^2}{k_2[O_2] + k_1'[NO]}[O] \approx 2\left(k_1[N_2] - \frac{k_1'k_2'[NO]^2}{k_2[O_2]}\right)[O] \quad (3.3.7)$$

Since the constants k_1' and k_2' which determine the rate of the exothermic reaction of an atom with a molecule, are of the same order of magnitude and since the concentration of nitric oxide is much lower than the concentration of molecular oxygen (the equilibrium concentrations of nitric oxide at temperatures from 3000 K up to 10000 K and $P = 1$ atm do not exceed a few percent), we neglect the term $k_1'[NO]$ in the denominator of Eq. (3.3.7).

Using the principle of detailed balancing for the reactions (3.3.2), (3.3.3), and $O_2 \leftrightarrow 2O$ (k_3, k_3'), which yields

$$K_1 = \frac{k_1}{k_1'} = \frac{(NO)(N)}{(O)(N_2)}, K_2 = \frac{k_2}{k_2'} = \frac{(NO)(O)}{(N)(O_2)}, K_3 = \frac{k_3}{k_3'} = \frac{[O]^2}{[O_2]}$$

(the conventional parentheses denote equilibrium concentrations), we can reduce Eq. (3.3.7) to the form

$$\frac{d[NO]}{dt} = k'[N_2][O_2] - [NO]^2 = k\{(NO)^2 - [NO]^2\}, \qquad (3.3.8)$$

where

$$k = \frac{k'}{K^1K^2} = \frac{2k_1'k_2'K_3^{1/2}}{k_2[O_2]^{1/2}} \frac{53}{[O_2]^{1/2}}\exp\{-92000/RT\},$$

$$k' = 2\frac{k_1K_3^{1/2}}{[O_2]^{1/2}} = \frac{1.1 \cdot 10^3}{[O_2]^{1/2}}\exp\{-135000/RT\}$$

are taken from experiment (k' and k in (cm^3/sec) $[O_2]$ in cm^{-3}).

Equation (3.3.8) for the reaction rate differs from the kinetic equation for the bimolecular reaction of N_2 with O_2 in that the rate constants depend on the concentration of one of the reagents, oxygen. This is a consequence of a chain reaction. The experimental value of the activation energy for the production of nitric oxide is 135 kcal/mol. As equation for k' shows, the temperature dependence of this reaction is determined by the temperature variations of the equilibrium constant K_3, which varies with the temperature as $\exp(-61000/RT)$ (61 kcal/mol is the effective energy of formation of an oxygen atom at temperatures of 2000–5000 K), and of the reaction rate k_1. Thus, the activation energy for the first step (3.3.2) in the oxidation of nitrogen equals $E_1 = 135 - 61 = 74$ kcal/mol, which is the same as the heat of the endothermic reaction (the reverse reaction to (3.3.2) occurs with practically zero activation energy).

3.4 Chain-Branching Reactions: Explosions

Compounds, which are highly stable at normal temperatures, usually emulate the noble gases (helium, argon, etc.) in the configuration of their electrons. That is, these highly inert gases have outer electron shells empirically filled by 2 and 8 electrons in the case of He and Ne, respectively. Compounds, which attempt to mimic this situation are usually quite stable and are hard to break apart. An example is a hydrogen molecule, H_2 which looks like He in the sense that each hydrogen atom shares its electron with the other atom so that there are two electrons in the configuration. Similarly, a water molecule, H_2O, looks like Ne where there are total of 8 electrons being shared, 6 from the outer shell in oxygen and 2 from each of the hydrogen atoms.

Any atom or compound with an unfilled outer electron configuration is highly reactive. Examples are the hydroxyl radical, OH, and the oxygen and hydrogen atoms, O and H. Species such as H, O, OH and HO_2 are called *radicals*. Radicals have unpaired valence electrons, which make them very reactive and short lived. They try to partner with other radicals to form covalent bonds. While an equilibrium calculation at usual temperatures shows that radicals are only present in small amounts, they play a central role in chemical kinetics reaction schemes. In chain branching reactions there is a net production of radicals. Chain branching reactions lead to rapid production of radicals, which causes the overall reaction to proceed extremely fast an explosively. The reaction comes to completion through chain termination reactions, where the radicals recombine to form final products.

Radicals usually have positive heats of formation, which means that energy is required for their formation from the standard state elements. They soak up collision energy in their formation and wish to give up this energy in attacking other species in the path toward reaction completion. Hungry for electrons, radicals are efficient intermediates in reaction schemes for taking the reaction path forward. It should be pointed out that the above reasoning concerning the electron configuration is not always straightforward. For example, a complex electron sharing occurs with CO, and it is not a radical, although it would appear to be one. Radicals are extremely important in the kinetic schemes relevant to atmospheric pollution. Examples of reactions producing nitric oxide (NO) as a harmful air pollutant, which is called the thermal mechanism (or Zel'dovich's mechanism) of NO for formation, are

$$O + N_2 \rightarrow NO + N, \quad N + O_2 \rightarrow NO + O.$$

Some chemical reactions develop in time smoothly while others "explosively" and are usually accompanied by some unexpected phenomena such as a flash, acoustic effects, etc. The important features of explosions are that the rate of the reaction varies extremely rapidly with temperature. Thus, for example, at room temperature and atmospheric pressure hydrogen and oxygen hardly react over a period of many years. However, if the pressure or temperature are raised sufficiently high, or being kept at temperature just slightly higher 450 C and the pressure in the vessel at a few torr, a violent exothermic reaction will be initiated with the product, water being formed. The reaction is very fast, and though it is difficult to define precisely, such a fast, exothermic reaction is called an explosion. As the temperature is raised the reaction rate remains extremely small up to some critical value, which depends on the experimental conditions. For a stoichiometric mixture of hydrogen and oxygen, the so-called "detonating mixture", at atmospheric pressure the critical temperature is about 550 C. At higher temperatures, even those only a few degrees above critical, the detonating mixture reacts very rapidly, the pressure increases suddenly, and the vessel may be broken apart. Unlike an ordinary reaction, an explosive reaction is characterized by the existence of a temperature at which the reaction rate changes very rapidly, almost discontinuously. This temperature is called the ignition (or self-ignition) temperature. An equally rapid change in the reaction rate can be obtained by changing the pressure at a given temperature. At some pressures the reaction either does not proceed or it proceeds very slowly, but sometimes even a very small pressure change is sufficient to cause the reaction to take place over a short time. In observing an explosive reaction the first impression is that for temperatures and pressures below critical nothing is happening in the mixture, but when the critical parameters are reached everything reacts all at once. This is the reason for calling the reaction explosive.

3.5 Hydrogen-Oxygen Reactions: Explosion Limits

Explosions are conventionally classified into two different categories: *chain-branching explosions*, in which the reaction rate increases spontaneously without limit due to chain branching, and *thermal explosions* in which the reaction rate increases exponentially due to exothermic chemical reaction resulting in heating of the reactants and an increase of the reaction rate constant.

To what extend a reaction is accelerated by the branching process can be demonstrated by numerical example. Let a 1 cm^3 container consisting reactant molecules at normal conditions, i.e. the number density is $n = 10^{19}\,cm^{-3}$, and one active radical. The average collision frequency is $n\sigma\bar{v}$, where σ is the cross section of collision, $\bar{v}$ is the average thermal velocity of molecules. For normal conditions, and $n = 10^{19}\,cm^{-3}$, we have $n\sigma\bar{v} \approx 10^8\,s^{-1}$. If the reaction in the volume is chain-carrying, i.e., one active radical generates another active radical, then the time required for all the molecules to react will be $10^{19}/10^8 = 10^{11}s \approx 30.000$ years. If the reaction is chain-branching, i.e. one active radical generates two new chain radicals, then the number of generation needed for reproduction of the active radical up to 10^{19} will be defined by geometric progression

$$\sum_{i=0}^{N} 2^i = \frac{2^{N+1} - 1}{2 - 1} = 10^{19}.$$

Thus, the number of generation is $N = 62$, and the time of the reaction becomes $62 \cdot 10^{-8}\,s \approx 1\,\mu s$.

In the case of hydrogen and oxygen, the explosion can be spontaneous, without requiring any ignition source, such as a spark. Though it might seem reasonable that the reaction scheme

$$H_2 + H_2 + O_2 \rightarrow H_2O + H_2O$$

would explain the formation of water, it never occurs in reality. First of all, it would require a three-body collision that has the three bodies perfectly aligned in the proper orientation to form the two water molecules. Second, it would require that all diatomic bonds be broken. Third, since this is a highly exothermic reaction, where would the energy go? Note that though total heat of the reaction $2H_2 + O_2 = 2H_2O$ is 116 kcal/mol, it require to broke bonds of two hydrogen molecules (206 kcal/mol) and one oxygen molecule (118 kcal/mol), which means that the activation energy for such reaction should be very high and it will never proceed in this way. It is presumed plausible that in a mixture of hydrogen and oxygen the bulk reaction $2H_2 + O_2 = 2H_2O$ contains a series of elementary steps involving a free valence radicals – active centers: the OH radical, the hydrogen and oxygen atoms. The minimum number of individual reactions, which can explain realistic reaction of hydrogen-oxygen depends on temperature and pressure. The global hydrogen-oxygen reaction proceeds via the following elementary reactions, known as a reaction mechanism:

$$\left.\begin{array}{l}H_2 + M \rightarrow H + H + M\end{array}\right\} \text{chain initiation}$$

$$\left.\begin{array}{l}H + O_2 + M \rightarrow HO_2 + M \\ HO_2 + H_2 \rightarrow H_2O + OH \\ OH + H_2 \rightarrow H_2O + H\end{array}\right\} \text{chain propagation}$$

$$\left.\begin{array}{l}H + O_2 \rightarrow OH + O \\ H_2 + O \rightarrow OH + H\end{array}\right\} \text{chain branching}$$

$$\left.\begin{array}{l}H + OH + M \rightarrow H_2O + M \\ H + H + M \rightarrow H_2 + M \\ O + O + M \rightarrow O_2 + M\end{array}\right\} \text{chain termination}$$

Some of these reactions are bimolecular reactions and their rates will scale with pressure squared. Some of these reactions involve three particles collision and will increase its rate in proportion to pressure cubed. Besides, at very low pressure a radical formation will be terminated by wall destruction, and no explosions may occur. The wall reactions scale with pressure to approximately of the first power.

In experimental studies of the reaction of hydrogen with oxygen three explosion limits have been observed. These limits can be conveniently represented in the form of a "pressure-temperature" diagram. The explosion limits of a stoichiometric hydrogen-oxygen mixture are plotted in Fig. 3.1, which shows kinetic curves for the combustion of a stoichiometric hydrogen-oxygen mixture. If the initial pressure and temperature of the mixture correspond to a point lying to the right of the curve ABCD, then ignition occurs. The segment AB corresponds to the first explosion limit, the segment BC to the second, and the segment CD to the third. The kinetic curves for the hydrogen-oxygen in the explosion region have the following appearances: over a certain time interval t_i, known as the induction period (or delay time), there is practically no reaction (the reaction cannot be observed) and the concentration is practically constant; then the reaction takes place rapidly and ends in a very short time. Using the reaction of hydrogen with oxygen as an example, we can note two characteristic features of explosive reactions: the induction period and the existence of critical conditions, that is, system parameters, which separate the explosion region from the non-explosion region in which the reaction hardly occurs.

Let us now consider the oxidation of hydrogen, which is one of the most widely studied chain reactions. The bulk reaction $2H_2 + O_2 = 2H_2O$ contains a series of elementary steps, which are chain initiation, chain propagation, chain branching and chain termination, listed above. The chain initiation reaction is

$$H_2 + M = H + H + M. \tag{3.5.1}$$

Consider the following reactions involving active centers (the OH radical, the hydrogen (H) and oxygen (O) atoms), which play an important role:

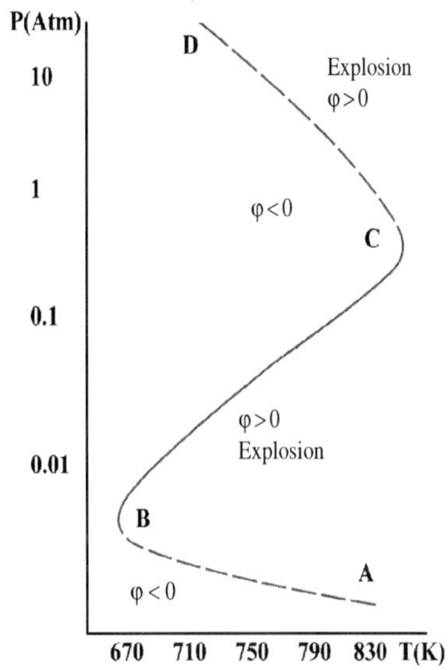

Fig. 3.1 P-T diagram of kinetic curves for the combustion of a stoichiometric hydrogen-oxygen mixture

$$OH + H_2 = H_2O + H \ (6\ \text{kcal/mol}) \tag{3.5.2}$$

$$H + O_2 = OH + O \ (17\ \text{kcal/mol}) \tag{3.5.3}$$

$$O + H_2 = OH + H \ (10\ \text{kcal/mol}) \tag{3.5.4}$$

Reactions (3.5.2) and (3.5.4) have activation energies of roughly 6 and 10 kcal/mol, respectively, and therefore proceed much more rapidly than reaction (3.5.3) whose activation energy is 17 kcal/mol. Therefore, reaction (3.5.3) is the limiting step for the entire process and determines its overall rate. The hydrogen atoms that participate in these reactions are the main active centers governing the reaction. In the cycle of reactions (3.5.2)–(3.5.4) a hydrogen atom entering reaction (3.5.3) is regenerated in the other reactions. A single H atom that enters reaction (3.5.3) generates active radicals, which in its turn regenerate two H atoms that is, chain branching occurs. Since the rate of reaction (3.5.3) is much larger than the rate of the reactions in which active centers are created (the dissociation of hydrogen molecules, $H_2 = 2H$ requires 103 kcal/mol!), the concentration of active centers in the branching chain reaction will be many times larger than that of the equilibrium value. This is the fundamental difference between these reactions and non-branching chain reactions in which the concentration of active particles cannot become larger than the equilibrium level. From an energetic standpoint this means that the heat released during the basic reaction $2H_2 + O_2 = 2H_2O + 116\ \text{kcal/mol}$ goes almost entirely into the dissociation of hydrogen molecules, $H_2 = 2H - 103\ \text{kcal/mol}$.

If the initial temperature of the mixture corresponds to a point lying to the right of the curve ABCD in Fig. 3.1, then ignition occurs, which results in thermal explosion. The self-acceleration of a branching chain reaction occurs because the rate of loss of active particles is lower than their rate of formation.

At lower temperature and pressures below the pressure, corresponding point B (below the first explosion limit in Fig. 3.1), active particles can diffuse to and be destroyed at the vessel walls. In order to describe the phenomena, reactions (3.5.2)–(3.5.4) must be supplemented by equations describing diffusion, absorption and distortion of H, O, and OH, rate of which is proportional to the first power of pressure. As the pressure increases the diffusion rate decreases and the critical conditions may be reached. When the pressure is raised further, the rate of production of active particles exceeds their rate of loss owing to diffusion to the walls and an explosion occurs (above curve AB). In the initial stage of the reaction, when the concentration of H atoms is rising, little heat is released. The thermal yield of the reaction is released only when the concentration of hydrogen atoms is close to the maximum and they begin to recombine with a large thermal yield. In the initial stage of the reaction the chemical energy of the reagents goes into chemical energy of the active particles rather than into thermal energy, thereby causing the rapid progress of this reaction.

At still higher pressures the mechanism for loss of active particles changes. Three body reactions such as

$$H + O_2 + M = HO_2 + M \tag{3.5.5}$$

begin to play an important role (this reaction competes with $H + O_2 = OH + O$). Reaction (3.5.5) leads to formation of the HO_2 radical, which is very inert from a chemical standpoint and is able to diffuse to the vessel walls before it reacts. Thus, this reaction begins to play as a chain termination reaction. It should be noted that the loss of the HO_2 radical depends strongly on the material of which the wall is made. With some materials all radicals that reach the wall are lost, while with others a substantial fraction is reflected back into the vessel volume without change. With increasing pressure the frequency of three body collisions according to reaction (3.5.5) increases more rapidly than the frequency of binary collisions yielding reactions (3.5.2)–(3.5.4), and the rate of loss of active centers exceeds their rate of production after a certain pressure. There is no explosion to the left from curve BCD and this explains the existence of the second explosion limit in Fig. 3.1.

As the pressure is raised further diffusion becomes difficult and before radicals reach the walls, the HO_2 radicals begin to participate in another phase of chain propagation and branching reactions of following types:

$$HO_2 + H_2 = H_2O_2 + H - 15\ \text{kcal/mol} \tag{3.5.6}$$

$$HO_2 + H = OH + OH + 37\ \text{kcal/mol} \tag{3.5.7}$$

$$HO_2 + H = H_2O + O + 54\ \text{kcal/mol.} \tag{3.5.8}$$

A substantial amount of heat is released in reactions (3.5.7) and (3.5.8) which also leads to an increase in the reaction rate that, in turn, causes further heating of the mixture. If the rate of heat release exceeds the rate of cooling owing to heat transfer to the vessel walls, then a thermal explosion takes place. This is the origin of the third explosion limit above curve CD.

These are the basic concepts involved in the $H_2 - O_2$ reaction, which may be viewed as a model reaction for the study of chain reactions. At present the mechanism of this reaction has been studied quite thoroughly for a wide range of reaction conditions with detailed discussions given in a number of textbooks. Most fuel and oxidizer combinations have more complicated chemical kinetics as compared with the limits in the $H_2 - O_2$ system.

Up to this point in explaining the existence of critical conditions for explosive reactions, we have mainly proceeded from the idea of branching chain reactions. However, the process that occurs at the third explosion limit of a hydrogen-oxygen explosive mixture is associated with heating of the reacting mixture by the energy released as a result of chemical change and corresponds to thermal explosion.

Problems

3.1. Write the stoichiometric coefficients for the product in the reactions below:

$$H_2 + 3O_2 = ?H_2O + O_2,$$

$$H_2 + 0.5F_2 = ?HF + ?H_2,$$

$$CH_4 + 3O_2 = ?CO_2 + ?H_2O + ?O_2,$$

$$C_2H_5OH + 6O_2 = ?H_2O + ?CO_2 + ?H_2O.$$

3.2. Determine the heat formation of gaseous methane CH_4 if bond energies C-C(85.5 kcal/mol), C-H(98.1 kcal/mol), C(s) = C(g) + 171.5 kcal/mol, $H_2 = 2H + 103$ kcal/mol.

3.3. Calculate distribution of the products in the reaction $N_2 + 0.5O_2 \rightarrow aN_2 + bO_2 + cNO$, if the equilibrium constant for final pressure 1 atm and final temperature 4000 K for the equilibrium reaction $0.5N_2 + 0.5O_2 = NO$ is $K_P = 10^{-0.5241} \approx 0.3$.

3.4. Determine the heat formation of gaseous butane C_4H_{10} if bond energies C-C (85.5 kcal/mol) and C-H (98.1 kcal/mol), C(s) = C(g)+171.3 kcal/mol, $H_2 = 2H + 104.2$ kcal/mol. Result: $\Delta H^0_{f,C_4H_{10}} \approx -31.3$ kcal/mol.

3.5. Determine heat of formation of CO_2 from the known heat of formation of CO –26.4 kcal/mol, and the experimentally know heat of reaction $CO_{(g)} + \frac{1}{2}O_{2(g)} = CO_{2(g)} - 67.6\,\text{kcal}$ (298 K).

3.6. Calculate distribution of the products in the reaction hydrogen-fluorine, assuming that the final pressure is 1 bar and final temperature 3500 K: $H_2 + F_2 \rightarrow n_{HF}HF + n_H H + n_F F$.

3.7. Calculate distribution of the products in the reaction $CO_2 + 0.5O_2 + 0.5N_2 \rightarrow aCO + bNO + cCO_2 + dO_2 + eN_2$, assuming that the final pressure is 1 bar and final temperature 3000 K. The equilibrium constants for the standard reactions are:

$$CO_2 \leftrightarrow CO + 0.5O_2 \rightarrow K_1(3000\,K) = 0.3273,$$

$$0.5O_2 + 0.5N_2 \leftrightarrow NO \rightarrow K_2(3000\,K) = 0.1222$$

3.8. Consider the rate of production for the hydrogen-bromine reaction, taking into account chain character of the $H_2 - Br_2$ reaction:

$$\left.Br_2 + M \rightarrow 2Br + M \quad (k_1)\right\} \quad \text{chain initiation}$$

$$\left.\begin{array}{ll} Br + H_2 \rightarrow HBr + H & (k_2) \\ Br_2 + H \rightarrow HBr + Br & (k_3) \\ HBr + H \rightarrow H_2 + Br & (k_4) \end{array}\right\} \quad \text{chain propagation}$$

$$\left.M + 2Br \rightarrow Br_2 + M \quad (k_5)\right\} \quad \text{chain termination}$$

What is the overall order of this reaction?

Chapter 4
Self-Accelerating Reactions, Explosions

4.1 Self-Accelerating Reactions

Some chemical reactions develop in time very slowly while others develop extremely fast – "explosively", and such phenomena as a flash, strong sound, etc. usually accompany them in the later case. The important features of explosions are that the rate of the reaction varies extremely rapidly with temperature. As it was discussed in previous chapter, very often the explosive paths of a reaction is explained by its chain character, that is, by the dependence of its rate on the concentration of active centers (atoms, radicals) formed during the reaction.

Chain reactions take place through the formation of active particles. During a reaction the amount of these particles may increase for two reasons. First, active particles may be formed as a result of thermal motion, independently of the chain reaction, since reagent molecules may dissociate upon colliding with one another; the rate for this process can be high enough only for high temperatures. Second, branching of the chain may occur, that is, an elementary process involving one active atom or radical that yields two atoms (or free radicals). Branching always occurs due to the reaction energy. For example, the reaction $H + H_2 = 3H$ is possible from the standpoint of the material balance. However from the standpoint of energy this reaction it is not possible, but the reaction $H + 3H_2 + O_2 = 3H + 2H_2O$, on the other hand, is possible. The rate of formation of active particles in this fashion is proportional to their concentration. The presence of a source of active particles is extremely important and determines the main features of chain kinetics. The appearance of active particles can initiate branching that is formation of multiple numbers of active particles if it is possible energetically. Besides, always there are processes, which cause their loss. The rate of removal of active centers from the reaction by collisions with stable molecules or by diffusion to the vessel walls also is proportional to the concentration of active centers. When the rate of chain branching becomes greater than the rate of chain termination, there is an explosion conditions; when the contra conditions exist, the explosion is impossible. Essentially, the same type of concept holds for thermal explosion (or thermal ignition). When the rate of thermal energy release is greater than the

M.A. Liberman, *Introduction to Physics and Chemistry of Combustion*,
DOI: 10.1007/978-3-540-78759-4_4, © Springer-Verlag Berlin Heidelberg 2008

rate of thermal energy loss, an explosive condition exists; when the contra condition exists, the explosion is impossible.

Thus, the equation for the time variation of the concentration of active centers can be written in the form

$$\frac{dn}{dt} = W_0 + fn - gn, \tag{4.1.1}$$

where W_0 is the rate of creation of active centers and f and g are the rate constants for chain branching and chain termination. Writing $f - g = \varphi$ we obtain

$$\frac{dn}{dt} = W_0 + \varphi n \tag{4.1.2}$$

Note that the rate of creation of active centers, W_0 is very small, almost equal to zero, so that in experiment we can't measure either change in the concentration of active centers or heat of the reaction. A change in the external conditions (temperature, pressure) causes changes in both f and g. Since no activation energy is required for loss of an active center, the constant characterizing the formation of new active centers by chain branching typically have much stronger temperature dependence than the rate of chain termination. Thus, the difference $f - g = \varphi$ changes sign as the temperature is raised. At low temperatures it is negative and at high temperatures it is positive. An examination of Eq. (4.1.2) for the reasonable initial condition $n(t = 0) = 0$ shows that when $\varphi < 0$ (at low temperatures), the concentration of active centers approaches the limit $W_0/(-\varphi)$, still a small quantity, and $dn/dt \to 0$. On the other hand, when $\varphi > 0$, the rate of formation and the concentration of active centers rise continuously. Thus, depending on the sign of φ the type of solution changes drastically. The temperature at which φ equals zero is the critical temperature below which an explosion is impossible. These qualitative considerations explain the existence of an explosion region for an explosive mixture at fixed initial pressure when the initial temperature passes through a critical value corresponding to $\varphi = 0$.

The solution of Eq. (4.1.2) satisfying the initial condition $n(t = 0) = 0$ is

$$n = \frac{W_0}{\varphi}(e^{\varphi t} - 1). \tag{4.1.3}$$

The rate of formation of the end products, that is the reaction rate, is

$$W = \nu \cdot (fn) = \frac{\nu f W_0}{\varphi}(e^{\varphi t} - 1), \tag{4.1.4}$$

where ν is an integer of the order unity that tells how many molecules of the end product are formed as the result of the reaction of one active center. When

$\varphi > 0$, i.e., chain branching occurs more rapidly than termination, n increases with time exponentially as does the reaction rate. In the early stages of the reaction, while there is a build up of active centers, the rate remains lower than the limit of measurability since W_0 is very small for explosive reactions and the change is hardly noticed. After a certain time the rate reaches a measurable value and continues to rise without limit. This time corresponds to the induction or delay period t_i. Since W_0 is always much smaller than W_{min} for an explosive reaction, $\exp(\varphi t)$ is much greater than unity for $t > t_i$ and with sufficient accuracy we can set

$$W \approx \nu \frac{W_0 f}{\varphi} e^{\varphi t}$$

The induction time is therefore equal to

$$t_i = \frac{1}{\varphi} \ln\left\{\frac{W_{min}\varphi}{\nu W_0 f}\right\} \tag{4.1.5}$$

The variation in the logarithmic term can be neglected, so that

$$t_i = \text{const}/\varphi.$$

For large φ, when $f >> g$, $\varphi = f - g \approx f$,

$$W \approx \nu W_0 \exp(\varphi), \text{ and } t_i = \frac{1}{\varphi} \ln\left\{\frac{W_{min}}{\nu W_0}\right\} \approx \text{const}_2/\varphi.$$

A change in φ for constant W_0 causes a sharp change in the induction period and reaction rate. Thus, when φ is reduced there is a noticeable increase in t_i and the reaction rate curve is expanded in time. As φ is decreased, the self-acceleration of the reaction becomes less and less distinct and, finally, when $\varphi = 0$, the kinetic curve becomes a straight line given by the equation

$$W = \nu W_0 f t.$$

Now, the induction time corresponding to detectable rate W_{min} is reached after a time

$$t_i = \frac{W_{min}}{\nu f W_0}.$$

In this limiting case the reaction rate increases without bound until burn up of the reactants begins to have an effect. If φ is decreased still further and becomes negative $\varphi < 0$, then the character of the reaction changes substantially. It follows from Eq. (4.1.4) that for $\varphi < 0$ the reaction rate will approach a limit $\nu f W_0/|\varphi|$. Since this quantity is less than W_{min}, being extremely small, we conclude that when $\varphi < 0$ the reaction does not proceed. For very small negative φ, only a small change in the external conditions (pressure or temperature) leading to a tiny increase in f is needed to cause φ to change sign and to make the reaction undergo self-acceleration.

4.2 Thermal Self-Ignition

The self-acceleration of a chemical reaction may also be associated with the build up of heat in the system. Two conditions must be fulfilled for this to occur: first, the mixture must release a certain amount of energy as heat during the reaction and, second, the reaction rate must rise with temperature. The overwhelming majority of reactions between stable molecules satisfy the second condition. These reactions do not proceed at low temperatures while at high temperatures they proceed rapidly. If the first condition is satisfied, that is, the reactions are exothermic, then the heat of reaction alone may cause the reaction to speed up and lead to an explosion, unlike chain reactions in which the rate of the initial thermal reaction W_0 is very small.

Let us consider the simple case when the reaction rate depends on the concentration of only one reacting component. The kinetic equation is

$$\frac{da}{dt} = -W(a,t) = -k_0 a^n e^{-E/RT}, \tag{4.2.1}$$

where n is the order of the reaction, for example, $n = 2$ for the bimolecular reactions.

Let Q is the heat of the reaction, T_0 is the initial temperature of the mixture consisting of N_0 moles; $a = N/N_0$ is the concentration. From the first law of thermodynamics we can write down condition for the enthalpy balance for adiabatic process

$$c_P N_0 (T - T_0) = Q(N_0 - N), \text{ or } c_P (T - T_0) = Q(a_0 - a), \tag{4.2.2}$$

where N is the number of the reacted moles.

For the highest temperature T_b reached during complete combustion of the reagent under adiabatic conditions in a vessel we obtain

$$c_P (T_b - T_0) = Q a_0. \tag{4.2.3}$$

If reaction proceeds in a vessel of constant volume then we obtain formulas similar to (4.2.2) and (4.2.3) but with the specific heat at constant volume c_V instead of c_P.

Using (4.2.2) and (4.2.3) we can express concentration via temperature as

$$a = a_0 \frac{T_b - T}{T_b - T_0}, \tag{4.2.4}$$

and rewrite equation for (4.2.1) for the only variable, for example, for temperature

$$\frac{dT}{dt} = k_0 a_0^{n-1} (T_b - T_0)^{1-n} (T_b - T)^n e^{-E/RT} = W_1(T) \tag{4.2.5}$$

or for concentration

$$\frac{da}{dt} = -k_0 a^n \exp\left\{ - \frac{E/RT_0}{1 + (Qa_0/c_p \rho_0 T_0)(1 - a/a_0)} \right\} = -W_2(a), \tag{4.2.6}$$

where we used relation between adiabatic temperature and concentration for isobaric process $T_b - T_0 = Qa_0/c_P\rho_0$.

Equations (4.2.5) and (4.2.6) determine the time variations of the temperature, concentration and reaction rate. The reaction rate increases very rapidly with the reduction in the amount of reactants that have been consumed, and only when very little of the reactants are left does the reaction rate reach a maximum and begins to fall. Such dependence of the reaction rate is typical feature of explosive reactions that are catalyzed by the heat released during the chemical change with the rate given by the Arrhenius law with large activation energy.

For typical reactions, the activation energy is of the order of several tens of kilocalories so that, $E/RT >> 1$ over the entire temperature range of interest (300–4000 K). Calculating the maximum of the functions $W_1(T)$ and $W_2(a)$, we find expressions for a_* and T_*, corresponding to maximum of the reaction rate (for $RT_b/E << 1$)

$$T_*/T_b = 1 - n(RT_b/E) \tag{4.2.7}$$

$$a_*/a_0 = nRT_b^2/E(T_b - T_0). \tag{4.2.8}$$

For $RT_b/E << 1$ the maximum of the functions $W_1(T)$ and $W_2(a)$ at the points for a_* and T_*, is very sharp. Using (4.2.7) and (4.2.8) we can obtain

$$\frac{W(a_*, T_*)}{W(a_0, T_0)} = \frac{n}{e} \left(\frac{RT_b^2}{E(T_b - T_0)} \right)^n e^{\frac{E}{RT_0}\left(1 - \frac{T_0}{T_b}\right)}. \tag{4.2.9}$$

It is interesting to examine the effect of burn up of the reactant on the peak value of the reaction rate. For this we use the ratio

$$\frac{W(a_*, T_*)}{W(a_0, T_*)} = \left(\frac{a_*}{a_0}\right)^n = n\left(\frac{RT_b^2}{E(T_b - T_0)}\right)^n. \qquad (4.2.10)$$

A numerical solution of Eqs. (4.2.5) and (4.2.6) in Fig. 4.1 shows the time variations of the temperature, concentration and reaction rate for $E/RT_0 = 12.5$, $T_b = 2.5T_0$, $T_0 = 800\,K$, $n = 1$. It is seen that from the beginning of the reaction until the reagents are almost fully consumed, the entire process has two characteristic stages, which are similar to the case of branching chain reactions. From the beginning, the temperature of the mixture and the concentration of the reagent change slowly over a comparatively long time interval – the induction period. The second stage is a rapid heating of the mixture accompanied by fast burn up of the reacting substance. If we estimate the induction period for this example to be the time over which the reaction reaches 5% of its maximum value, then the induction period occupies 85% of the duration of the entire process. At higher activation energies there is a still more distinct difference in duration of the induction stage, during which almost none of the reactant is consumed, followed by practically instantaneous burn up.

Taking this into account we can for a high activation energies use a model with the time dependence of the reaction rate being approximated by a delta-function as

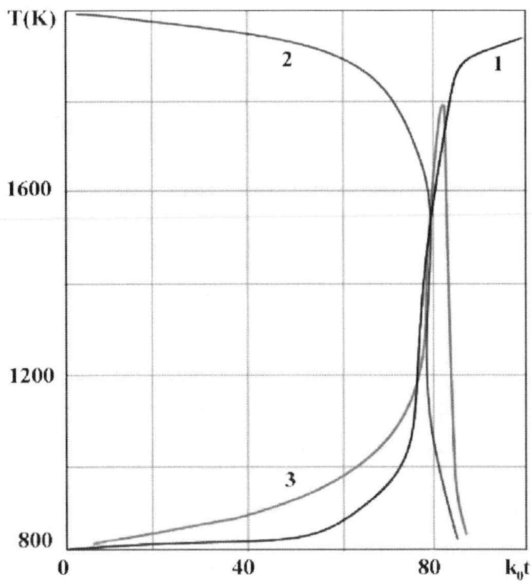

Fig. 4.1 The time variations of temperature (1), concentration (2) and reaction rate (3) in the adiabatic thermal explosion. [From Zel'dovich Ya. B., Barenblatt G. I., Librovich V. B., Makhviladze G. M. "*The Mathematical Theory of Combustion and Explosion*", Consultants Bureau, NY, 1985.]

$$W(t) = A_0 \delta(t - t_i) \tag{4.2.11}$$

where the normalizing constant A_0 is determined from the condition that all the fuel be used up in the reaction,

$$\int_0^\infty W(t)dt = A_0, \tag{4.2.12}$$

and the induction period t_i can be defined as the time over which the reaction rate reaches its maximum value.

The induction period can be calculated by integrating Eq. (4.2.5) with the initial condition $T(0) = T_0$

$$k_0 a_0^{n-1} (T_b - T_0)^{1-n} t = \int_{T_0}^{T_b} \frac{dT}{(T_b - T)^n} e^{\frac{E}{RT}}. \tag{4.2.13}$$

For the reaction orders $n \geq 1$ the integral in (4.2.13) diverges as $T \to T_b$. This means that the final state, characterized by complete consumption of the reagent, attained asymptotically over an infinite time. Therefore, the time for completion of the reaction is usually estimated as the time over which the concentration falls to an amount of the order of 1% of the initial value a_0. The integral in (4.2.13) can be calculated analytically using expression for the exponential integral and its asymptotic for large argument. The result for the adiabatic induction time is

$$t_{ia} = \frac{1}{k_0} \frac{RT_0^2}{E(T_b - T_0)} \exp(E/RT_0), \text{ for } n = 1, \tag{4.2.14}$$

$$t_{ia} = \frac{1}{k_0 a_0} \frac{RT_0^2}{E(T_b - T_0)} \exp(E/RT_0), \text{ for } n = 2. \tag{4.2.15}$$

4.3 The Frank-Kamenetskii Transformation

The fact that the rate of a chemical reaction depends strongly on the temperature according to Arrhenius law for $E/RT >> 1$ makes possible to simplify all calculations considerably, and to investigate analytically the basic properties of explosions and combustion due to the function $W(T) = C\exp(-E/RT)$ can be replaced by $W'(T) = C' \exp\{(T - T_1)/\Theta\}$ which is very close to the Arrhenius temperature dependence over a temperature interval in the vicinity of some given temperature T_1 for suitable values of the constants C' and Θ. To make this replacement, we expand the exponent in the Arrhenius formula in a Taylor series to the first linear term near T_1

$$W(T) = Ce^{-\frac{E}{RT}} \approx Ce^{-\frac{E}{RT_1}} e^{\frac{E}{RT_1^2}(T-T_1)}, \tag{4.3.1}$$

which gives

$$C' = Ce^{-\frac{E}{RT_1}} \text{ and } \Theta = RT_1^2/E.$$

The functions W(T) and W′(T) and their first derivatives are equal at $T = T_1$. This approximation for the Arrhenius′ law temperature dependence of the reaction rate was proposed by D.A. Frank-Kamenetsky (the Frank-Kamenetsky transformation) and it is widely used in combustion theory. Since the basic equations of the temperature dependence of the reaction rate are nonlinear, it is much easier to work with the function exp (Ax) than with exp (B/x), explicitly introducing into the analysis the characteristic temperature interval $\Theta = RT_1^2/E$ and representing a small parameters in asymptotic solutions with a good accuracy. Formally, the characteristic temperature interval is the temperature range over which the rate of a chemical reaction varies by a factor of *e*. Table below shows the values of the Arrhenius law temperature dependence of the reaction rate, W(T), and the corresponding approximation W′(T) for $E/RT_1 = 2000$ K at various temperatures (we use the values $T_1 = 2000$ K and $E = 80000$ kcal/mol, which yield $\Theta = 100$ K).

T	1000	1800	1900	2000	2100	2200	4000
T/T_1	0.5	$1 - 2\Theta/T_1$	$1 - \Theta/T_1$	1	$1 + \Theta/T_1$	$1 + 2\Theta/T_1$	2
$W(T)/W(T_1)$	$2\ 10^{-9}$	$1.1\ 10^{-1}$	$3.49\ 10^{-1}$	1	2.59	6.16	$2.2\ 10^4$
$W'(T)/W(T_1)$	$4.5\ 10^{-5}$	$1.35\ 10^{-1}$	$3.67\ 10^{-1}$	1	2.72	7.39	$4.85\ 10^4$

As can be seen in the table, within a characteristic interval of T_1 the error in using the approximation is less than 5%. Though the relative error increases rapidly for large differences between T and T_1 this difference, is not important since the reaction rate at temperatures far from T_1 is very small and may be neglected compared with the rates at temperatures near T_1. The peak temperature, which typically is the adiabatic flame temperature, is usually well known.

4.4 Semenov's Theory of Thermal Explosions

N. N. Semenov (1927) first considered the progress of the reaction of a combustible gaseous mixture that has initial temperature T_0 in a vessel, which walls kept at a fixed temperature T_0 throughout the process. We assume that at the beginning the gas temperature is also T_0. The amount of heat release in the reaction can be represented in simplified form as

$$q_+ = QW = Qa^n A \exp(-E/RT). \tag{4.4.1}$$

The rate of heat transfer to the vessel walls can be estimated, taking into account that the heat flux through the surface element dS is $dS\frac{\partial T}{\partial n}$, and total heat flux from the vessel of volume V and its surface S is

$$q_- = \kappa \frac{S}{V}(T - T_0), \tag{4.4.2}$$

where κ is the coefficient of thermal conduction.

The temperature change for the system with simultaneous heat generation and heat losses is defined by the overall energy conservation

$$\rho c_V \frac{dT}{dt} = (q_+ - q_-), \tag{4.4.3}$$

where c_V is the specific heat at constant volume.

Neglecting by burn up of the initial reactants we obtain the condition for stationary state of the reaction by equating the heat production to the heat loss ($q_+ = q_-$) and determine the values of the steady-state temperature in the vessel. The corresponding solutions are shown in Fig. 4.2, where q_+ and q_- are plotted as functions of the temperature. Note, that q_+ rises with the temperature in the same way as the reaction rate (Arrhenius law, curve 1) while the rate of heat loss rises linearly (curve 2).

Let the heat transfer be large enough so that the heat loss line intersects the heat influx curve. Since the temperature of the mixture at the initial time is equal to the wall temperature T_0, at first there is no heat loss. As soon as heat is released in the reaction, the temperature of the mixture begins to rise. Then the temperature difference $T - T_0$ between the gas and vessel walls rises and causes the rate of heat loss to increase. At some gas temperature $T = T_F$ the rates of heat production and heat loss become equal (point F of Fig. 4.2) and the gas will not be heated further. If by some reason the temperature becomes slightly greater than T_F, then the temperature falls to T_F since at that point the heat loss is more intense than the heat release. This indicates that the steady-state regime corresponding to point F is stable. Thus, if the heat transfer is large enough that the heat loss line intersects the heat production curve, then a gas initially at temperature T_0 will be heated to temperature T_F, close to T_0 and will react for a long time at this temperature.

Fig. 4.2 The dependence of the rate of heat release (curve 1) and the heat loss (curves 2, 3) on the temperature

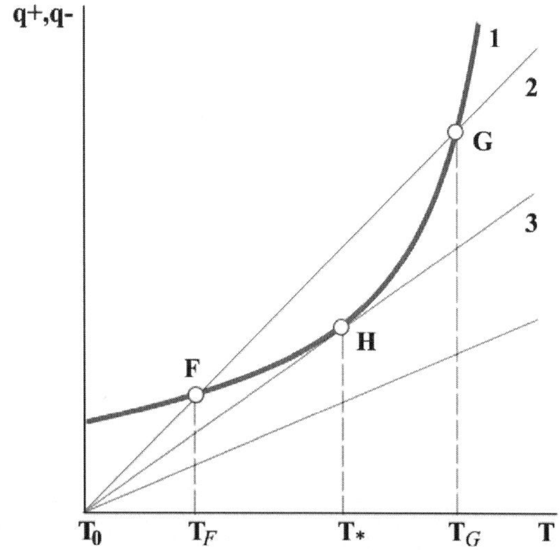

The second intersection point, G, existing at the temperature $T = T_G$, corresponds to the unstable regime. If the temperature is reduced below than T_G, the rate of heat loss exceeds the rate of heat production and the temperature falls to T_F. If the gas is heated to $T > T_G$, then the heat release rate will rise further. For an initial gas temperature below T_G the gas cools to T_F and reacts at a constant rate, while for an initial gas temperature above T_G the mixture ignites and temperature will rise further. Ignition occurs only if we heat the mixture to a temperature $T > T_G$ much higher than the vessel temperature. A mixture with an initial temperature $T < T_G$ cannot ignite, which means that this is not self-ignition.

Let us assume that the heat losses decreased, for example, owing to the decrease in S/V for larger vessel, while the wall temperature T_0 remains the same. The heat release curve is the same but the slope of the heat loss line is smaller. If the q_+ curve is above the q_- for all temperatures (the critical slope corresponds to curve 3 in Fig. 4.2), then temperature of the reacting mixture will rise from initial temperature T_0 unboundedly. The reaction rate will also rise unboundedly; the stable exothermal reaction regime cannot exist and self-ignition will occur. Also if the initial pressure is increased, the heat release curve shifts to higher values, which are proportional to $\rho^n \propto P^n$ according to Eq. (4.4.1). If the value of pressure is such that q_+ is everywhere greater than q_- then self-ignition will occur.

The same conclusion can be obtained by changing the wall temperature T_0 for the same vessel. In this case the slope of the heat loss line is unchanged but the q_- line shifted parallel to itself in the direction of the temperature change (Fig. 4.3). A steady regime corresponding to temperature T_F exists for a wall temperature $T_{01} < T_{0*}$, but for a wall temperature $T_{02} > T_{0*}$, the lines q_+ and q_- do not intersect and an explosion occurs (Fig. 4.3).

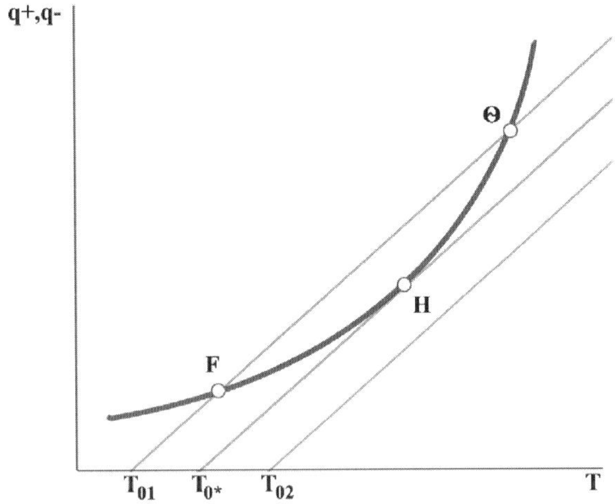

Fig. 4.3 The temperature dependences of the heat release and loss rates for various initial gas temperatures

The highest wall temperature at which a stationary regime is possible corresponds to $T_0 = T_{0*}$; when the heat production curve is tangent to the heat loss curve. From these considerations it is clear that the ability of a fuel mixture to explode i.e. a self-ignition of the mixture depends on the shape and dimensions of the vessel, the thermal conductivity of the gas, and so on, as well as on the rate of the chemical reaction. The point H in the Figures 4.2 and 4.3 represents the ignition temperature $T_0 = T_{0*}$.

A fundamental feature of Semenov's theory is that a steady-state approach is used in the analysis of what might seem to be an essentially unsteady proccss. The condition of an explosion (nonstationary regime) is formulated as the loss of the steady-state regime, that is, as the impossibility of a balance between heat production and heat losses. Semenov showed that, although the reaction takes place at a higher temperature than the initial wall temperature, but the wall temperature does determine many important characteristics of the process, such as the condition for ignition, the reaction time, etc. If, for example, at a temperature of 580 C self-ignition does not occur while at 600 C it does, this does not mean that the reaction rate is high enough at 600 C to produce the observable external manifestations of an explosion. Actually, at this initial temperature the conditions are such that the reaction rate rises rapidly and the high temperatures (above 2000 K) are reached at which the reaction rate is high enough to produce effects accompanied an explosion. The initial temperature at which an explosion occurs is not some singular point where the reaction rate experiences a discontinuity, but ignition is a consequence of the fact that, beginning with this temperature, the heat loss cannot balance the heat production. Contrary, for a chain explosion with chain termination in the vessel volume, the critical ignition condition $\varphi = 0$ is a consequence solely of the properties of the mixture.

4.5 Critical Conditions for a Thermal Explosion

We now apply method of the Frank-Kamenetskii transformation to analyze the critical conditions for a thermal explosion. Consider the system parameters at the boundary separating the two possible regimes of a stationary slow reaction and an explosion. Substituting the expressions for q_+ and q_-, and W in the heat balance Eq. (4.4.3) and applying the Frank-Kamenetsky transformation near the wall temperature $T = T_0$, which equals the initial gas temperature, we obtain

$$\frac{d\theta}{d\tau} = e^{\theta} - \gamma\theta \tag{4.5.1}$$

where we introduced the dimensionless variables

$$\theta = \frac{T - T_0}{\Theta}, \tau = \frac{t}{t_{ia}}, t_{ia} = \frac{\Theta c_V \rho}{Q a_0^n}\frac{1}{k_0}\exp\left\{\frac{E}{RT_0}\right\} \tag{4.5.2}$$

and the parameter

$$\gamma = t_{ia}/t_a, t_a = c_V \rho V/\kappa S, \tag{4.5.3}$$

which is the ratio of the adiabatic induction time to the characteristic time of the heat loss. The magnitude $\gamma = t_{ia}/t_a$ depends on the volume, surface area of the vessel, the thermal conductivity of the mixture and the wall temperature T_0. If the heat losses are small so that $t_a >> t_{ia}$, then $\gamma = t_{ia}/t_a << 1$ and we can neglect by the second term on the right side of Eq. (4.5.1). For small γ the characteristic heat transfer time is much greater than the characteristic heat production time; the heat cannot be removed to the vessel walls and, therefore, an explosion must occur. This means that for sufficiently small γ the right hand side of the Eq. (4.5.1) must be positive for arbitrary θ; that is $d\theta/dt > 0$ for all t, and the equation

$$e^{\theta} - \gamma\theta = 0 \tag{4.5.4}$$

has no roots, so that the temperature increases unboundedly.

If γ is such that the roots of Eq. (4.5.4) are real, then a stationary regime is possible: the temperature initially rises to a value such that $e^{\theta} - \gamma\theta$ goes to zero and then the reaction proceeds at a constant temperature.

The condition for a transition from a stationary burning regime to an explosion is the same as we obtained above in the Semenov's theory (see Figs. 4.2 and 4.3); this is the tangency of the curves $\exp\theta$ and $\gamma\theta$. At the point separating the two reaction regimes

$$e^{\theta} = \gamma\theta, \text{ and } \frac{d}{d\theta}e^{\theta} = \frac{d}{d\theta}\gamma\theta, \tag{4.5.5}$$

that gives the ignition conditions in the form

$$\gamma = e, \text{ and } \theta = 1. \tag{4.5.6}$$

The equality $\theta = 1$, which can be rewritten in the form

$$T_* - T_0 = \Theta, \tag{4.5.7}$$

defines the maximum pre-explosion heating of the mixture.

Writing the condition of tangency of $q_+(T)$ and $q_-(T)$ we obtain for the pre-explosion temperature

$$T_* = \frac{E}{2R}\left[1 - \sqrt{1 - 4\frac{RT_0}{E}}\right],$$

which coincides with (4.5.7) for $RT_0 << E$.

Substituting the expression for γ from Eq. (4.5.3) in the first of conditions (4.5.6), we obtain

$$\kappa\frac{S}{V}\frac{\Theta\exp(E/RT_0)}{k_0 a_0^n Q} = e. \tag{4.5.8}$$

This equation gives the relation between the initial temperature T_0, the reaction energy Q, the concentration of the reactant a_0, the volume and shape of the vessel (V and S), the heat conductivity coefficient of the mixture, and the initial pressure of the fuel mixture at the explosion limit ($a_0^n \propto \rho^n \propto P^n$).

The induction time of thermal explosion can be obtained integrating Eq. (4.5.1) and neglecting by consumption of the reagents

$$\tau_i = \int_0^\infty \frac{d\theta}{e^\theta - \gamma\theta}. \tag{4.5.9}$$

The upper limit in the integral in (4.5.9) was set at infinity assuming that the condition $E(T_b - T_0)/RT_0^2 >> 1$ holds. If in addition it is possible to neglect by the heat losses, then the induction time calculated from (4.5.9) will coincide with the adiabatic induction time, which was calculated before, Eqs. (4.2.12) and (4.2.13).

4.6 Spark Ignition and Minimum Ignition Energy

Contrary to the spontaneous ignition the spark is the most prevalent form of forced ignition. In the automobile cylinder it initiates a flame that propagates through the engine cylinder. Typically the spark is fired before the piston reaches top dead center and as flame is traveling the combustible mixture ahead of the flame is compressed. Forced ignition can also be brought about by a shock wave, which is strong enough to raise temperature of the mixture above the ignition point. The most commonly used spark systems are those developed from the electric discharge with the duration of about 0.01/100 μs.

Let us estimate minimum energy of a spark required for ignition, assuming that the spark can be modeled as a point heat source, which releases a quantity of heat. In this model the distribution of the heat is given by equation

$$c_P\rho\frac{\partial T}{\partial t} = \kappa\,\Delta T \tag{4.6.1}$$

with the boundary conditions $T(r = \infty) = T_0$.

The input energy at any time must also obey the equality

$$Q/c_P\rho = \int T(x, y, z)\,d^3x = 4\pi\int_0^\infty (T(r) - T_0)r^2dr. \tag{4.6.2}$$

where Q is the energy released at constant pressure, or at constant volume.

Two Eqs. (4.6.1) and (4.6.2) are equivalent to the Eq. (4.6.1) with the sources of energy in the form of delta-function, $q = Q\delta(r)\delta(t)$. The solution of Eq. (4.6.1) is (see Sect. 7.3)

$$T(\mathbf{r}, t) - T_0 = \frac{Q}{8(\pi\chi t)^{3/2}}\exp\left(-\frac{r^2}{4\chi t}\right), \chi = \kappa/\rho c_P \tag{4.6.3}$$

It is seen that temperature is maximum at the point $r = 0$ and decreases with time as $1/t^{3/2}$, and simultaneously temperature of the surrounding gas is gradually increasing. The condition for ignition can be specified assuming that the cooling time associated with $T(0)$ is greater than the reaction duration time for a laminar flame. A simple estimate of the ignition condition can be obtained considering stability of the solutions of stationary equations associated with Eq. (4.6.1) with the energy source in the right hand part. If right hand part is positive, then positive is time derivation of the temperature, which can be view as instability and ignition. In such consideration cooling is associated with energy loss due to thermal conduction. As larger temperature gradient as larger thermal losses are. This leads to condition for ignition as condition for the size of the spark where energy was released

$$r_0 > 4.2\,L_f, \tag{4.6.4}$$

where L_f is the width of the laminar flame defined by thermal conduction of the combustible mixture.

Problems

4.1. Explain the explosion limits for stoichiometric hydrogen-oxygen reaction.

4.2. Analyze possible regimes of thermal explosion of exothermal reaction for a mixture in container, using steady state Semenv's approach.

4.3. Explain the difference between chain explosion and the thermal explosion.

4.4. Derive equation for concentration of the intermediate product, B in a consecutive reactions of the first order given by the equation: $A(k_1) \rightarrow B\,(k_2) \rightarrow C + D$.

4.5. Explain what is the order of the reaction $dC_a/dt = -kC_a^2C_0$ and how the reaction rate depends on pressure?

Chapter 5
Velocity and Temperature of Laminar Flames

5.1 Reaction Waves Propagation Through a Combustible Mixture

Under ordinary conditions, i.e. atmospheric pressure and room temperature, the rate of chemical reactions in most combustible mixtures is negligible. A perfectly premixed fuel-oxidant mixture ignites only under conditions when either there is a spark, which can heat considerably the mixture, or the walls of the vessel can heat the mixture or some seed active centers are introduced. Let these conditions are realized at a single place inside the vessel, for example, a spark is used to produce local heating of the fuel mixture to create either a sufficient number of active centers, or very high local temperature. After local ignition a reaction wave starts moving through the mixture.

A self-sustained wave of the exothermic chemical reaction spreading through a homogeneous combustible mixture is known to occur either as a subsonic *deflagration* (premixed flame) or supersonic *detonation* waves. From the mathematical point of view, both deflagration and detonation appear to be stable attractors each being linked to its own base of initial data. A detonation wave is caused by rapid compression of the substance in a shock wave, which rapidly compresses and heats the substance so the reaction can proceed at a high rate. In turn, the heat released in the chemical reaction keeps the intensity of the shock wave constant, thereby ensuring that it will propagate over large distances. The second mode of propagation of a chemical reaction wave is at velocities substantially below that of sound and is associated with molecular thermal conductivity and diffusion. This is the flame propagation regime. During thermal propagation of a wave, the heat released in the chemical reaction is transmitted to neighboring parts of the unburned reactant gas by thermal conductivity, heats this gas, and initiates the chemical reaction in the neighboring parts. In chain propagation, the reaction of new layers is driven by diffusion of active centers. Also, it is possible that diffusion and thermal conductivity may act together.

Typically, during the propagation of the reaction, the burning proceeds in a layer, which is very thin compared to length scales of the problem, for example, the size of the combustion chamber. This layer is known as the reaction zone or the flame thickness. In the first approximation the propagation of a flame can

M.A. Liberman, *Introduction to Physics and Chemistry of Combustion*,
DOI: 10.1007/978-3-540-78759-4_5,

thus be treated as propagation of a geometrical surface of negligible thickness, which separates by unburned reagents and combustion products, and which is moving relative to the unburned gas with a known velocity. This velocity is called the normal propagation velocity of the flame or also referred to as the laminar flame speed. The thermal theory of flame propagation was first suggested by Mallard and Le Chatelier (1883), who postulated that the heat transfer controls the flame propagation, and the flame consists of two zones separated at the point where the next layer of combustible mixture ignites. Father development of the theory of flame propagation is due to Lewis and von Elbe (1934) and Zel'dovich and Frank-Kamentskii (1939).

The normal propagation velocity of a flame determines the volume of fuel mixture consumed per unit time per unit surface of the flame surface has the dimensions of linear velocity (cm/s). For a curved flame front, the normal combustion velocity characterizes the velocity of the flame front relative to the unburned mixture in the direction normal to the surface of the flame front. The magnitude of the normal propagation velocity of a flame is determined by the kinetics of the chemical reaction and by mass transport processes inside the flame front: thermal conductivity and diffusion. Because of poor knowledge of the reaction kinetic coefficients, for practical applications usually the magnitude of the normal flame velocity is taken from experimental measurements rather than calculated theoretically. Such experimental measurements can be easily performed with a quite high precision. When the tube is opened at both ends, the velocity of combustion wave is in the range from a few cm/s to 10 m/s. The normal flame velocity for most hydrocarbon-air mixtures is about 40 cm/s.

Considering a flame front as a geometrical surface of zero thickness it is convenient to choose a coordinate system co-moving with the flame front, so the flame front is at rest in this co-ordinate system. The unburned fuel mixture moves into the flame front with velocity u_n (upstream flow) and the combustion products flow out of the flame front with velocity u_b (downstream flow). Clearly, the upstream velocity u_n is equal to the normal velocity of the flame relative to the motionless gas. The relationship between u_n and u_b for a plane flame front steams from the condition of conservation of the total mass flux. The total mass of gas entering the flame per unit area of the front must be equal to the mass of combustion products leaving this surface downstream:

$$\rho_u u_n = \rho_b u_b, \tag{5.1.1}$$

where ρ_u and ρ_b are the densities of the initial unburned mixture and combustion products. Thus,

$$u_b = \frac{\rho_u}{\rho_b} u_n. \tag{5.1.2}$$

Obviously the velocity u_b of the burned products is greater than u_n since the gas is heated and expands during combustion, which means that the flame front propagating in the motionless gas mixture in a laboratory co-ordinate system will inevitably involve gas in the motion. While gas is at rest ahead of the flame front, the gas of the burned products will be involved in a motion. The overall picture of the gas dynamics ahead and behind the flame depends on the boundary conditions of the particular problem. For example, it is different for a flame propagating through the tube with the closed or open ends. It is important feature of the gaseous burning that the burning mixture undergoes strong expansion. Because of the expansion the burning process becomes strongly coupled to the gaseous flow; which leads to different flame instabilities, generation of pressure waves, etc.

Since normal velocity of a flame is much smaller than sound velocity, the pressure is practically constant over the gas volume. As we shall see, the pressure varies as $\Delta P/P \cong M^2$, where $M \propto u_n/a_s << 1$ is the Mach number. Thus, taking into account that pressure is constant, we can rewrite (5.1.1) as

$$\frac{u_b}{u_n} = \frac{\rho_u}{\rho_b} = \frac{\mu_u}{\mu_b}\frac{T_b}{T_u}, \tag{5.1.3}$$

where T_u and T_b are the temperatures of the initial unburned mixture and the adiabatic flame temperature – the temperature of combustion products, and μ_u and μ_b are the molecular weight of the initial mixture and combustion products. Since for ordinary combustion reactions $\mu_u \approx \mu_b$ and the temperature changes by a factor of 6/12, we find that $\rho_u/\rho_b = u_b/u_n \approx 6/12$. The expansion coefficient, $\Theta = \rho_u/\rho_b$ is one of the most important flame parameters. It is convenient to introduce the combustion mass flux $\rho_u u_n$, which is the product of the gas density and velocity. This quantity has the dimensions g/cm^{2}s and represents the mass of material burned per square centimeter per second in the flame front. It is also convenient to refer the rate of heat release to unit surface area of the flame. The corresponding quantity has the dimensions cal/cm^{2}s.

5.2 Velocity and Thickness of Laminar Flames

In this section we shall discuss the physical basis of the thermal flame propagation, which serves as the foundation of the concept of how reaction waves propagate.

For the sake of simplicity we shall consider premixed flames propagating in gaseous combustible mixture, which is the opposite case to diffusion flames. This means that all components necessary for the reaction are present in the fuel mixture from the very beginning and in order to start the reaction one has only to heat the mixture. In the case of premixed flames the burning rate is not controlled by the diffusion process, still even for premixed flames the fuel diffusion is an important factor of propagation together with thermal conduction. Particularly, when the fuel diffusion is stronger than the thermal conduction, then the thermal-diffusion instability of a planar stationary flame may develop. We shall consider mostly the situations of equally strong fuel diffusion and thermal conduction, when the hydrodynamic flame properties can be elucidated.

A remarkable feature of premixed combustion is a very strong dependence of the reaction rate on temperature expressed by the Arrhenius law for the reaction rate $\propto \exp(-E/RT)$, for high activation energy, $E/RT >> 1$. The activation energy of many reactions is so large, that the reaction rate at the room temperature may be taken zero. On the contrary, increase of the fuel temperature even by a factor 2 may lead to increase of the reaction rate by 10–12 orders of magnitude. In the case of a strongly exothermic reaction when a considerable amount of energy releases, relatively slight increase of the temperature at some region ignites the reaction, which eventually extends over the whole gas. It is obvious, that because of the strong dependence of the reaction rate on the temperature, the chemical reaction will proceed nonuniformly inside a flame front, and the reaction will be mainly concentrated in the portion of the flame front adjacent to the hot combustion products. Where the temperature is lower the reaction proceeds at a much lower rate and can generally be neglected.

In the flame the transport processes dominate because of the rapid drop in the temperature and concentrations of the reactants through the flame front. Once a reaction is ignited it can propagate in a self-supporting regime. The burnt matter has larger temperature and thermal conduction transports energy from the hot burned matter to the cold fuel: the temperature of the fuel close to the burned matter increases, the reaction in this fuel develops faster until another portion of the fuel is burned. The released energy is transported by thermal conduction to the next fuel layer and so on, resulting in propagation of the reaction front.

Flame velocity and thickness may be estimated on the basis of a simple dimensional analysis. Let the burning process is characterized by some characteristic time τ_b, which is the time of the heat release, then the only combination of velocity dimension that may be constructed out of the thermal diffusivity coefficient $\lambda/\rho_u C_P$ and the reaction time τ_b is

$$U_f \propto \sqrt{\frac{\lambda}{\rho_u C_P \tau_b}}, \tag{5.2.1}$$

where λ is the coefficient of thermal conduction and C_P is the specific heat of the fuel at constant pressure. In a similar way, the dimensional analysis gives the formula for the flame thickness. The distance heated by thermal conduction can be evaluated as $\Delta x \cong L_f = \sqrt{\frac{\lambda}{\rho_u C_P}\tau_b}$, and the reaction wave velocity is of the order of $U_f \cong \Delta x/\tau_b$, so that taking into account (5.2.1) we have

$$L_f = \frac{\lambda}{\rho_u C_P U_f}. \tag{5.2.2}$$

The exact magnitude in these formulas depends on the definition since both a thermal conduction and density are depending on the gas temperature. We can choose the value $\lambda = \lambda_u(T_b)$ as the definition for L_f. The first important conclusion from these simple estimates is that as shorter the reaction time and stronger the thermal conduction, as faster flame propagates and thinner the flame front.

Let us compare the flame velocity to the sound speed. From kinetic theory the thermal diffusivity coefficient can be expressed through the sound speed a_s (that is about the thermal velocity of molecules) and the mean free time τ_{coll}, which is the average time between molecules collisions, $\lambda\rho_u/C_P \approx a_s^2\tau_{coll}$. Then we obtain $U_f/a_s \propto \sqrt{\tau_{coll}/\tau_b}$. Since only a very small fraction of colliding molecules react because of the large potential barrier of a reaction (because of a large activation energy), then $\tau_{coll} << \tau_b$ and this means that flame velocity is much smaller than the sound speed, $U_f << a_s$. Typical velocities of flame at the normal laboratory conditions are given in Table 5.1.

Table 5.1 Velocities of flames at normal conditions ($P = 1$ atm, $T_u = 300$ K)

Fuel	Combustible mixture	U_f(cm/s)	T_b(K)	$\Theta = \rho_u/\rho_b$
Methane CH_4	6% CH_4+94%Air	5	1850	6.3
Methane CH_4	6.84% CH_4-Air	15	1900	6.3
Methane CH_4	7.76% CH_4-Air	27	2000	6.7
Methane CH_4	9.5% CH_4-Air	43	2250	7.5
Methane CH_4	11.6% CH_4-Air	25	2100	7.5
Methane CH_4	10%CH_4-O_2	80	2200	7.3
Methane CH_4	25%CH_4-O_2	304	3000	10
Methane CH_4	40%CH_4-O_2	305	3000	12.7
Methane CH_4	50%CH_4-O_2	112	2650	13.3
Propane C_3H_8	2.86% C_3H_8-Air	28	1870	6.4
Propane C_3H_8	3.64% C_3H_8-Air	35	2170	7.5
Propane C_3H_8	4.02% C_3H_8-Air	40	2240	7.8
Propane C_3H_8	5.5% C_3H_8-Air	17	2030	7.5
Propane C_3H_8	5.9% C_3H_8-Air	5.9	1830	6.9
Propane C_3H_8	7.4%C_3H_8 -O_2	240	2500	9
Propane C_3H_8	19.3%C_3H_8 -O_2	320	3000	13.6
Propane C_3H_8	21.9%C_3H_8 -O_2	235	2800	14.3
Propane C_3H_8	22.7%C_3H_8 -O_2	190	2750	14.5
Ethane C_2H_6	C_2H_6+3.5 O_2	44	3086	10.3
Butane C_4H_{10}	C_4H_{10}+6.5 O_2	45	3101	10.3
Benzene C_6H_6	C_6H_6+7.5 O_2	48	3136	
Hydrogen H_2	20% H_2-Air	100	1910	5.7
Hydrogen H_2	30% H_2-Air	195	2300	6.5
Hydrogen H_2	40% H_2-Air	265	2240	6.5
Hydrogen H_2	57% H_2-Air	190	1850	5.5
Hydrogen H_2	H_2+0.5O_2	900	3080	7.8
Hydrogen H_2	H_2+0.5O_2+0.5 N_2	520		
Acetylene C_2H_2	C_2H_2+2.5O_2	1330	3342	
Carbon Monoxide	CO+0.5O_2	29	2977	

Let l will be molecule mean free path. The reaction time τ_b is the product of the average number of collisions Z experienced in that time by a reactive molecule and the mean collision time: $\tau_b = Z(l/a_s)$. Substituting this dependence in the expression for flame velocity, we find

$$U_f \approx a_s/Z \tag{5.2.3}$$

and the characteristic width of the flame to be

$$L_f \approx l\sqrt{Z} \tag{5.2.4}$$

Equation (5.2.3) is similar to the Einstein's expression for the mean diffusion velocity in a given direction with Z being the total number of collisions experienced by a particle during the diffusion time and Eq. (5.2.4) is similar to the Einstein's formula for the mean distance covered by a molecule after Z collisions. It is clear from the molecular picture of flame propagation: a fuel molecule undergoes Z collisions in the combustion zone and during this time it covers a distance L_f at mean velocity U_f. The difference between a diffusion process and the propagation of a flame is that the mean velocity of diffusing particles falls with time (as $1/\sqrt{t}$) while the velocity of a flame is constant in time. The point is that the motion of a flame does not involve displacement of the molecules along the entire path followed by the flame front. A given molecule moves within the combustion zone for only an extremely short time and undergoes Z collisions on the average, after which it reacts and, thereby, creates the conditions necessary for the next molecule to pass the same way.

Let us consider qualitatively propagation of the reaction wave from another standpoint. In the co-moving co-ordinate system flame is at rest and we consider the reaction zone as a surface located at $x = 0$ with fixed temperature T_b. Then, the temperature distribution is defined by equation of thermal conduction

$$\rho u C_p \frac{dT}{dx} = \lambda \frac{d^2T}{dx^2}, \tag{5.2.5}$$

with the boundary conditions $T(-\infty) = T_u$ and $T(0) = T_b$. For simplicity, all the material properties are assumed to be constant and independent of the temperature. Obviously Eq. (5.2.5) is applicable only to the region $x < 0$ where there is no source of heat release and the reaction is very slow due to strong dependence on temperature. For $x > 0$ the combustion products are at a constant temperature T_b. The Eq. (5.2.5) can be integrated introducing new variable $z = dT/dx$; the equation becomes

$$z = \frac{\lambda}{\rho u c_p} \frac{dz}{dx},$$

which is easily integrated, and the solution of (5.2.5) satisfying the boundary conditions is

$$T = T_u + (T_b - T_u)e^{u\,x/\chi}, \tag{5.2.6}$$

where $\chi = \lambda/C_p\rho$ is the thermal diffusivity (the ratio of the thermal conductivity λ to the product of the density and specific heat at constant pressure).

The solution (5.2.6) shows that we can choose the distance over which the heating rises by a factor of e as the scale length for the heating zone with the velocity of the flow entering the flame front being equal to the flame velocity U_f:

$$L_f = \frac{\chi}{U_f} = \frac{\lambda}{\rho C_p U_f}. \tag{5.2.7}$$

We now calculate this length for typical laboratory flames: a slowly burning mixture (6% methane in air) and one of the fastest the hydrogen-oxygen flame – stoichiometric ($2H_2 + O_2$) mixture. Taking velocities from Table 5.1, and setting $\chi = 0.3\,\text{cm}^2/\text{s}$, we obtain $L_f = 0.06\,\text{cm}$ (CH_4) and $L_f = 0.00033\,\text{cm}$ ($2H_2 + O_2$). In both cases the width of the heating zone – thickness of the flame is quite small. The typical thickness of combustion zone ranges from $5 \cdot 10^{-2}$ to $5 \cdot 10^{-4}\,\text{cm}$.

The reaction time is of the same order as the residence time of the reactive mixture in the heating zone. An order of magnitude of the residence time of the reactive mixture in the flame can be evaluated by dividing the width of the heating zone by the flame speed: $t_{L_f} = 4 \cdot 10^{-3}\text{s}$ for CH_4 and $t_{L_f} = 4 \cdot 10^{-7}\text{s}$ for $2H_2 + O_2$. This time is many times greater than the mean collision time of the molecules in the gas. Because of the large activation energy a very large number of molecular collisions takes place in a flame for each effective collision that leads to a chemical reaction. We emphasize once again that the temperature and composition of the mixture in a flame vary for other reasons besides the chemical reaction. They also vary in the heating zone where there is no chemical reaction: the temperature changes because of the thermal conductivity of the gas and the composition may changes because of diffusion. In a flame the chemical reaction takes place in a mixture which has already been heated and whose composition has already been changed. Typical structure of a premixed flame is shown in Fig. 5.1, compared to a structure of diffusion flame shown in Fig. 5.2. In the later case the main chemical transformation is focused in a narrow zone near the peak temperature. Diffusion continuously supplies reagents to the reaction zone and redistributes them there. Completion of the reaction and thermodynamic equilibrium is achieved asymptotically, where the diffusive fluxes are practically become equal to zero.

Flame is not the only possible self-supporting regime of reaction propagation. A reaction can propagate also in a fast supersonic regime of detonation, which will be discussed in Chap. 8. In case of a detonation the reaction is

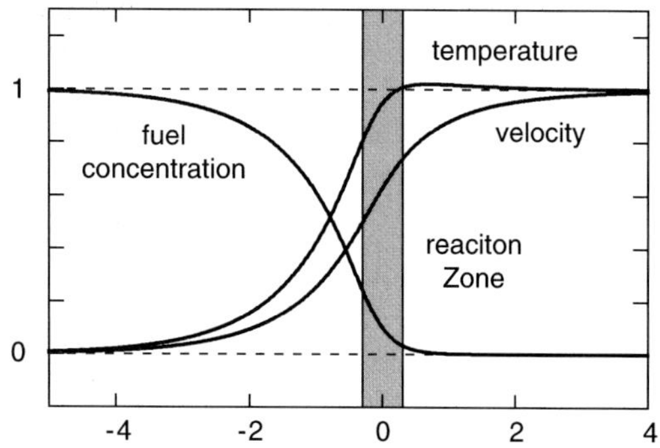

Fig. 5.1 Structure of the flame front in premixed combustion

induced by a shock wave compressing and heating the fuel. The burning mixture expands and acts like a piston pushing a leading shock and supporting the detonation. Transitions from the slow deflagration regime of flame propagation to a detonation regime are observed quite often in experiments. In industrial combustion and for safety reasons this can be a very undesirable phenomenon.

If initial temperature (or induction time) distribution is a non-uniform one, then subsequent (though independent) development of the reaction in the neighboring fuel layers may be interpreted as propagation of a reaction front with the phase velocity depending on the initial temperature distribution. This is specific regime known as spontaneous reaction (Ya. B. Zel'dovich, 1980).

We can equally consider another type of flames instead of chemical one, where energy release may be supplied from other sources. An example is thermonuclear burning of white dwarfs, which results in Supernova explosion. In this case thermonuclear flame expands from the center of the star, and the energy source is thermonuclear reactions. Another example is the laser radiation absorbed by plasma layers close to the critical surface of a target used for experiments on inertial confined fusion. For all these examples of "flame" the

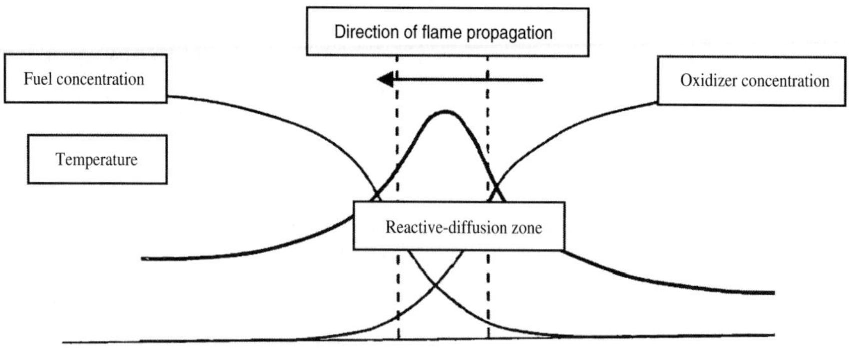

Fig. 5.2 Structure of a diffusion flame

released energy is transported by thermal conduction and the flame propagates relatively slow in comparison with speed of sound. Therefore a flame may also be defined as a subsonic regime of reaction propagation, called a *deflagration*.

5.3 Temperature and Concentration Distributions in Flames

We saw that due to velocity of a flame is much smaller than the velocity of sound; the conservation of momentum flux yields pressure to be nearly constant. Also, conservation of the energy yields constancy of the enthalpy, so that we have

$$P_u = P_b, H_u = H_b. \tag{5.3.1}$$

For a simple case of constant specific heats we obtain

$$H_u = H_{u0} + c_{pu} T_u = H_b = H_{b0} + c_{pb} T_b, \tag{5.3.2}$$

where c_{pu} and c_{pb} are specific heats of the fuel and combustion product at constant pressure. Since $Q = H_{u0} - H_{b0}$ is a heat released in the reaction, this implies that the final temperature of the combustion products is determined only by the initial enthalpy of the fuel mixture ($Q = H_{u0} - H_{b0}$).

$$T_b = \frac{c_{pu}}{c_{pb}} T_u + \frac{1}{c_{pb}} Q. \tag{5.3.3}$$

With account of (5.1.3) we also obtain

$$\frac{\rho_u}{\rho_b} = \frac{\gamma_u(\gamma_b - 1)}{\gamma_b(\gamma_u - 1)} \left(1 + \frac{Q}{c_{pu} T_u}\right), \tag{5.3.4}$$

where γ_u and γ_b is adiabatic constants for fuel and combustion products.

Structure of a laminar flame is described by a system of thermal conductivity and diffusion equations. We shall examine the simplest case of a plane flame front and a chemical reaction described by a single stoichiometric equation. In a coordinate system co-moving with the flame front we have:

$$\frac{d}{dx} J_T = QW(a_i, T), \tag{5.3.5}$$

$$\frac{d}{dx} J_i = -\frac{\nu_i}{\nu_1} W(a_i, T) \tag{5.3.6}$$

Here Q is the heat of reaction, a_i denotes the relative mass concentrations of the reagents, and $W(a_i, T)$ is the rate of the reaction. The chemical reaction rate is calculated relative to one mole of the substance denoted by the subscript 1. We shall assume that this substance is deficient in the initial mixture. The fluxes of heat and mass are made up of convective and molecular transport terms:

$$J_T = -\lambda \frac{dT}{dx} + \rho u c_p T, \tag{5.3.7}$$

$$J_i = -\rho D_i \frac{da_i}{dx} + \rho u a_i, \tag{5.3.8}$$

where D_i is the diffusion coefficient. Using the equation of continuity of the mass flux, $\rho u = \text{const}$, the system (5.3.7) and (5.3.8) can be rewritten as

$$\rho u c_p \frac{dT}{dx} = \frac{d}{dx} \lambda \frac{dT}{dx} + QW(a_i, T), \tag{5.3.9}$$

$$\rho u \frac{da_i}{dx} = \frac{d}{dx} \rho D_i \frac{da_i}{dx} - \frac{\nu_i}{\nu_1} W(a_i, T), \tag{5.3.10}$$

We assume for simplicity that the diffusion coefficients are the same form both sides of the flame, for example, a mixture of gases has similar molecular weight. In addition, we assume that the Lewis number (the ratio of the thermal diffusivity to the diffusion coefficients) $Le = D/\chi = 1$ over the entire temperature interval, which is true for many gases, i.e. $\lambda \approx \rho c_p D_i$.

Multiplying the diffusion Eq. (5.3.10) for the concentration a_i by $1/\nu_i$, and the diffusion equation for a_j by $1/\nu_j$, and subtracting one from the other, we obtain a linear diffusion equation (without term with rate function) for the linear combination $(a_i/\nu_i) - (a_j/\nu_j)$ of the densities. Its solution with the conditions in the initial fuel mixture, can be written in the form

$$\frac{a_{i0} - a_i}{\nu_i} = \frac{a_{j0} - a_j}{\nu_j}, \tag{5.3.11}$$

where the subscript 0 denotes the concentrations of the components in the initial mixture.

Thus, the concentrations of all the reagents can be calculated using (5.3.11) if the concentration distribution of one of them is known. Therefore, the Eqs. (5.3.9) and (5.3.10) can be written as two equations for only one reagent (we omit the subscript),

$$\rho u c_p \frac{dT}{dx} = \frac{d}{dx}\lambda \frac{dT}{dx} + QW(a, T), \tag{5.3.12}$$

$$\rho u \frac{da}{dx} = \frac{d}{dx}\rho D \frac{da}{dx} - W(a, T), \tag{5.3.13}$$

in which the reaction rate is written as a function of the temperature and of only one of the concentrations with the aid of (5.3.11).

Multiplying the diffusion Eq. (5.3.13) by the constant Q, adding it to the thermal conductivity equation, and taking into account that $\chi = D$, we obtain equation for enthalpy $H = Qa + \int_0^T c_p dT$

$$\rho u \frac{dH}{dx} = \frac{d}{dx}\left(\frac{\lambda}{c_p}\frac{dH}{dx}\right), \tag{5.3.14}$$

which does not include the reaction rate. The total enthalpy of the system is the sum of the chemical and thermal energy at constant pressure per unit mass of gas. The only solution of Eq. (5.3.14) that is bounded for all x is

$$H = \text{const}, \tag{5.3.15}$$

where the constant is defined by the initial fuel mixture conditions:

$$\text{const} = H_0 = Qa_0 + \int_0^{T_0} c_p T. \tag{5.3.16}$$

At the beginning of this section we have already received the same result, Eqs. (5.3.2). The idea that the sum of the chemical and thermal energy is constant in a flame was first proposed as a hypothesis (B. Lewis and G. von Elbe, 1934) for a chain reaction initiated by the diffusion of active centers. However, as we have shown, the constancy of the total enthalpy in a stationary flow depends on the relationship between the coefficients of diffusion and thermal conductivity in the medium, rather than on the mechanism of the reaction.

For constant specific heat, the constancy of the total enthalpy within the flame can be written in the form

$$Qa + c_p T = Qa_0 + c_p T_0 = c_p T_b, \tag{5.3.17}$$

that is, the temperature and concentration distributions are similar:

$$\frac{a_0 - a}{a_0} = \frac{T - T_0}{T_b - T_0}. \tag{5.3.18}$$

The temperature distribution and concentration distributions of the other reactants, as well as the concentrations of the combustion products are similar in exactly the same way since concentration of one of the reagent is linearly related to others by Eq. (5.3.11). The similarity also exists when the rate of a simple reaction depends on the concentrations of the end products as in a reversible reaction or autocatalysis by the end products. When the reaction is complex, stoichiometric relations do not relate the concentrations of intermediate products and these similarities do not exist. However, even in this case, when all the transport coefficients are equal ($D_i = \chi$), the total enthalpy of the system is still constant, but this fact can no longer be used to derive unique temperature dependence for each of the substances involved in the reaction.

The simplest case, in which the propagation of a flame is described by the system of Eqs. (5.3.9) and (5.3.10), is that when the reaction rate depends on the concentration of only one of the reacting substances. This situation is often encountered during the propagation of flames in fuel mixtures whose composition deviates strongly from stoichiometric. In such mixtures the reaction rate is limited by the concentration of one substance that is deficient. When the diffusion and thermal conductivity coefficients for this substance are equal, its concentration and the temperature have similar distributions. The deficient substance may be viewed as the carrier of chemical energy since the ultimate temperature reached by the mixture during combustion depends on its concentration.

As we already saw, the conservation of enthalpy

$$H_0 = Qa_0 + \int_0^{T_0} c_p T = \int_0^{T_b} c_p T \tag{5.3.19}$$

implies that the final temperature of the combustion products is determined only by the initial enthalpy of the fuel mixture Eq. (5.3.3). When $D \neq \chi$ within the flame the sum of the chemical and heat energies is not conserved, but it is possible to predict the direction of the shift with the aid of some general considerations. Thus in a lean mixture of hydrogen with air the diffusion coefficient of the hydrogen (the carrier of chemical energy) is an order of magnitude greater than the thermal conductivity of the mixture. Hence, diffusion brings hydrogen from the gas layer into the heating zone more rapidly than the gas is heated by molecular thermal conductivity. As a result, in a hydrogen-air mixture (i.e., when $D > \chi$) the total enthalpy has a minimum in the heating zone. On the other hand, in a lean mixture of a high-molecular-weight hydrocarbon (for example, benzene and air), the thermal conductivity exceeds the diffusion coefficient of benzene vapor ($D > \chi$) and heating of the mixture in the heating zone is more intense than diffusion of benzene into the reaction zone. Therefore, the total enthalpy has a maximum in the heating zone. Typical

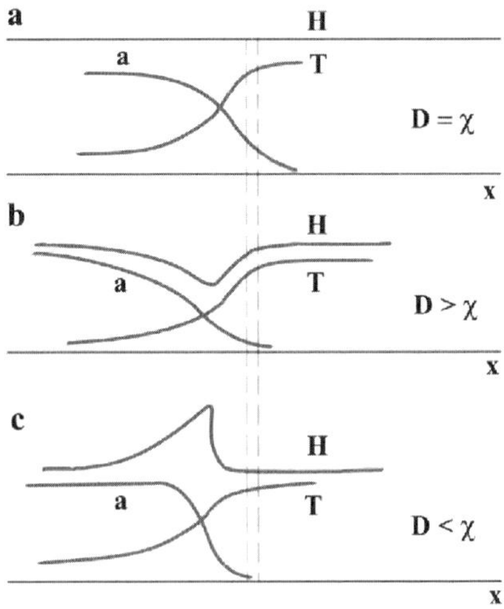

Fig. 5.3 Schematic distributions of reagent concentration, temperature, and total enthalpy in a flame front for various relationships between the diffusion coefficient and the thermal conductivity

distributions of reagent concentration, temperature, and total enthalpy in a flame for various relationships between the coefficients of diffusion and thermal conductivity are shown in Fig. 5.3.

5.4 Normal Velocity of Flame Propagation. Zel'dovich – Frank-Kamentskii Theory

In a simple case, when the diffusion and thermal conductivity coefficients are equal and the concentration and temperature are expressed by the linear relation between them, we can reduce the two differential equations for the fuel balance and energy to a single differential equation. To do this let us express the concentration in terms of the temperature in the formula for the chemical reaction rate and write the reaction rate $W(a, T)$ as a function of only the temperature $W[a(T), T] = W(T)$. Then we arrive to the energy equation alone:

$$\rho u c_p \frac{dT}{dx} = \frac{d}{dx}\left(\lambda \frac{dT}{dx}\right) + QW(T). \tag{5.4.1}$$

This is the second order equation for function $T(x)$, where the mass velocity of combustion ρu enters, which means that the normal flame velocity is an *eigenvalue* of the problem which must be found in solving the problem.

Here we shall obtain an approximate solution for the flame using the fact of very strong dependence of the reaction rate on temperature, that is a rapid chemical reaction takes place over a narrow range of temperatures close to the final temperature achieved in the flame. The heat released in the reaction is mainly carried from the reaction zone to the heating zone by thermal conduction for heating the unburned mixture from the initial temperature to the temperature of the reaction zone. In the narrow chemical reaction zone the derivative of the temperature changes from a finite value on the side facing the reactive mixture to zero on the side facing the combustion products where thermal and chemical equilibrium is reached. This rapid change in the temperature gradient over a short distance means that the contribution of the thermal conductivity is large. Because of this, the contribution of the convective heat flux may be neglected compared to that of the thermal conductivity in the heat conduction equation for the reaction zone where the temperature drop is small. Hence, in the reaction zone we can take the "truncated" equation without the convective heat flux

$$\frac{d}{dx}\left(\lambda\frac{dT}{dx}\right) + QW(T) = 0. \tag{5.4.2}$$

This equation can be integrated using the independent variable T and setting $\lambda dT/dx = z$

$$\frac{d}{dx}\lambda\frac{dT}{dx} = \frac{z}{\lambda}\frac{dz}{dT} \tag{5.4.3}$$

$$z\frac{dz}{dT} + \lambda QW(T) = 0. \tag{5.4.4}$$

In the reaction products $a = 0$ and the temperature is nearly constant, being equal to the combustion temperature T_b, so that the boundary conditions behind the flame are

$$T = T_b, \lambda\frac{dT}{dx} = z = 0. \tag{5.4.5}$$

Integrating Eq. (5.4.4) with respect to the variable temperature T up to T_b (the thermal conductivity is constant over the reaction zone, $\lambda = \lambda_b = \text{const}$), we obtain the thermal flux from the reaction zone into the heating zone

$$\lambda_b\left.\frac{dT}{dx}\right|_{T_b} = \sqrt{2\lambda_b Q\int_T^{T_b} W(T)dT}. \tag{5.4.6}$$

The integration in Eq. (5.4.6) can be extended to the initial temperature $T_0 = T_u$ since at low temperatures the reaction rate is negligibly small and the integral can be regarded as equal to zero in the low temperature region: the heat release function will contribute to the integral only at high temperatures. The actual integration region is on the order of the characteristic temperature interval (for an Arrhenius heat release function). The function $W(T)$ includes the concentration dependence $a(T)$, so that $W(T_b) = 0$. The amount of heat released per unit time by the chemical reaction and removed by thermal conductivity is equal to the chemical energy stored in the flowing fuel mixture (in co-moving with the flame coordinate system the flame is at rest while the fuel mixture moves at a speed equal to the normal propagation velocity of the flame). Thus, we can write

$$\sqrt{2\lambda_b Q \int_{T_0}^{T_b} W(T)dT} = \rho_u c_p u_n Q a_0. \tag{5.4.7}$$

This equation gives formula for the normal propagation velocity of a laminar flame

$$U_f = u_n = \frac{1}{\rho_0 c_p a_0}\sqrt{\frac{2\lambda_b}{Q}\int_{T_0}^{T_b} W(T)dT}. \tag{5.4.8}$$

We used here subscript "0" for the variables in the unburned mixture.

Formula (5.4.8) shows that the flame speed depends on the integral of the heat release function and is, therefore, related to the order of the chemical reaction, its activation energy, and other kinetic and physical-chemical characteristics of the fuel mixture. The only assumption used in deriving it was that about strong temperature dependence of the rate of heat release, which is true for large activation energy.

Since W depends strongly on the temperature according to the Arrhenius law, the integral in Eq. (5.4.8) can be estimated as the product of the maximum value $W = W_{max}$ and the effective width of the temperature interval equal to $RT_b^2/E \equiv ZT_b$, where $Z = RT_b/E << 1$ is the Zel'dovich number. We thus obtain putting $a_0 = 1$

$$U_f = u_n \approx \frac{1}{\rho_0 c_p}\sqrt{\frac{2\lambda_b W_{max} Z T_b}{Q}}. \tag{5.4.9}$$

Introducing the reaction time τ_R as the time over which all the heat contained in the initial mixture is released at the maximum heat production rate, with the aid of the heat balance equation

$$Q W_{max} \tau_R \simeq \rho_0 c_p T_b, \tag{5.4.10}$$

we obtain

$$u_n \approx \sqrt{\frac{2\lambda_b Z}{\rho_0 c_p T_b}\frac{1}{\tau_R}} = \sqrt{\frac{2Z}{T_b}} \cdot \sqrt{\frac{\chi_b}{\tau_R}} \tag{5.4.11}$$

The relationship $U_f = u_n \propto \sqrt{\chi_b/\tau_R}$ was obtained earlier, see Eq. (5.2.1) from dimensional considerations.

It is useful to obtain expression for the normal velocity of a laminar flame in a slightly different fashion. Integrating (5.4.2) we obtain similar to Eq. (5.4.6)

$$\lambda_b \left.\frac{dT}{dx}\right|_{T_b} = \sqrt{2\lambda_b Q \int_T^{T_b} W(T)dT}. \tag{5.4.12}$$

At the same time integrating Eq. (5.4.1) in the thermal conduction zone, where the reaction term can be dropped, we consider integral of the equation

$$\rho u c_p \frac{dT}{dx} = \frac{d}{dx}\left(\lambda \frac{dT}{dx}\right), \tag{5.4.13}$$

which is

$$\lambda \frac{dT}{dx} = \rho c_p u T + \text{const}. \tag{5.4.14}$$

With the boundary condition $T = T_0$ and $dT/dx = 0$ at $x = -\infty$, we obtain

$$\lambda \frac{dT}{dx} = \rho c_p u(T - T_0). \tag{5.4.15}$$

The continuity of the heat flux leads to the equation similar to (5.4.12)

$$\rho c_p u(T - T_0) = \sqrt{2\lambda Q \int_T^{T_b} W(T)dT}, \tag{5.4.16}$$

where T is the temperature at the end of the thermal conduction zone.

Taking into account that for large activation energy maximum reaction is very close to $T = T_b$ and the reaction zone is very narrow, we can take $T = T_b$ in the left-hand part of (5.4.16). To estimate integral in the right-hand part of (5.4.16) we expand in the power series argument at the Arrhenius exponent in the expression for $W(T) = A\exp(-E/RT)$.

$$\exp(-E/RT) \simeq \exp\left[-\frac{E}{RT_b}\left(1 + \frac{T_b - T}{T_b}\right)\right]$$
$$= \exp\left(-\frac{E}{RT_b}\right) \cdot \exp\left(-\frac{E}{RT_b}(T_b - T)\right).$$

Then integral in the right part can be easily calculated, and the result is

$$\int_T^{T_b} W(T)dT \simeq A\frac{RT_b^2}{E}\exp\left(-\frac{E}{RT_b}\right). \tag{5.4.17}$$

Finally, substituting (5.4.17) to (5.4.16) we obtain for the normal flame velocity

$$U_f = u_n \approx \sqrt{\frac{2\lambda}{\rho_0 c_P}\frac{A \cdot RT_b^2}{E(T_b - T_0)}} \cdot e^{-E/2RT_b}. \tag{5.4.18}$$

If we allow to remove the restricting assumptions for Lewis number not to be equal to unity and allow arbitrary for the ratio of number of moles of reactant and products (n_r/n_p), then the result for a first order reaction becomes

$$U_f = u_n = \sqrt{\frac{2\lambda c_{Pb}}{\rho_0 \langle c_P \rangle^2}\frac{T_0}{T_b}\frac{n_r}{n_p}(Le)\frac{A \cdot R^2 T_b^2}{E^2(T_b - T_0)^2}} \cdot e^{-E/2RT_b}, \tag{5.4.19}$$

and for a second order reaction

$$U_f = u_n = \sqrt{\frac{2\lambda c^2{}_{Pb}}{\rho_0 \langle c_P \rangle^3}\frac{T_0}{T_b}\left(\frac{n_r}{n_p}\right)^2 (Le)^2 \frac{A \cdot R^3 T_b^3}{E^3(T_b - T_0)^3}} \cdot e^{-E/2RT_b}, \tag{5.4.20}$$

where c_{Pb} is specific heat at $T = T_b$, and $\langle c_P \rangle$ is the average specific heat in the interval between T_0 and T_b.

The structure of the reaction zone is a consequence of the nonlinearity of the rate of heat release, which is zero everywhere except of a narrow region near the final temperature. It is precisely this circumstance, which makes it possible both to find the temperature distribution inside the flame and to find the conditions for yet another parameter of the equation – the flame velocity.

The constant characteristic velocity of the normal flame propagation is associated with the structure of the zone in which the chemical transformation takes place. The temperature distribution within this zone in the used approximation of this section, and the resulting rate of change of the fuel mixture into combustion products are independent on the velocity of the flame. The reaction zone thus "dictates" its condition, the magnitude of the heat flux that goes into heating the fuel mixture, which in turn determines the flame propagation velocity.

5.5 Consequences of the Formula for Normal Flame Velocity

It might seem surprising that the heat of reaction Q of the mixture appears in the denominator of the formula (5.4.8) for the flame velocity. This seems to be obvious that as the heat of reaction is increased the flame velocity rises too. The paradox is explained by the fact that a rise in Q also causes a rise in the combustion temperature, so that the increase in the integral of the reaction rate even larger than the increase in Q in the denominator. It should be noticed also that the initial fuel concentration enters formula (5.4.8) both in the denominator as well as through the fact that as a_0 increases, T_b increases and the rate of reaction also becomes larger in the integration region.

Since the density is proportional to the pressure and the reaction rate depends on the pressure exponentially as P^n (where n is the order of the reaction), we find that

$$u_n \propto P^{\frac{n}{2}-1}. \tag{5.5.1}$$

Thus, for bimolecular reactions the flame velocity is independent of the pressure, and it rises with pressure for trimolecular reactions, and for the first order reactions it falls with pressure.

For the second order reactions the rate of chemical change is determined by the number of binary collisions between reactive molecules, which is independent of the pressure. The mean free path and the collision time decrease in inverse proportion to the pressure while the speed of the molecules is independent of the pressure. This conclusion steams also from the Eq. (5.2.3), which says that the flame velocity is proportional to a_s and, thus, is independent of *P*. Furthermore, it can be stated that the propagation of a flame in which the chemical change involves an arbitrary number of bimolecular reactions will be self-similar as the pressure is changed. This is because as the pressure is increased, all the spatial scales (the dimensions of the heating and chemical reaction zones) and temporal scales (mean collision times of the molecule, chemical reaction time) are decreased in proportion to $1/P$ while the number of collisions, the thermal velocity of the molecules, and the flame velocity are unchanged. This similarity is destroyed in the case of monomolecular or three-molecular reactions for which it is necessary to take into account the characteristic decay time of the molecule or the lifetimes of metastable chemical complexes. On the other hand, the combustion mass flux $\rho_0 u_n$ increases with pressure. Since $\rho_0 u_n \propto P^{n/2}$, this means that the reaction rate increases with pressure. According to experimental data, the dependence on pressure of the flame velocity varies from $u_n \propto P^{-0.5}$ ($\rho_0 u_n \propto P^{0.5}$) for slowly burning mixtures (hydrocarbons) to $u_n \propto P^{0.5}$ ($\rho_0 u_n \propto P^{1.5}$) for rapidly burning mixtures (rich carbon monoxide-oxygen mixtures, hydrogen-oxygen mixtures).

Problems

5.1. Estimate width of a flame propagating in 6% CH_4+air ($\chi = \lambda/\rho C_p = 0.3\,\text{cm}^2/\text{s}$; $U_f = 5\,\text{cm/s}$) and in $2\,H_2+O_2$, ($\chi = 0.3\,\text{cm}^2/\text{s}$, $U_f = 1000\,\text{cm/s}$).

5.2. Proof that the velocity of a reaction propagating due to thermal conduction or diffusion is less than the sound speed.

5.3. Assuming that for a given combustion mixture the characteristic time of the reaction is $\tau_R = 3 \cdot 10^{-11} \exp((180\,\text{kcal/mol})/RT)\text{s}$, and $\chi = 0.3(T/298\,\text{K})^{1.7}\,\text{cm}^2/\text{s}$ calculate the values of the laminar flame velocity and its thickness, if the adiabatic flame temperature is 2500 K.

5.4. How the laminar flame velocity depends on initial temperature and pressure of the combustible mixture? Explain.

5.5. For a flame propagating in the two-dimensional channel the quenching width of the channel can be defined as the distance between the walls such that the rate of heat production is equal to the rate of heat loss to the walls. Assuming that the reaction rate, heat of reaction and thermal conduction is known derive formula for the minimal width of 2D channel at which the flame can propagate. How the quenching width depends on the flame velocity?

Chapter 6
Introduction to Hydrodynamics of Ideal Fluids

The combustion of a gas mixture is necessarily accompanied by motion of the gas. The process of combustion, flame propagation, detonation and explosion, are therefore not only a chemical phenomena but also one of gas dynamics. The goal of this chapter is to provide the basic knowledge in gas dynamics, which is necessary to study dynamics of combustion processes. Those who had taken course in fluid mechanics may find in this chapter some materials, which are missed in university courses.

6.1 The Fluid Dynamics

Fluid dynamics concerns itself with the study of the motion of fluids. When we are talking about fluids we keep in mind liquids or gases. Since the phenomena considered in fluid dynamics are macroscopic, a fluid is regarded as a continuous medium. This means that any small volume element in the fluid is always supposed so large that it still contains a very great number of molecules. Accordingly, when we speak of infinitely small elements of volume, we shall always mean those, which are "physically" infinitely small, i.e. very small compared with the volume of the body under consideration, but large compared with the distances between the molecules. The expressions *fluid particle* and *point in a fluid* are to be understood in a similar sense. If, for example, we speak of the displacement of some fluid particle, we mean the displacement of a volume element containing many molecules, though still regarded as a point.

The mathematical description of the state of a moving fluid is effected by means of functions, which give the distribution of the fluid velocity $\mathbf{u} = \mathbf{u}(\mathbf{r}, t)$ and of any two thermodynamic quantities pertaining to the fluid, for instance the pressure $P(\mathbf{r}, t)$ and the density $\rho(\mathbf{r}, t)$. All the thermodynamic quantities are determined by the values of any two of them, together with the equation of state. Therefore, if we are given five quantities, namely the three components of the velocity $\mathbf{u}(\mathbf{r}, t)$, the pressure $P(\mathbf{r}, t)$ and the density $\rho(\mathbf{r}, t)$, the state of the moving fluid is completely determined.

M.A. Liberman, *Introduction to Physics and Chemistry of Combustion*,

DOI: 10.1007/978-3-540-78759-4_6,

The situation is opposite to that we deal in mechanics when we can describe motion of a single particle if we do know how the radius vector changes in time $\mathbf{r} = \mathbf{r}(t)$. When the problem of interest involves many particles, we, in principle, can also introduce a radius vector $\mathbf{r}_i = \mathbf{r}_i(t)$ for each particle. However, for a large number of particles not only analytical but also numerical solution of the problem becomes hopeless. For example, 1 cm^3 of a gas under the normal conditions (room temperature, atmospheric pressure) contains 10^{19} particles. There is no and there will not be computers in the nearest future which can calculate such huge number of equations of motions. Besides, for any practical applications it is sufficient to know only averaged magnitudes characterizing macroscopic properties of the particles flow.

Thus, in fluid dynamics approach we need to know the velocity of the flow $\mathbf{u} = \mathbf{u}(\mathbf{r}, t)$, the density $\rho = \rho(\mathbf{r}, t)$, the pressure $P = P(\mathbf{r}, t)$, etc. in some fixed reference frame. All these quantities are, in general, functions of the coordinates $\mathbf{r} = (x, y, z)$ and of the time t. We emphasize that any function, for example $\mathbf{u}(\mathbf{r}, t)$, is the velocity of the fluid at a given point $\mathbf{r} = (x, y, z)$ in space and at a given time t, i.e. it refers to fixed points in space and not to specific particles of the fluid. In the course of time, the latter move in space.

We first introduce the notion of a fluid element, with a size of an element much smaller than the typical length scale of the flow $L_V \ll L_{flow}$, so that any part of the fluid element has the same velocity, density, pressure, etc. as the other parts of the element. From the microscopic point of view a fluid element contains huge number of particles (atoms or molecules). This means that the size of the fluid element is much larger than the mean free path – the average distance that molecules or atoms pass between collisions. We also assume that the fluid or gas is in local thermodynamic equilibrium for every fluid element. Otherwise we cannot define the thermodynamic parameters of a fluid element ρ, P, T, etc.

The velocity of the fluid element with coordinate $\mathbf{r} = \mathbf{r}(t)$ can be calculated as

$$\mathbf{u} = \frac{d\mathbf{r}}{dt}. \tag{6.1.1}$$

In general, thermodynamic function of this fluid element, for example, density changes in time and it is also function of space coordinates, which can be written as $\rho = \rho(\mathbf{r}(t), t)$. This means that the time derivative of density consists of two parts; one is related to the partial time derivative of density, and another is due to space coordinate time dependence

$$\frac{d\rho}{dt} = \frac{\partial\rho}{\partial t} + \frac{d\mathbf{r}}{dt} \cdot \frac{d\rho}{d\mathbf{r}} = \frac{\partial\rho}{\partial t} + \frac{d\mathbf{r}}{dt} \cdot \nabla\rho = \frac{\partial\rho}{\partial t} + \mathbf{u} \cdot \nabla\rho. \tag{6.1.2}$$

The combination $\frac{\partial}{\partial t} + \mathbf{u} \cdot \nabla$ is called the substantive derivative. If we have a vector value $\mathbf{u} = \left(u_x; u_y; u_z\right)$, then the substantive derivative for any component of the vector is

$$\frac{du_x}{dt} = \frac{\partial u_x}{\partial t} + (\mathbf{u} \cdot \nabla)u_x, \tag{6.1.3}$$

and for the whole vector value it becomes

$$\frac{d\mathbf{u}}{dt} = \frac{\partial \mathbf{u}}{\partial t} + (\mathbf{u} \cdot \nabla)\mathbf{u}, \tag{6.1.4}$$

where

$$(\mathbf{u} \cdot \nabla)\mathbf{u} = \left(u_x \frac{\partial}{\partial x} + u_y \frac{\partial}{\partial y} + u_z \frac{\partial}{\partial z}\right)\mathbf{u} = u_x \frac{\partial \mathbf{u}}{\partial x} + u_y \frac{\partial \mathbf{u}}{\partial y} + u_z \frac{\partial \mathbf{u}}{\partial z}. \tag{6.1.5}$$

6.2 The Equation of Continuity

The complete set of hydrodynamic equations corresponds to the conservation laws of mass, momentum and energy together with the equation of state describing thermodynamic properties of a fluid. In hydrodynamics we assume that all thermodynamic properties of the fluid under consideration are known. In this chapter we will discuss the hydrodynamic equations of an ideal fluid neglecting the transport phenomena related to the thermal motion of particles such as viscosity, thermal conduction, etc.

We shall now derive the fundamental equations of fluid dynamics. Let us begin with the equation, which expresses the conservation of matter. Consider first a simple example of a tank with two tubes of cross sections S_1 and S_2 as shown in Fig. 6.1. A fluid of density ρ_1 flows into the tank through the first tube with a uniform velocity U_1, and it moves out through the second tube with another uniform velocity U_2 and with another density ρ_2.

The fluid mass that enters the tank in a time interval dt is $\rho_1 S_1 U_1 dt$ and the mass of the fluid leaving the tank is $\rho_2 S_2 U_2 dt$, so that the resulting change of mass in the tank is

$$dM = (\rho_1 S_1 U_1 - \rho_2 S_2 U_2)dt$$

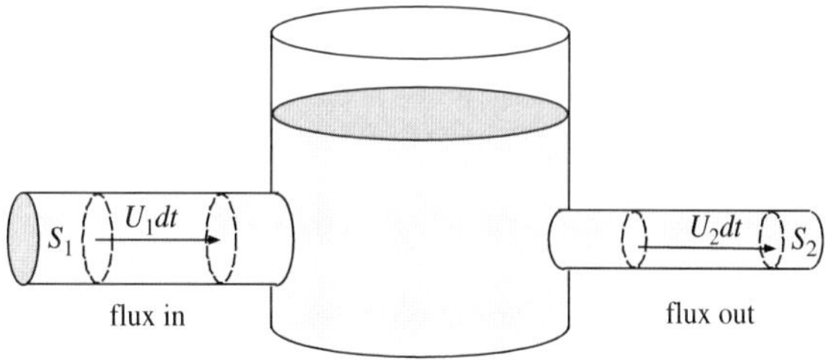

Fig. 6.1 A fluid flows through the tank

or

$$\frac{dM}{dt} = \rho_1 S_1 U_1 - \rho_2 S_2 U_2.$$

Obviously, if the fluid is an incompressible one, which means that its density does not change, $\rho_1 = \rho_2 = \text{const}$, flowing in a tube of a changing cross section, then the fluid velocity is related to the cross section as $U_1 S_1 = U_2 S_2$. Which means that the larger the cross section of a tube, the slower the velocity of the flow being inversely proportional to the cross section area.

Consider a volume V_0, which is fixed in the laboratory reference frame with the fluid flowing in and out of the volume (Fig. 6.2). The mass of fluid in this volume is $\int \rho dV$, where ρ is the fluid density, and the integration is taken over the volume V_0. The mass of fluid flowing in unit time through an element $d\mathbf{S} = \mathbf{n} dS$ of the surface bounding this volume is $\rho \mathbf{u} d\mathbf{S}$, with the magnitude of the vector $d\mathbf{S} = \mathbf{n} dS$ being equal to the area of the surface element and directed along the normal vector.

The total mass in the volume is the integral

$$M = \int_{V_0} \rho dV. \tag{6.2.1}$$

The change per unit time in the mass of fluid in the volume V_0 is

$$\frac{\partial M}{\partial t} = \int_{V_0} \frac{\partial \rho}{\partial t} dV. \tag{6.2.2}$$

Thus, the law of mass conservation may be written by equating the two expressions

$$(\text{Change of mass per unit time}) = -(\text{Total mass flux outwards}). \tag{6.2.3}$$

The mass flux through an elementary surface area $d\mathbf{S} = \mathbf{n} dS$ is

$$\frac{\rho \mathbf{u} \cdot d\mathbf{S} dt}{dt} = \rho \mathbf{u} \cdot d\mathbf{S}. \tag{6.2.4}$$

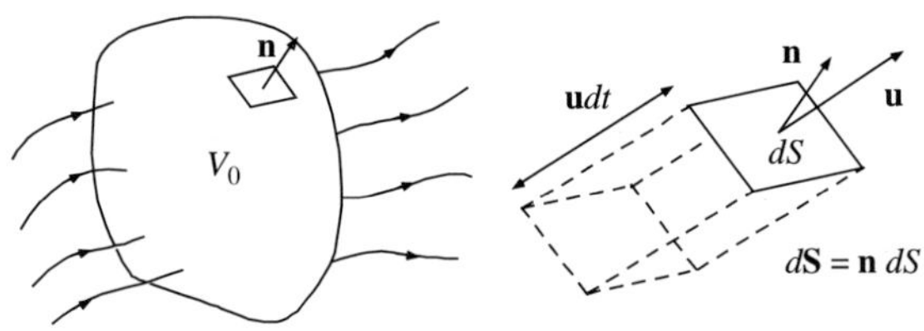

Fig. 6.2 Calculation of mass flux through volume V_0; **n** is a normal unit vector to the volume surface

The total flux is given by the surface integral, which can be transformed into a volume integral with the help of the Gauss theorem:

$$(\text{Total mass flux outwards}) = \int_{S_0} \rho\mathbf{u} \cdot d\mathbf{S} = \int_{V_0} \nabla \cdot (\rho\mathbf{u})dV. \tag{6.2.5}$$

Then the conservation of mass Eq. (6.2.3) takes the form

$$\int_{V_0} \frac{\partial \rho}{\partial t} dV = -\int_{V_0} \nabla \cdot (\rho\mathbf{u})dV, \tag{6.2.6}$$

or

$$\int_{V_0} \left[\frac{\partial \rho}{\partial t} + \nabla \cdot (\rho\mathbf{u})\right] dV = 0. \tag{6.2.7}$$

Since the Eq. (6.2.7) must hold for any volume, the integrand in (6.2.7) must vanish, and we come to the equation of mass conservation, which is called *the equation of continuity*:

$$\frac{\partial \rho}{\partial t} + \nabla \cdot (\rho\mathbf{u}) = 0. \tag{6.2.8}$$

Using $\nabla \cdot (\rho\mathbf{u}) = \rho\nabla \cdot \mathbf{u} + \mathbf{u} \cdot \nabla\rho$, we can rewrite (6.2.8) as

$$\frac{\partial \rho}{\partial t} + \rho\nabla \cdot \mathbf{u} + \mathbf{u} \cdot \nabla\rho = 0. \tag{6.2.9}$$

The vector $\mathbf{j} = \rho\mathbf{u}$ is called the *mass flux density*. Its direction is that of the motion of the fluid, while its magnitude equals the mass of fluid flowing in unit time through unit area perpendicular to the velocity.

6.3 The Euler Equation

The Newton law for a fluid element of a small fixed mass $\delta M = \rho\delta V$ is

$$\delta M \frac{d\mathbf{u}}{dt} = \delta\mathbf{F}. \tag{6.3.1}$$

The force acting on an elementary surface $d\mathbf{S} = \mathbf{n}dS$ is the product of pressure and surface area $d\mathbf{S} = \mathbf{n}dS$

$$d\ \delta\mathbf{F}_{\text{pressure}} = -P\mathbf{n}dS = -Pd\mathbf{S}. \tag{6.3.2}$$

We take the negative sign in (6.3.2), since the pressure force acts inward, while the normal vector **n** and the elementary surface area d**S** conventionally point outwards. The total force acting on the volume is equal to the surface integral of (6.3.2), which can be transformed into the integral over the volume:

$$d\,\delta\mathbf{F}_{\text{pressure}} = \oint_{S} P d\mathbf{S} = -\int_{\delta V} \nabla P dV \approx -\nabla P \delta V. \qquad (6.3.3)$$

Hence we see that the fluid surrounding any volume element δV exerts on that element a force $-\nabla P = -\text{grad}P$. In other words, we can say that a force $-\nabla P$ acts on unit volume of the fluid. In general, the gravity force acting on the fluid element is

$$d\,\delta\mathbf{F}_{\text{gravity}} = \rho\mathbf{g}\delta V. \qquad (6.3.4)$$

Substituting the obtained expressions for the forces into (6.3.1), we come to the equation

$$\rho\delta V\frac{d\mathbf{u}}{dt} = (-\nabla P + \rho\mathbf{g})\,\delta V. \qquad (6.3.5)$$

The derivative $d\mathbf{u}/dt$ must be expressed in terms of quantities referring to points fixed in space. Notice that the change in the velocity of the given fluid particle during the time dt consists of two parts, namely the change during dt in the velocity at a point fixed in space, and the difference between the velocities (at the same instant) at two points $d\mathbf{r}$ apart, where $d\mathbf{r}$ is the distance moved by the given fluid particle during the time dt. The first part is $(\partial\mathbf{u}/\partial t)dt$ where the derivative $\partial\mathbf{u}/\partial t$ is taken for constant x,y,z i.e. at the given point in space. The second part is $(d\mathbf{r}\cdot\nabla)\mathbf{u}$. Thus, taking into account the expression for the substantive derivative we find

$$\rho\frac{\partial\mathbf{u}}{\partial t} + \rho(\mathbf{u}\cdot\nabla)\mathbf{u} = -\nabla P + \rho\mathbf{g}. \qquad (6.3.6)$$

This is the equation of motion of the fluid, which is called *Euler equation* (L. Euler, 1755).

In deriving the equations of motion we have taken no account of processes of energy dissipation, which may be consequence of internal friction (viscosity) in the fluid and heat exchange between different parts. Such fluids are said to be *ideal.* Below in this section we will be dealing mainly with *ideal fluids* and gases. The absence of heat exchange between different parts of the fluid means that the process is adiabatic throughout the fluid. Thus the motion of an *ideal fluid* must necessarily be *adiabatic* and the entropy of any particle of fluid remains constant. Thus the condition for adiabatic motion is

$$dS/dt = 0, \tag{6.3.7}$$

where the total derivative with respect to time denotes the rate of change of entropy for a given fluid particle as it moves. So that the equation describing adiabatic motion of an ideal fluid is

$$\frac{\partial S}{\partial t} + (\mathbf{u} \cdot \nabla)S = 0. \tag{6.3.8}$$

Using the continuity equation, we can rewrite it as "continuity equation" for entropy:

$$\frac{\partial(\rho S)}{\partial t} + \nabla \cdot (\rho S\mathbf{u}) = 0. \tag{6.3.9}$$

The product $(\rho S\mathbf{u})$ is the *entropy flux density*.

If the entropy is constant throughout the volume of the fluid at some initial instant, it retains everywhere the same constant at all times and for any subsequent motion of the fluid. In this case a motion is called *isentropic* and the adiabatic equation is simply

$$S = \text{constant}. \tag{6.3.10}$$

Using the thermodynamic relation $dH = TdS + VdP$, where H is enthalpy per unit mass of fluid, $V = 1/\rho$ is the specific volume, and T is the temperature, we obtain for $S = \text{const}$

$$dH = VdP = dP/\rho$$

and therefore

$$\frac{1}{\rho}\nabla P = \nabla H. \tag{6.3.11}$$

Taking into account Eq. (6.3.11), equation of motion takes the form

$$\frac{\partial \mathbf{u}}{\partial t} + (\mathbf{u} \cdot \nabla)\mathbf{u} = -\nabla H \tag{6.3.12}$$

(for the sake of simplicity, we did not consider gravity force in the last equation).

With help of known formula of vector analysis,

$$\frac{1}{2}\nabla(\mathbf{u}^2) = \mathbf{u} \times \nabla \times \mathbf{u} + (\mathbf{u} \cdot \nabla)\mathbf{u}. \tag{6.3.13}$$

we can rewrite the (6.3.12) as

$$\frac{\partial \mathbf{u}}{\partial t} - \mathbf{u} \times \nabla \times \mathbf{u} = -\nabla\left(H + \frac{1}{2}\mathbf{u}^2\right). \tag{6.3.14}$$

Acting by operation $(\nabla\times)$ to both sides of the last equation, we obtain the Euler equation for isentropic flow, which involves only velocities

$$\frac{\partial}{\partial t}(\nabla \times \mathbf{u}) = \nabla \times (\mathbf{u} \times \nabla \times \mathbf{u}). \tag{6.3.15}$$

The equations of motion have to be supplemented by the boundary conditions that must be satisfied at the surfaces bounding the fluid. For an ideal fluid, the boundary condition is simply that the fluid cannot penetrate a solid surface. This means that the component of the fluid velocity normal to the bounding surface must vanish if that surface is at rest:

$$(\mathbf{u} \cdot \mathbf{n}) = u_n = 0. \tag{6.3.16}$$

In the case of a moving surface, u_n must be equal to the corresponding component of the velocity of the surface. At a boundary between two ideal fluids, the condition is that the pressure and the velocity component normal to the surface of separation must be the same for both fluids, and each of these velocity components must be equal to the corresponding component of the velocity of the surface.

A complete system of equations of fluid dynamics should be five in number. For an ideal fluid these are the Euler equations, the continuity equation, and the adiabatic equation.

6.4 Conservation of Energy

Similar to the derivation of the continuity equation we consider a volume, which is fixed in the laboratory reference frame. Let us choose some volume element fixed in space. The total energy E in the volume is given by the integral of the thermal energy per unit volume $\rho\varepsilon$, where ε is the thermal (internal) energy per unit mass, and the kinetic energy per unit volume $\rho\mathbf{u}^2/2$

$$E = \int_{V_0} \left(\rho\varepsilon + \rho\frac{u^2}{2}\right) dV. \tag{6.4.1}$$

The equation of energy conservation is

$$\begin{aligned}&(\text{change of energy per unit time})\\&\quad = -(\text{energy flux outwards}) + (\text{work of external forces per unit time}).\end{aligned} \tag{6.4.2}$$

The change of energy per unit time in a fixed volume is

$$\frac{\partial E}{\partial t} = \int_{V_0} \frac{\partial}{\partial t}\left(\rho\varepsilon + \rho\frac{u^2}{2}\right)dV, \tag{6.4.3}$$

and the energy flux may be calculated similar to the mass flux. Using transformation from integral over the surface into a volume integral with the help of the Gauss theorem (compare with Eq. (6.2.5)) we obtain:

$$\int_{S_0} \left(\rho\varepsilon + \rho\frac{u^2}{2}\right)\mathbf{u}\cdot d\mathbf{S} = \int_{V_0} \nabla\cdot\left(\rho\mathbf{u}\left[\varepsilon + \frac{u^2}{2}\right]\right)dV. \tag{6.4.4}$$

The external forces acting on the fluid are the pressure force applied to the fluid surface of the volume and the gravity force that acts on all fluid particles inside the volume. The work of the pressure force per unit time performed on the fluid is the product of the elementary force and the fluid velocity: $-Pd\mathbf{S}\cdot\mathbf{u}$. The total work of the pressure force per unit time is the surface integral of the elementary work

$$\int_{S_0} (-P\mathbf{u})\cdot d\mathbf{S} = -\int_{V_0} \nabla\cdot(\mathbf{u}P)dV. \tag{6.4.5}$$

The elementary work of the gravity force per unit time is $\rho\mathbf{g}\cdot\mathbf{u}dV$ and the total work is

$$\int_{V_0} \rho\mathbf{g}\cdot\mathbf{u}dV. \tag{6.4.6}$$

Similar to the derivation of continuity equation, substituting all the terms into (6.4.2) we obtain equation of energy conservation for an ideal fluid

$$\frac{\partial}{\partial t}\left(\rho\varepsilon + \rho\frac{u^2}{2}\right) + \nabla\cdot\left(\rho\mathbf{u}\left[\varepsilon + \frac{P}{\rho} + \frac{u^2}{2}\right]\right) = \rho\mathbf{g}\cdot\mathbf{u}. \tag{6.4.7}$$

The term with the partial derivative $\partial/\partial t$ can be transformed as

$$\frac{\partial}{\partial t}\left(\rho\frac{u^2}{2}\right) = \frac{1}{2}u^2\frac{\partial\rho}{\partial t} + \rho\mathbf{u}\cdot\frac{\partial\mathbf{u}}{\partial t}.$$

Using the continuity equation for $\partial\rho/\partial t$, and the equation of motion (Euler equation for $\partial\mathbf{u}/\partial t$), we rewrite this as

$$\frac{\partial}{\partial t}\left(\rho\frac{u^2}{2}\right) = -\frac{1}{2}u^2\nabla(\rho\mathbf{u}) - \rho\mathbf{u}\cdot(\mathbf{u}\cdot\nabla)\mathbf{u} - \mathbf{u}\cdot\nabla P. \tag{6.4.8}$$

The second term in the right hand side of (6.4.8) we shall transform using vector analysis formula $\frac{1}{2}\nabla(\mathbf{u}^2) = \mathbf{u}\times\nabla\times\mathbf{u} + (\mathbf{u}\cdot\nabla)\mathbf{u}$ (see Eq. (6.3.13)), so that it will be

$$\rho\mathbf{u}\cdot(\mathbf{u}\cdot\nabla)\mathbf{u} = \frac{1}{2}\rho\mathbf{u}\cdot\nabla(\mathbf{u}^2).$$

Let us consider this equation without gravity force and transform it using the continuity equation and the thermodynamic relation $dH = TdS + \frac{1}{\rho}dP$

$$\mathbf{u}\cdot\nabla P = \rho\nabla H - \rho T\nabla S.$$

So that (6.4.8) becomes

$$\frac{\partial}{\partial t}\left(\rho\frac{u^2}{2}\right) = -\rho\mathbf{u}\cdot\nabla\left(\frac{1}{2}\mathbf{u}^2 + H\right) - \frac{1}{2}u^2\nabla\cdot(\rho\mathbf{u}) + \rho T\mathbf{u}\cdot\nabla S. \tag{6.4.9}$$

In order to transform the derivative $\partial(\rho\varepsilon)/\partial t$, we use the thermodynamic relation

$$d\varepsilon = TdS - PdV = TdS + \frac{P}{\rho^2}d\rho.$$

Since $\varepsilon + PV = \varepsilon + P/\rho = H$ is the enthalpy per unit mass, we obtain

$$d(\varepsilon\rho) = \varepsilon d\rho + \rho d\varepsilon = \varepsilon d\rho + \rho TdS + \frac{P}{\rho}d\rho = Hd\rho + \rho TdS, \tag{6.4.10}$$

and for $\partial(\rho\varepsilon)/\partial t$ we obtain

$$\frac{\partial(\rho\varepsilon)}{\partial t} = H\frac{\partial\rho}{\partial t} + \rho T\frac{\partial S}{\partial t}. \tag{6.4.11}$$

Using for the terms in the right hand side of this equation the continuity equation for time variation of density and adiabatic motion equation for an ideal fluid, $\frac{\partial S}{\partial t} + (\mathbf{u}\cdot\nabla)S = 0$, we can write (6.4.11) as

$$\frac{\partial(\rho\varepsilon)}{\partial t} = -H\nabla\cdot(\rho\mathbf{u}) - \rho T\mathbf{u}\cdot\nabla S. \tag{6.4.12}$$

Combining (6.4.9) and (6.4.12), we find the change in the energy

$$\frac{\partial}{\partial t}\left(\rho\frac{u^2}{2}+\rho\varepsilon\right) = -\rho\mathbf{u}\cdot\nabla\left(\frac{1}{2}\mathbf{u}^2+H\right) - \left(\frac{1}{2}\mathbf{u}^2+H\right)\nabla\cdot(\rho\mathbf{u}), \qquad (6.4.13)$$

or

$$\frac{\partial}{\partial t}\left(\rho\frac{u^2}{2}+\rho\varepsilon\right) = -\nabla\cdot\left[\rho\mathbf{u}\left(\frac{1}{2}\mathbf{u}^2+H\right)\right]. \qquad (6.4.14)$$

Let us integrate the last equation over some volume V:

$$\frac{\partial}{\partial t}\int_V\left(\rho\frac{u^2}{2}+\rho\varepsilon\right)dV = -\int_V\nabla\cdot\left[\rho\mathbf{u}\left(\frac{1}{2}\mathbf{u}^2+H\right)\right]dV.$$

Transforming the volume integral on the right side into a surface integral with help of Gauss theorem, we can rewrite this equation as

$$\frac{\partial}{\partial t}\int_V\left(\rho\frac{u^2}{2}+\rho\varepsilon\right)dV = -\oint_S\rho\mathbf{u}\left(\frac{1}{2}\mathbf{u}^2+H\right)\cdot d\mathbf{S}. \qquad (6.4.15)$$

The left-hand side in Eq. (6.4.15) is the rate of change of the energy of the fluid in some given volume, while the right-hand side is the amount of energy flowing out of this volume per unit time. Hence we see that the expression

$$\rho\mathbf{u}\left(\frac{1}{2}\mathbf{u}^2+H\right) \qquad (6.4.16)$$

is nothing but the *energy flux density* vector. Its magnitude is the amount of energy transporting in unit time through unit area perpendicular to the direction of the velocity.

The expression (6.4.16) shows that any unit mass of fluid carries with it during its motion an amount of energy $(H + u^2/2)$. The fact that the enthalpy (H) appears here, and not the internal energy (ε) has a simple physical significance. Putting $H = \varepsilon + P/\rho$, we can write the flux of energy through a closed surface in the form

$$-\oint_S\rho\mathbf{u}\left(\frac{1}{2}\mathbf{u}^2+\varepsilon\right)\cdot d\mathbf{S} - \oint_S P\mathbf{u}\cdot d\mathbf{S}. \qquad (6.4.17)$$

The first term is the sum of kinetic and internal energy transported through the surface in unit time by the mass of fluid, and the second term is the work done by pressure forces on the fluid within the surface.

Summarizing, we write down the total set of hydrodynamic equations for an ideal fluid

$$\frac{\partial \rho}{\partial t} + \nabla \cdot (\rho \mathbf{u}) = 0, \qquad (\textit{Continuity})$$

$$\rho \frac{\partial \mathbf{u}}{\partial t} + \rho(\mathbf{u} \cdot \nabla)\mathbf{u} = -\nabla P + \rho \mathbf{g} \qquad (\textit{Euler})$$

$$\frac{\partial}{\partial t}\left(\rho\varepsilon + \rho\frac{u^2}{2}\right) + \nabla \cdot \left(\rho \mathbf{u}\left[\varepsilon + \frac{P}{\rho} + \frac{u^2}{2}\right]\right) = \rho \mathbf{g} \cdot \mathbf{u} \qquad (\textit{Energy})$$

As has been said in earlier, the state of a moving fluid is determined by five quantities: three components of the velocity and by two thermodynamic functions, for example, the pressure and the density. Accordingly, a complete system of equations of fluid dynamics should be five in number. For an ideal fluid these are the equation of continuity, the Euler equations, and the equation of energy, which in the case of ideal fluid is the adiabatic equation.

6.5 The Equation of State

The set of hydrodynamic equations includes two scalar equations (*continuity* and *energy*) and one vector equation (*Euler*), which makes totally five partial differential equations. These equations involve three unknown scalar functions: density ρ, pressure P, internal energy ε, and one vector function $\mathbf{u}$, which imply total 6 unknown functions. The missing equation is not a hydrodynamic equation, but it is the equation of state describing thermodynamic properties of a fluid (or gas). It is supposed in fluid mechanics that thermodynamic properties of the fluid are known. In general, the equation of state can be written as an algebraic equation

$$P = P(\rho, \varepsilon). \tag{6.5.1}$$

More often it is convenient to introduce temperature T of the substance and present the equation of state in the form

$$P = P(\rho, T), \qquad \varepsilon = \varepsilon(\rho, T). \tag{6.5.2}$$

For example, the equation of state for an ideal gas is

$$P = (\gamma - 1)C_v \rho T, \qquad \varepsilon = C_v T, \tag{6.5.3}$$

where $\gamma = C_P/C_V$ is the adiabatic exponent of the gas, and C_P and C_V are the specific heat at constant pressure and constant volume, respectively. As a rule, both these values can be considered as constants for a perfect gas. Particularly $\gamma = 5/3$ for a one-atomic gas and $\gamma = 7/5$ for a two-atomic gas. Since the Earth atmosphere consists mostly of nitrogen and oxygen molecules, then for the air one can also take $\gamma \approx 7/5$.

Two Eqs. (6.5.3) can be combined into one that does not contain temperature

$$\varepsilon = \frac{1}{\gamma - 1}\frac{P}{\rho}. \tag{6.5.4}$$

Sometimes we do not need to use the complete set of the hydrodynamic equations. For example, for certain flows it is possible to specify in advance an unambiguous relation between pressure and density $P = P(\rho)$. The most typical examples are considered below.

Adiabatic Flow of an Ideal Gas The adiabatic flow takes place, when exchange of internal energy between fluid (gas) elements due to the random thermal motion of fluid particles is negligible. The approximation of negligible thermal motion implies that viscosity, thermal conduction, diffusion, etc. are zero. For the adiabatic approximation pressure is related to density as $P/\rho^\gamma = \text{const}$. The adiabatic approximation is applicable when we consider sound waves.

Isothermal Flow The isothermal flow corresponds to the case when temperature is constant everywhere, for example, when thermal conduction is very large. The equation of state is $P/\rho = \text{const}$.

Incompressible Fluids In many cases of the density of the gas flow may be supposed invariable, i.e. constant throughout its motion. This is approximation of incompressible flow. The condition of incompressible flows is justified when variations of density are much smaller than the density itself, i.e. $\Delta\rho/\rho \ll 1$. The equations of fluid dynamics are considerably simplified for an incompressible fluid. The equation of continuity for constant density becomes

$$\nabla \cdot \mathbf{u} = 0. \tag{6.5.5}$$

The Euler equation is not much simplified if we put $\rho = \text{constant}$, except that ρ can be taken under the gradient operator:

$$\frac{\partial \mathbf{u}}{\partial t} + (\mathbf{u} \cdot \nabla)\mathbf{u} = -\nabla\left(\frac{P}{\rho}\right) + \mathbf{g}. \tag{6.5.6}$$

Since the density is no longer an unknown function, the equations in fluid dynamics for an incompressible fluid can be taken to be equations involving the velocity only. These may be the equation of continuity (6.5.5) and Eq. (6.3.15):

$$\frac{\partial}{\partial t}(\nabla \times \mathbf{u}) = \nabla \times (\mathbf{u} \times \nabla \times \mathbf{u}). \tag{6.5.7}$$

As a matter of fact, an incompressible flow may be a particular case of adiabatic or isothermal flows.

Let us consider the conditions under which the fluid may be regarded as incompressible. When the pressure changes adiabatically by ΔP the density variation may be expressed through the pressure variation as

$$\Delta\rho = \left(\frac{\partial P}{\partial \rho}\right)_{ad}^{-1} \Delta P = \frac{1}{a_s^2}\Delta P, \tag{6.5.8}$$

where a_s is the sound speed. Let us estimate the pressure variations ΔP using the Euler equation. For characteristic length scale and velocity of the flow being L and U, we have and $\nabla P \propto \Delta P/L$, i.e. $\Delta P \propto \rho U^2$. Substituting this estimate in (6.5.8) we obtain $\Delta P \propto \rho U^2 \propto \Delta\rho a_s^2$, i.e. condition of small density variation is

$$\frac{\Delta\rho}{\rho} \propto \frac{U^2}{a_s^2} << 1,\ U^2 << a_s^2 \tag{6.5.9}$$

which means that condition for incompressible flow is that the fluid velocity is small compared with that of sound speed, $U^2 << a_s^2$.

In fact condition $U^2 << a_s^2$ is sufficient only for steady flow. In non-steady flow, a further condition must be fulfilled. Let L and τ be a length scale and time over which the fluid velocity undergoes significant changes. Estimating in the Euler equation terms $d\mathbf{u}/dt \propto u/\tau$ and $(1/\rho)\nabla P \propto \Delta P/\rho L$, we find, in order of magnitude, $u/\tau \approx \Delta P/L\rho$, or $\Delta P \approx Lu\rho/\tau$, and the corresponding change in ρ then can be estimated as $\Delta\rho \approx Lu\rho/a_s^2\tau$. Comparing the terms $\partial\rho/\partial t \approx \Delta\rho/\tau$ and $\rho\nabla \cdot \mathbf{u} \approx \rho u/L$ in the equation of continuity, we find that the derivative $\partial\rho/\partial t$ may be neglected if $\Delta\rho/\tau << \rho u/L$, or

$$L/a_s\tau << 1. \tag{6.5.10}$$

The condition (6.5.10) has an obvious meaning: the time L/a_s required for sound to pass the distance L must be small compared with the time τ during which the flow changes appreciably, so that sound propagates through the flow "instantaneously".

6.6 Hydrostatics

The simplest solutions of the hydrodynamic equations correspond to the case of a fluid at rest. In this case $\mathbf{u} = 0$, all time derivatives are zero, $\frac{\partial}{\partial t}(\ldots) = 0$, and the continuity equations of mass and energy conservation are satisfied identically

$$\frac{\partial \rho}{\partial t} + \nabla \cdot (\rho \mathbf{u}) = 0 \quad \Rightarrow \quad 0 = 0.$$

The Euler equation now expresses balance of the pressure force and the gravity force on any fluid element, and describes the mechanical equilibrium of the fluid

$$\rho \frac{\partial \mathbf{u}}{\partial t} + \rho(\mathbf{u} \cdot \nabla)\mathbf{u} = 0, \quad \Rightarrow \quad \nabla P = \rho \mathbf{g}, \tag{6.6.1}$$

If there is no external force, the equation of equilibrium is simply $\nabla P = 0$, i.e.; the pressure is the same at every point in the fluid.

If in addition the fluid (gas) can be regarded as incompressible, then the Eq. (6.6.1) can be integrated immediately. For the z-axis directed upward, and gravity acceleration, $\mathbf{g} = (0;\ 0;\ -g)$ the equation of hydrostatics becomes

$$\frac{\partial P}{dx} = \frac{\partial P}{dy} = 0, \quad \frac{dP}{dz} = -\rho g, \tag{6.6.2}$$

which expresses balance of the pressure force and the gravity force on any fluid element, so that

$$P = -\rho g z + \text{const.} \tag{6.6.3}$$

Constant is defined here from the boundary conditions. For example, if the fluid at rest has a free surface at height $z = h$, where an external pressure $P = P_0$, then, $P = P_0 + \rho g(h - z)$.

Let us find pressure distributions for the case of constant gravity acceleration in the models of an adiabatic, isothermal and incompressible gas. Let the pressure and the density at $z = 0$ are P_0 and ρ_0. Substituting the relation between pressure and density for an *adiabatic gas,* $P = P_0(\rho/\rho_0)^\gamma$, $T = T_0(\rho/\rho_0)^{\gamma-1}$ into Eq. (6.6.2) we come to the differential equation

$$\gamma P_0 \left(\frac{\rho}{\rho_0}\right)^{\gamma-1} \frac{d}{dz}\left(\frac{\rho}{\rho_0}\right) = -\rho g, \ \text{or} \ \left(\frac{\rho}{\rho_0}\right)^{\gamma-2} \frac{d}{dz}\left(\frac{\rho}{\rho_0}\right) = -\frac{\rho_0 g}{\gamma P_0}.$$

Integrating this equation, we obtain the density, pressure and temperature distributions

$$\rho = \rho_0\left(1 - \frac{\gamma - 1}{\gamma}\frac{\rho_0 g}{P_0}z\right)^{\frac{1}{\gamma-1}}, \qquad P = P_0\left(\frac{\rho}{\rho_0}\right)^{\gamma} = P_0\left(1 - \frac{\gamma - 1}{\gamma}\frac{\rho_0 g}{P_0}z\right)^{\frac{\gamma}{\gamma-1}},$$

$$T = T_0\left(1 - \frac{\gamma - 1}{\gamma}\frac{\rho_0 g}{P_0}z\right).$$

It is seen that, the characteristic length scale of the adiabatic atmosphere is

$$L \propto \frac{P_0}{\rho_0 g}.$$

Assuming air to be an ideal gas with the adiabatic exponent $\gamma = 7/5$, we find the air density, pressure and temperature changes with height. Taking the air density, pressure and temperature at the surface of the Earth, $\rho_0 = 1.3\,\mathrm{kg/m^3}$, $P_0 = 10^5\,\mathrm{Pa}$, $T_0 = 300\,\mathrm{K}$ and the gravity acceleration, $g = 9.8\,\mathrm{m/s^2}$, we obtain the temperature decrease with height, $dT/dz \approx -10^{-2}\,\mathrm{K/m}$. According to this estimate at the top of Everest ($z = 8.7\,\mathrm{km}$), the temperature is about $-50°$ when it is $40°$ at the sea level, and the level of snow at $0°$ in mountains starts at height of about 3 km. By this reason snow is always at the peaks of high mountains, for example, at Mont-Blanc, which is about 5 km. We see that the adiabatic approximation is a quite a good model for the Earth atmosphere.

In the case of an *isothermal gas* $P = P_0\frac{\rho}{\rho_0}$, and integrating the equation of hydrostatics we find the distributions of density and pressure

$$\rho = \rho_0 \exp\left(-\frac{\rho_0 g}{P_0}z\right) \qquad P = P_0 \exp\left(-\frac{\rho_0 g}{P_0}z\right).$$

For *incompressible gas* density is constant, $\rho = \rho_0$, and pressure changes as

$$P = P_0 - \rho_0 g z.$$

Let us consider equilibrium of a very large mass of gas like a star atmosphere, which is kept by the gravitational forces. The Newtonian gravitational potential, ϕ satisfies the equation

$$\Delta\phi = 4\pi G\rho, \tag{6.6.4}$$

where G is the Newtonian constant of gravitation, the gravitational acceleration is $-\nabla\phi$, and the force acting on a mass ρ is $-\rho\nabla\phi$. Then the condition of equilibrium is

$$\nabla P = -\rho\nabla\phi. \tag{6.6.5}$$

For large masses of a gas or liquid, the approximation of incompressibility cannot be used, the density ρ cannot be supposed constant. Dividing both sides (6.6.5) by ρ, taking the divergence of both sides, and using Eq. (6.6.4), we obtain

$$\nabla\left(\frac{1}{\rho}\nabla P\right) = -4\pi G\rho. \tag{6.6.6}$$

It must be emphasized that the present discussion concerns only mechanical equilibrium; Eq. (6.6.6) does not presuppose the existence of complete thermal equilibrium. If the body is not rotating, its shape is spherical when in equilibrium, and the density and pressure distributions will be spherically symmetrical. In spherical polar coordinates then Eq. (6.6.6) takes the form

$$\frac{1}{r^2}\frac{d}{dr}\left(\frac{r^2}{\rho}\frac{dP}{dr}\right) = -4\pi G\rho.$$

If fluid is not only in mechanical equilibrium but also in thermal equilibrium, then the temperature is the same at every point. For constant temperature, introducing the Gibbs free energy G per unit mass we have

$$dG = -SdT + VdP = VdP = \frac{1}{\rho}dP,$$

Since $\mathbf{g} = -\nabla\phi$, the equation of equilibrium (6.6.1) takes the form

$$\nabla\Phi = \mathbf{g}. \tag{6.6.7}$$

For a constant vector $\mathbf{g}$ directed along the negative z-axis we can write $\mathbf{g} = -\nabla(gz)$, and (6.6.7) becomes

$$\Phi + gz = \text{constant}. \tag{6.6.8}$$

This is condition for thermodynamic equilibrium of a system in an external gravity field.

A simple consequence of Eq. (6.6.1) is that if a fluid is in mechanical equilibrium in a gravitational field, the pressure can be function only of the altitude z. Indeed, if the pressure were different at different points with the same altitude, motion would result. From (6.6.1) it follows that the density is also a function of z only, as well as the temperature is therefore a function of z only. Thus, in mechanical equilibrium in a gravitational field, the pressure, density and temperature distributions depend only on the altitude. If, for example, the temperature is different at different points with the same altitude, then mechanical equilibrium is not possible.

6.7 A Stationary Flow. The Bernoulli Equation

The equations of fluid dynamics are much simplified in the case of *steady flow* with the velocity being constant in time. This means that $\mathbf{u}$ is a function of the coordinates only, so that $\partial \mathbf{u}/\partial t = 0$. We shall transform the Euler equation

$$\frac{\partial \mathbf{u}}{\partial t} + (\mathbf{u} \cdot \nabla)\mathbf{u} = -\frac{1}{\rho}\nabla P + \mathbf{g}$$

with the help of the vector formula

$$(\mathbf{u} \cdot \nabla)\mathbf{u} = \frac{1}{2}\nabla u^2 - \mathbf{u} \times [\nabla \times \mathbf{u}] = \frac{1}{2}\nabla u^2 - \mathbf{u} \times \boldsymbol{\omega},$$

where we introduced the definition of a flow vorticity as

$$\boldsymbol{\omega} = [\nabla \times \mathbf{u}]. \tag{6.7.1}$$

For an adiabatic flow, Eq. (6.3.14), obtained in Sect. 6.3 then reduces to

$$\frac{1}{2}\nabla(\mathbf{u}^2) - \mathbf{u} \times \nabla \times \mathbf{u} = -\nabla(H + \Phi), \tag{6.7.2}$$

or, using definition (6.7.1), we can write this equation as

$$\mathbf{u} \times \boldsymbol{\omega} = \nabla\left(\frac{1}{2}u^2 + H + \Phi\right). \tag{6.7.3}$$

We now introduce the concept of *streamlines*, which are lines such that the tangent to a streamline at any point gives the direction of the velocity at that point. The streamlines show direction of the fluid velocity, so that a streamline is a continuous line, the tangent of which at any point is in the direction of the velocity at this point $\vec{\tau} = \mathbf{u}/u$. According to the definition, the streamlines are determined by the system of differential equations:

$$\frac{dx}{u_x} = \frac{dy}{u_y} = \frac{dz}{u_z}. \tag{6.7.4}$$

The definition of a streamline is illustrated in Fig. 6.3, which also gives a typical sketch of the streamlines of a flow past a fixed cylinder. According to the definition, two streamlines cannot intersect, otherwise at the point of intersection the velocity would be directed in two different ways simultaneously. The only exception is the stagnation points of a flow, where the flow velocity is zero. A single streamline shows only the velocity direction in a flow, but provides no information about the velocity value. However, when several streamlines are drawn, then the distance between the streamlines tells us about the velocity value: the smaller the distance, the larger the velocity.

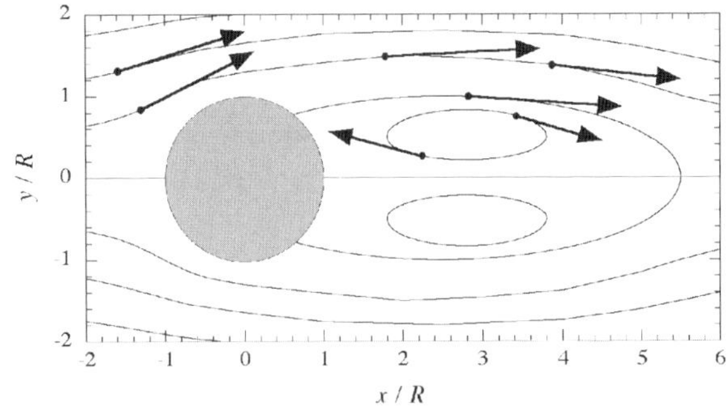

Fig. 6.3 Streamlines of a flow past a cylinder

If we choose a closed loop and draw all streamlines through all points of this loop, then we get a *stream-tube* (Fig. 6.4). When we consider a flow in a real tube then the tube plays the role of a stream tube too. The *particle path* shows motion of a fluid element in a flow.

In steady flow the streamlines do not vary with time, and coincide with the paths of the fluid particles. In non-steady flow this coincidence no longer occurs: the tangents to the streamlines give the directions of the velocities of fluid particles at various points in space at a given instant, whereas the tangents to the paths give the directions of the velocities of given fluid particles at various times.

Equation (6.7.3) can be integrated along a streamline. To do this we form the scalar product of Eq. (6.7.3) with the unit vector along a streamline $\boldsymbol{\omega}_l = \mathbf{u}/\mathrm{u}$, which is tangent to the streamline at each point. The projection of the gradient on any direction is the derivative in that direction, so that $\nabla \mathrm{H}$ is $\partial \mathrm{H}/\partial \mathbf{r}_l$. The vector $\mathbf{u} \times \boldsymbol{\omega}$ is perpendicular to $\mathbf{u}$, and its projection on the direction of $\vec{\tau} = \mathbf{u}/\mathrm{u}$ is zero. Thus we obtain from (6.7.3)

$$\frac{\partial}{\partial \mathrm{r}_l}\left(\frac{\mathrm{u}^2}{2} + \mathrm{H} + \Phi\right) = 0. \tag{6.7.5}$$

It follows from this equation that expression in bracket is constant along a streamline:

$$\frac{1}{2}\mathrm{u}^2 + \mathrm{H} + \Phi = \text{constant}. \tag{6.7.6}$$

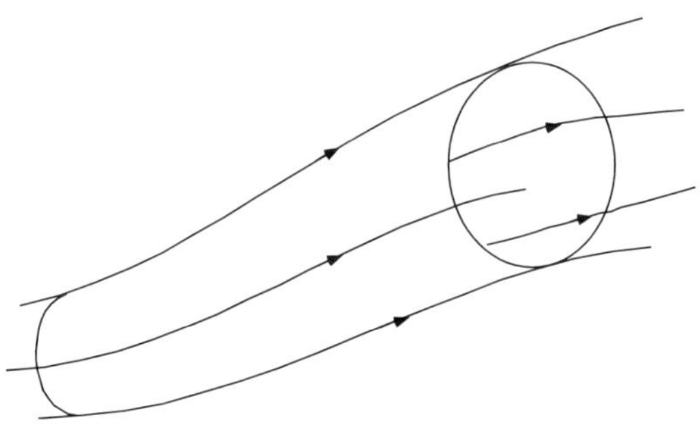

Fig. 6.4 A stream tube

Equation (6.7.6) is called the *Bernoulli equation* (D. Bernoulli, 1738). It is important to note that the constant is different for different streamlines. The Bernoulli integral expresses conservation of energy for a fluid element with the first term of Eq. (6.7.6) representing kinetic energy, the second term corresponding to internal energy and the last term describing potential energy of an element in a gravitational field.

Let us take the direction of gravity as the z-axis, with z directed upwards. Then the cosine of the angle between the directions of $\mathbf{g}$ and $\vec{\tau}_l$ is equal to the derivative $dz/d\vec{\tau}_l$, so that the projection of $\mathbf{g}$ on $\vec{\tau}_l$ is $(-gdz/d\vec{\tau}_l)$. Accordingly, we now have

$$\frac{\partial}{\partial r_l}\left(\frac{u^2}{2}+H+gz\right)=0,$$

and the Bernoulli equation states that along a streamline

$$\frac{1}{2}u^2+H+gz=\text{constant}. \qquad (6.7.7)$$

Consider examples illustrating application of the Bernoulli equation. Let us calculate pressure at the stagnation point in a flow around an obstacle shown in Fig. 6.5, where the fluid velocity $\mathbf{u}=0$.

Far ahead of the obstacle the flow is uniform with the flow velocity U_0 and pressure P_0, and we suppose that influence of gravity is negligible. Let us assume here that inner energy is constant so that $H=P/\rho$ and consider the Bernoulli integral that starts at infinity and ends at the stagnation point S:

$$\frac{u^2}{2}+\frac{P}{\rho}+\Phi=\frac{1}{2}U_0^2+\frac{P_0}{\rho}=\frac{P_s}{\rho}=\text{constant}$$

Thus, pressure at the stagnation point exceeds pressure in the flow ahead of the obstacle:

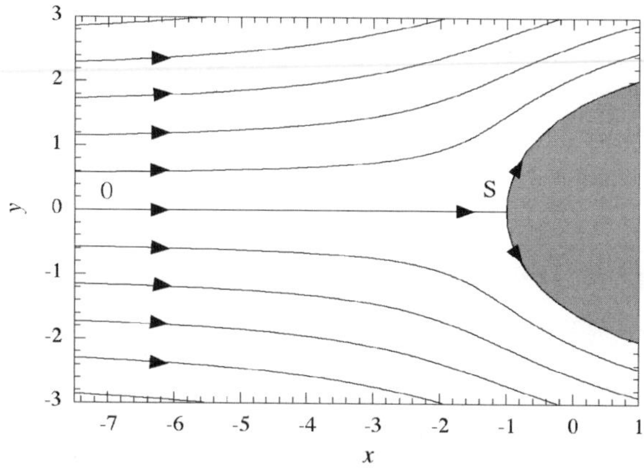

Fig. 6.5 A flow streamlines ahead of a cylinder

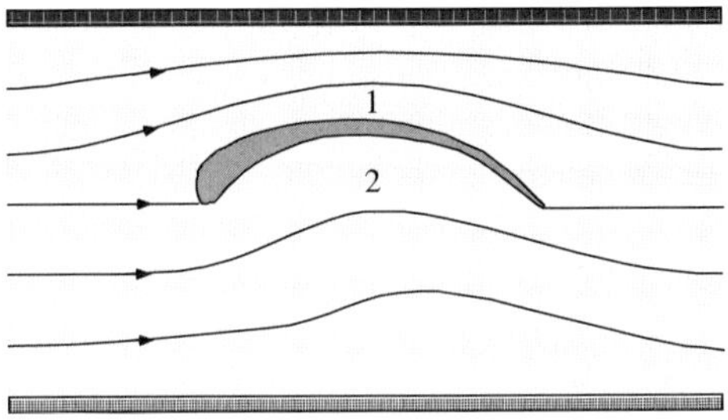

Fig. 6.6 Flow stream lines for asymmetric airfoil

$$\Delta P = P_s - P_0 = \frac{1}{2}\rho U_0^2. \tag{6.7.8}$$

A simple consequence of the Bernoulli integral is the lifting force acting on an asymmetric airfoil in a channel, shown in Fig. 6.6. Far ahead of the airfoil the flow is uniform with velocity U_0 and pressure P_0. Because of the asymmetric shape of the airfoil the effective cross-section (1) above the airfoil decreases, the fluid velocity increases in comparison with U_0 and pressure decreases $P_1 < P_0$.

On the contrary, the cross-section (2) increases in the flow below the airfoil, so that fluid velocity decreases and pressure increases in comparison with the initial one $P_2 > P_0$. As a result pressure below the airfoil is larger than pressure above it, and the resulting pressure force acts upwards.

6.8 The Conservation of Velocity Circulation

Let us consider a closed contour drawn in the fluid at some instant. We suppose it to be a "fluid contour", i.e. composed of the fluid particles that lie on it. In the course of time these particles move about, and the contour moves with them. The integral

$$\Gamma \equiv \oint \mathbf{u} \cdot d\mathbf{l}, \tag{6.8.1}$$

being calculated conventionally in the counter-clockwise direction taken along some closed contour C, is called the *velocity circulation*.

Let us calculate the time derivative

$$\frac{d}{dt}\oint \mathbf{u} \cdot d\mathbf{l}. \tag{6.8.2}$$

We have written here the total derivative with respect to time, since we are seeking the change in the circulation round a "fluid contour" as it moves about, and not round a contour fixed in space. To avoid confusion, we shall temporarily denote differentiation with respect to the coordinates by the symbol δ, retaining the symbol d for differentiation with respect to time. Next, we notice that an element dl of the length of the contour can be written as the difference $\delta\mathbf{r} = \mathbf{r}_2 - \mathbf{r}_1$ between the position vectors $\mathbf{r}$ of the points at the ends of the element. Thus we write the velocity circulation as $\oint \mathbf{u} \cdot \delta\mathbf{r}$. In differentiating this integral with respect to time, it must be kept in mind that not only the velocity but also the shape of contour changes. Hence, on taking the time differentiation under the integral, we must differentiate not only $\mathbf{u}$ but also $\delta\mathbf{r}$:

$$\frac{d}{dt}\oint \mathbf{u} \cdot \delta\mathbf{r} = \oint \frac{d\mathbf{u}}{dt} \cdot \delta\mathbf{r} + \oint \mathbf{u} \cdot \frac{d\delta\mathbf{r}}{dt}. \tag{6.8.3}$$

Since the velocity $\mathbf{u}$ is the time derivative of the position vector $\mathbf{r}$, we have

$$\mathbf{u} \cdot \frac{d\delta\mathbf{r}}{dt} = \mathbf{u} \cdot \delta\frac{d\mathbf{r}}{dt} = \mathbf{u} \cdot \delta\mathbf{u} = \delta\left(\frac{1}{2}u^2\right) \tag{6.8.4}$$

Thus, the second integral vanishes, since the integral of a total differential along a closed contour is zero, and we obtain from (6.8.3):

$$\frac{d}{dt}\oint \mathbf{u} \cdot \delta\mathbf{r} = \oint \frac{d\mathbf{u}}{dt} \cdot \delta\mathbf{r}. \tag{6.8.5}$$

Substituting for isentropic flow $d\mathbf{u}/dt = -\nabla H$ (Eq. (6.3.12)) using the Stokes formula, and taking into account that $\nabla \times (\nabla H) \equiv 0$, we obtain:

$$\oint \frac{d\mathbf{u}}{dt} \cdot \delta\mathbf{r} = \oint \nabla \times \left(\frac{d\mathbf{u}}{dt}\right) \cdot d\mathbf{S} = \oint (\nabla \times \nabla H) \cdot d\mathbf{S} = 0. \tag{6.8.6}$$

Thus, we found that:

$$\frac{d}{dt}\oint \mathbf{u} \cdot d\mathbf{l} = 0, \tag{6.8.7}$$

or

$$\oint \mathbf{u} \cdot d\mathbf{l} = \text{constant}. \tag{6.8.8}$$

This is the *Kelvin's* theorem (1869) or it is also known as the *law of conservation of circulation*, which says that in an ideal fluid the velocity circulation round a closed "fluid" contour is constant in time. It should be emphasized that this result has been obtained by using the Euler equation in the form of Eq. (6.3.12), and therefore involves the assumption that the flow is isentropic. The theorem does not hold for flows, which are not isentropic.

Applying Kelvin's theorem to an infinitesimal closed contour δC and transforming the integral according to Stokes' theorem, we obtain

$$\oint \mathbf{u} \cdot d\mathbf{l} = \oint \nabla \times \mathbf{u} \cdot d\mathbf{S} \cong \delta\mathbf{S} \cdot (\nabla \times \mathbf{u}) = \text{constant}, \qquad (6.8.9)$$

where $d\mathbf{S}$ is a fluid surface element spanning the contour δC. The vector $\boldsymbol{\omega} = \nabla \times \mathbf{u}$ is called the *vorticity* of the fluid flow at a given point. The constancy of the product (6.8.9) can be intuitively interpreted as meaning that the vorticity moves with the fluid.

Calculations above gives us the qualitative idea whether a flow goes with rotation of fluid elements or not, with the corresponding characteristic of a flow being vorticity, $\boldsymbol{\omega} = \nabla \times \mathbf{u}$. Flows with non-zero circulation are called rotational flows, and flows with zero circulation are called irrotational.

The circulation theorem holds also for adiabatic flows and remains valid if fluid elements experience any other potential force as well. On the contrary, a non-potential force like the Coriolis force, a viscous force, etc. breaks down the conservation of circulation: these forces may produce circulation in an initially irrotational flow. Viscous forces may also lead to dissipation of vorticity.

6.9 Potential Flow

There is an important consequence of the law of conservation of circulation. Let us at first suppose that the flow is steady, and consider a streamline where $\boldsymbol{\omega} = \nabla \times \mathbf{u}$ is zero at some point. We draw an arbitrary infinitely small closed contour to encircle the streamline at that point. In the course of time, this contour moves with the fluid, but always encircles the same streamline. Since the product $\delta\mathbf{S} \cdot (\nabla \times \mathbf{u})$ must remain constant (Sect. 6.8), it follows that $\boldsymbol{\omega} = \nabla \times \mathbf{u}$ must be zero at every point on the streamline. Thus we came to the conclusion: if at any point on a streamline the vorticity is zero, the same is true at all other points on that streamline. If the flow is not steady, the same result holds, except that instead of a streamline we must consider the path described in the course of time by some particular fluid particle. We recall that in non-steady flow these paths do not in general coincide with the streamlines. A flow for which $\boldsymbol{\omega} = \nabla \times \mathbf{u} = 0$ in all space is called a *potential flow* or *irrotational flow,* as opposed to *rotational flow,* in which the circulation of the velocity is not everywhere zero.

To avoid misunderstanding, we may mention here that this result has no meaning in turbulent flow. We may also remark that a non-zero vorticity may occur on a streamline after the passage of a shock wave and a flame front. We shall see that this is because the flow is no longer isentropic (see Sect. 6.10). Also the proof given above that $\boldsymbol{\omega} = \nabla \times \mathbf{u} = 0$ all along a streamline is, strictly speaking, invalid for a line which lies in the surface of a solid body past which the flow takes place, since in the presence of this surface it is not possible to draw

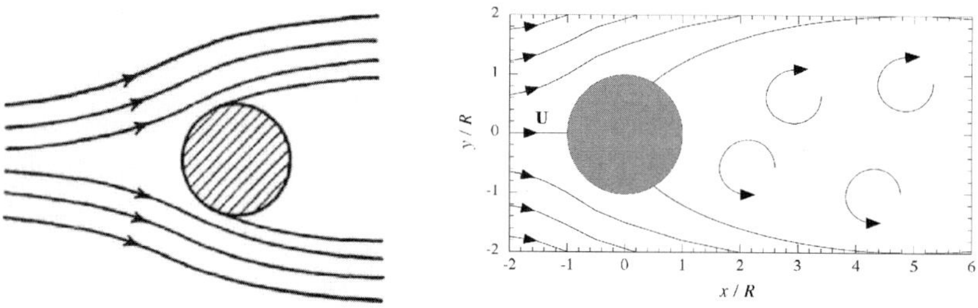

Fig. 6.7 Stream-lines of stationary flow ahead and behind the body

a closed contour in the fluid encircling such a streamline. A stationary flow ahead and behind the body is shown in Fig. 6.7. The equations of motion of an ideal fluid therefore admit solutions for which *separation* occurs at the surface of the body: the streamlines, having followed the surface for some distance, become separated from it at some point and continue into the fluid. The resulting flow pattern is characterized by the presence of a "surface of tangential discontinuity" proceeding from the body. On this surface the fluid velocity, which is everywhere tangential to the surface, has a discontinuity. From mathematical point of view, the discontinuity in the tangential velocity corresponds to a surface on which the circulation of the velocity is non-zero.

Any realistic fluid has a certain viscosity, even if the viscosity is small and has practically no effect on the motion of most of the fluid, but, no matter how small it is, it will be important in a thin layer of fluid adjoining the body (boundary layer). In the general case of flow past bodies will result in development of turbulence.

An important case of potential flow occurs for *small oscillations* of a body immersed in fluid. If the amplitude of the oscillations is small compared with the linear dimension of the body ($a << L$), the flow past the body will be potential flow. To show this, we estimate the order of magnitude of the various terms in the Euler equation

$$\frac{\partial \mathbf{u}}{\partial t} + (\mathbf{u} \cdot \nabla)\mathbf{u} = -\nabla H. \qquad (6.9.1)$$

The velocity $\mathbf{u}$ changes markedly, by an amount of the same order as the velocity $\mathbf{U}$ of the oscillating body, over a distance of the order of the dimension L of the body. Hence the derivatives of $\mathbf{u}$ with respect to the coordinates are of the order of U/L. The order of magnitude of $\mathbf{u}$ itself at fairly small distances from the body is determined by the magnitude of $\mathbf{U}$. Thus we have $(\mathbf{u} \cdot \nabla)\mathbf{u} \approx U^2/L$. The derivative $\partial \mathbf{u}/\partial t$ is of the order of ωU, where $\omega \propto U/a$ is the frequency of the oscillations, so that $\partial u/\partial t \propto U^2/a$. It follows that the term $(\mathbf{u} \cdot \nabla)\mathbf{u} \propto U^2/L$ is small compared with $\partial u/\partial t \propto U^2/a$ and can be neglected if $a << 1$, that so that the equation of motion of the fluid becomes

$$\frac{\partial \mathbf{u}}{\partial t} = -\nabla H. \tag{6.9.2}$$

The fact that for small oscillations we can neglect by nonlinear term in the Euler equation will be used below, when we will consider gravity and sound waves and problems of stability against small disturbances.

Applying the operation $[\nabla \times]$ of both sides of (6.9.2), we obtain

$$\frac{\partial}{\partial t}[\nabla \times \mathbf{u}] = 0, \text{ and, consequently, } [\nabla \times \mathbf{u}] = \text{constant}.$$

In oscillatory motion, however, the time average of the velocity is zero, and therefore $[\nabla \times \mathbf{u}] = \text{constant}$ implies that $[\nabla \times \mathbf{u}] = 0$. Thus, *the motion of a fluid executing small oscillations is potential flow to a first approximation.*

We recall that the derivation of the law of conservation of circulation, and therefore all its consequences, were based on the assumption that the flow is isentropic. If the flow is not isentropic, the law does not hold, and therefore, even if we have potential flow at some instant, the vorticity will in general be non-zero at subsequent instants. Thus only isentropic flow can in fact be potential flow. In rotational flow the velocity circulation is not in general zero. In this case there may be closed streamlines, but it must be emphasized that the presence of closed streamlines is not a necessary property of rotational flow.

Like any vector field having zero circulation, $[\nabla \times \mathbf{u}] = 0$, the velocity in potential flow can be expressed as the gradient of some scalar function. This scalar is called the *velocity potential*. Let us denote it by

$$\mathbf{u} = \nabla \phi. \tag{6.9.3}$$

Writing the Euler equation in the form (6.9.1), using formula from vector analysis

$$(\mathbf{u} \cdot \nabla)\mathbf{u} = \frac{1}{2}\nabla u^2 - \mathbf{u} \times [\nabla \times \mathbf{u}],$$

and substituting $\mathbf{u} = \nabla\phi$, we obtain the following equation

$$\frac{\partial}{\partial t}(\nabla\phi) + \nabla\left(\frac{u^2}{2} + H\right) = \nabla\left(\frac{\partial \phi}{\partial t} + \frac{u^2}{2} + H\right) = 0. \tag{6.9.4}$$

Integration of this equation gives

$$\frac{\partial \phi}{\partial t} + \frac{u^2}{2} + H = f(t), \tag{6.9.5}$$

where f(t) is an arbitrary function of time. This equation is a first integral of the equations of potential flow. The function f(t) in Eq. (6.9.5) can be put equal to zero without loss of generality, because the potential is not uniquely defined: since the velocity is the space derivative of ϕ, we can add to ϕ any function of the time.

For steady flow, taking the potential ϕ to be independent of time: $\partial\phi/\partial t = 0$, we have, $f(t) = \text{const}$ and (6.9.5) becomes the Bernoulli equation for a stationary flow

$$\frac{u^2}{2} + H = \text{constant}. \tag{6.9.6}$$

It must be emphasized here that there is an important difference between the Bernoulli equation for potential flow and that for other flows. In the general case, the "constant" on the right-hand side is a constant along any given streamline, but is different for different streamlines. In potential flow, however, it is constant throughout the fluid. This enhances the importance of the Bernoulli equation in the study of potential flow.

6.10 Linear Waves and Instabilities

Not every flow allowed by equations of fluid dynamics may exist in reality. Some flow created with especially elaborated initial conditions, may be destroyed by hydrodynamic instabilities. Therefore, one of the very important subjects of fluid dynamics is investigation of stability of a solution against infinitely small disturbances imposed on the solution. The basic idea of a stable or unstable configuration may be understood considering a simple example of a small ball at the top of a hill and at the bottom of a well (Fig. 6.8). In both cases the balls are in equilibrium: the net forces acting on the balls are zero and they can remain in their present positions forever. However, if we slightly shift the balls, then the ball at the bottom will return to its previous position, while the

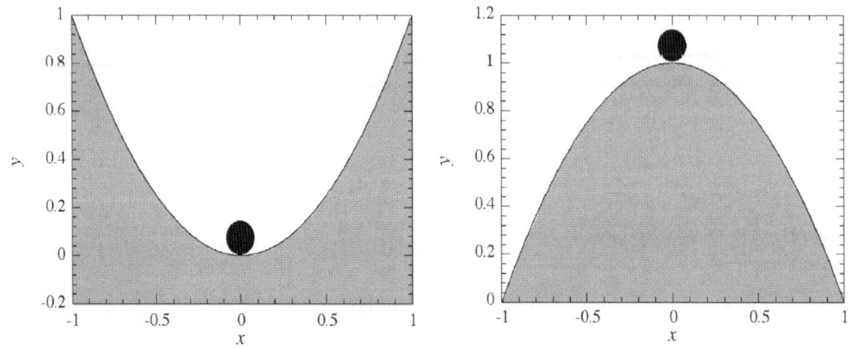

Fig. 6.8 The stable configuration of a ball in a well and the unstable configuration of a ball on a hill

ball at the top of the hill will leave its position going down. In the last case the initial configuration will be destroyed completely.

Let us consider stability of a flow of an ideal incompressible fluid. Suppose the velocity and pressure distributions, $\mathbf{u} = \mathbf{u}_0(\mathbf{r}, t)$, $P = P_0(\mathbf{r}, t)$, are solutions of the fluid equations:

$$\nabla \cdot \mathbf{u}_0 = 0, \tag{6.10.1}$$

$$\frac{\partial \mathbf{u}_0}{\partial t} + (\mathbf{u}_0 \cdot \nabla)\mathbf{u}_0 = -\frac{1}{\rho}\nabla P_0 + \mathbf{g}. \tag{6.10.2}$$

In order to investigate stability of the flow given by the solution $\mathbf{u} = \mathbf{u}_0(\mathbf{r}, t)$, $P = P_0(\mathbf{r}, t)$ we must study whether the flow being slightly disturbed will return to its initial configuration, or small disturbances will grow, and the initial configuration of the flow will be destroyed.

Let us assume that some small external disturbances perturb the initial flow leading to new velocity and pressure distributions

$$\mathbf{u} = \mathbf{u}_0 + \tilde{\mathbf{u}}(\mathbf{r}, t),\ P = P_0 + \tilde{P}(\mathbf{r}, t), \tag{6.10.3}$$

where $\tilde{\mathbf{u}}$ and $\tilde{P}$ are small perturbations, such that $\tilde{\mathbf{u}} << \mathbf{u}_0$, $\tilde{P} << P_0$. These new velocities and pressure for the perturbed flow must also satisfy the equations of fluid dynamics

$$\nabla \cdot (\mathbf{u}_0 + \tilde{\mathbf{u}}) = 0, \tag{6.10.4}$$

$$\frac{\partial}{\partial t}(\mathbf{u}_0 + \tilde{\mathbf{u}}) + [(\mathbf{u}_0 + \tilde{\mathbf{u}}) \cdot \nabla](\mathbf{u}_0 + \tilde{\mathbf{u}}) = -\frac{1}{\rho}\nabla\left(P_0 + \tilde{P}\right) + \mathbf{g}. \tag{6.10.5}$$

Extracting Eqs. (6.10.1) and (6.10.2) for the initial flow out of Eqs. (6.10.4) and (6.10.5) we come to the set of equations for small perturbations $\tilde{\mathbf{u}}$ and $\tilde{P}$

$$\nabla \cdot \tilde{\mathbf{u}} = \mathbf{0}, \tag{6.10.6}$$

$$\frac{\partial \tilde{\mathbf{u}}}{\partial t} + (\mathbf{u}_0 \cdot \nabla)\tilde{\mathbf{u}} + (\tilde{\mathbf{u}} \cdot \nabla)\mathbf{u}_0 + (\tilde{\mathbf{u}} \cdot \nabla)\tilde{\mathbf{u}} = -\frac{\mathbf{1}}{\rho}\nabla\tilde{P}, \tag{6.10.7}$$

where the coordinate and time dependence of the unperturbed flow, $\mathbf{u} = \mathbf{u}_0(\mathbf{r}, t)$ and $P = P_0(\mathbf{r}, t)$ is supposed to be known.

The nonlinear term $(\tilde{\mathbf{u}} \cdot \nabla)\tilde{\mathbf{u}}$ is of the second order and it is small in comparison with the linear term $(\mathbf{u}_0 \cdot \nabla)\tilde{\mathbf{u}}$, which of the first order in $\tilde{\mathbf{u}}$. Therefore, neglecting by $(\tilde{\mathbf{u}} \cdot \nabla)\tilde{\mathbf{u}}$ we obtain the linearized equation of momentum transfer in an incompressible ideal fluid

$$\frac{\partial \tilde{\mathbf{u}}}{\partial t} + (\mathbf{u}_0 \cdot \nabla)\tilde{\mathbf{u}} + (\tilde{\mathbf{u}} \cdot \nabla)\mathbf{u}_0 = -\frac{1}{\rho}\nabla\tilde{P}. \tag{6.10.8}$$

Stability analysis implies investigation of perturbation dynamics on the basis of the linearized Eqs. (6.10.6) and (6.10.8). If solution of these equations is found, and the time dependence of small perturbations $\tilde{\mathbf{u}} = \tilde{\mathbf{u}}(\mathbf{r}, t)$ shows that amplitude grows with no limit, then we can say that the initial flow $\mathbf{u} = \mathbf{u}_0(\mathbf{r}, t)$, $P = P_0(\mathbf{r}, t)$ is unstable. If the perturbation amplitude decreases with time, then the initial flow is stable.

In the case of stationary unperturbed flows, when velocity and pressure, $\mathbf{u} = \mathbf{u}_0(\mathbf{r})$, $P = P_0(\mathbf{r})$, are time independent, the time dependence of perturbations takes the form

$$\tilde{\mathbf{u}}(\mathbf{r}, t) = \tilde{\mathbf{u}}(\mathbf{r}) \exp(-i\Omega t), \tag{6.10.9}$$

where Ω is some complex number. Now, the unperturbed flow is unstable, if the imaginary part of the frequency is positive $\sigma = \mathrm{Im}\Omega > 0$, while in the opposite case of a negative imaginary part $\mathrm{Im}\Omega < 0$ the initial flow is stable. At the same time the real part of Ω may be zero or non-zero. The imaginary part of Ω is called the instability growth rate, while the real part is called frequency. If the flow is unstable then the amplitude of small perturbations grows exponentially

$$\tilde{\mathbf{u}}(\mathbf{r}, t) = \tilde{\mathbf{u}}(\mathbf{r}) \exp(\sigma t - i\omega t) = \tilde{\mathbf{u}}(\mathbf{r}) \exp(\sigma t)[\cos(\omega t) - i \sin(\omega t)] \tag{6.10.10}$$

Since stability problems are linear, it is often convenient to work with complex functions $\exp(-i\omega t) = \cos(\omega t) - i \sin(\omega t)$. If the instability growth rate is zero $\sigma = 0$, while the frequency is non-zero $\omega = \mathrm{Re}\Omega \neq 0$, then the perturbation amplitude oscillates in time

$$\tilde{\mathbf{u}}(\mathbf{r}, t) = \tilde{\mathbf{u}}(\mathbf{r}) \exp(-i\omega t) = \tilde{\mathbf{u}}(\mathbf{r})[\cos(\omega t) - i \sin(\omega t)]. \tag{6.10.11}$$

In case of uniform unperturbed fluid a solution for small perturbations takes the form

$$\tilde{\mathbf{u}}(\mathbf{r}, t) = \tilde{\mathbf{u}} \exp(i\mathbf{k} \cdot \mathbf{r} - i\omega t). \tag{6.10.12}$$

In this case the solution represents linear waves propagating in a fluid, with $\mathbf{k}$ being the wave number inversely proportional to the perturbation wavelength $k = 2\pi/\lambda$. Another important value that specifies propagation of waves is the phase velocity $U_p = \omega/k$, which is the propagation velocity of harmonic wave with a frequency ω and a wave number k. In the next sections we will consider several examples of linear waves and instabilities: gravitational waves on the liquid surface, the Rayleigh-Taylor instability and sound waves.

6.11 The Gravity Waves

The free surface of a liquid in equilibrium in a gravitational field is a plane surface. If, under the action of some external perturbation, the surface is moved from its equilibrium position at some point, motion will occur in the liquid. This motion will propagate over the whole surface in the form of waves, which are called *gravity waves,* since they are due to the action of the gravitational field. Suppose that some external force has created a hump on the fluid surface. Pressure below the hump is larger than the ambient pressure, which makes the fluid flow out of the hump. As a result the hump goes down and overshoots the stationary planar level because of inertia creating a well on the fluid surface with reduced pressure below the well. The fluid is pushed towards the well until a hump is created once more by inertia: the fluid level oscillates. Interaction of neighboring fluid elements makes these oscillations propagate in a form of a wave. Gravity waves, which appear on the surface of the liquid also affect the interior, but as less as greater depths.

We shall here consider gravity waves in which the velocity of the moving fluid is so small that we may neglect the term $(\mathbf{u}\cdot\nabla)\mathbf{u}$ in comparison with $d\mathbf{u}/dt$ in the Euler equation. The physical meaning of this is easily seen. During a time interval of the order of the period T of the oscillations of the fluid particles in the wave, these particles travel a distance of the order of the amplitude a of the wave. Their velocity u is therefore of the order of a/T. It varies noticeably over time intervals of the order of T and distances of the order of the wavelength λ in the direction of propagation. Hence the time derivative of the velocity is of the order of u/T, and the space derivatives are of the order of u/λ. Thus, $\partial u/\partial t \approx u/T$, $u \approx a/T$, $\nabla \approx 1/\lambda$, and the condition $(\mathbf{u}\cdot\nabla)\mathbf{u} << \partial\mathbf{u}/\partial t$ is equivalent to

$$\frac{1}{\lambda}\left(\frac{a}{T}\right)^2 << \frac{a}{T^2}, \quad \Rightarrow \quad a << \lambda, \tag{6.11.1}$$

i.e. the amplitude of the oscillations in the wave must be small compared with the wavelength.

Thus, the term $(\mathbf{u}\cdot\nabla)\mathbf{u}$ in the equation of motion may be neglected, as we have seen the motion of a fluid executing small oscillations is potential flow. Let us take the z-axis directed upward and the xy-plane (plane $z = 0$) is the equilibrium surface of the liquid. Assuming the fluid incompressible, we can therefore use Eq. (6.9.6), which with account gravity potential and the velocity potential $\mathbf{u} = \nabla\phi$ takes the form:

$$\frac{\partial\phi}{\partial t} + \frac{1}{2}(\nabla\phi)^2 + \frac{P}{\rho} + gz = f(t), \tag{6.11.2}$$

The term $u^2/2 = (\nabla\phi)^2/2$ in the Eq. (6.11.2) may be neglected, since it contains the square of the velocity. We have seen also in Sect. 6.9 that the function f(t) in the Eq. (6.11.2) can be put equal to zero, because the potential is defined with an arbitrary additive constant. Thus, we obtain from (6.11.2)

$$P = -\rho\frac{\partial\phi}{\partial t} - \rho gz. \tag{6.11.3}$$

According to what was said in Sect. 6.9, there is no vorticity in the fluid under consideration and small deviations of the fluid surface from the initial level produce no vorticity either. Therefore, the perturbed motion is potential, and using continuity equation $\nabla\cdot\mathbf{u} = 0$ we obtain equation for the velocity potential

$$\nabla\cdot\mathbf{u} = \nabla\cdot(\nabla\phi) = \nabla^2\phi = \Delta\phi = 0. \tag{6.11.4}$$

The Laplace Eq. (6.11.4) must be supplemented by boundary conditions on the free surface and at the bottom of the fluid. The boundary conditions on the free surface are the continuity condition related to the mass conservation and the condition of pressure balance. In the unperturbed state we suppose that the fluid is at rest, $\mathbf{u} = 0$, and the pressure balance follows from the Bernoulli equation $P_0 + \rho gz = \text{const} = P_{atm}$.

Let us denote by ς the z coordinate of a point on the surface. ζ is a function of x,y and t. In equilibrium $\zeta(x, y, t) = 0$ so that ζ gives the vertical displacement of the surface in its oscillations (see Fig. 6.9). Let a constant external (atmosphere) pressure P_0 acts on the surface. Then at the surface the Eq. (6.11.3) gives

$$P_0 = -\rho\frac{\partial\phi}{\partial t} - \rho g\zeta(x, y). \tag{6.11.5}$$

The constant P_0, can be eliminated by redefining the potential ϕ, adding to it a quantity $P_0 t/\rho$, independent of the coordinates. We then obtain the condition at the surface in the form

$$\left(\frac{\partial\phi}{\partial t}\right)_{z=\zeta} + g\zeta(x, y) = 0. \tag{6.11.6}$$

We can come to the same condition (6.11.6) considering the pressure balance at the free surface of the liquid. Let us consider forces acting on a surface element

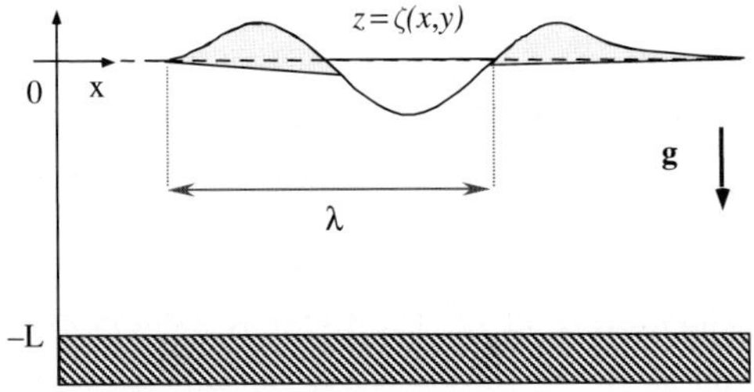

Fig. 6.9 The linear (gravitational) waves on a liquid surface

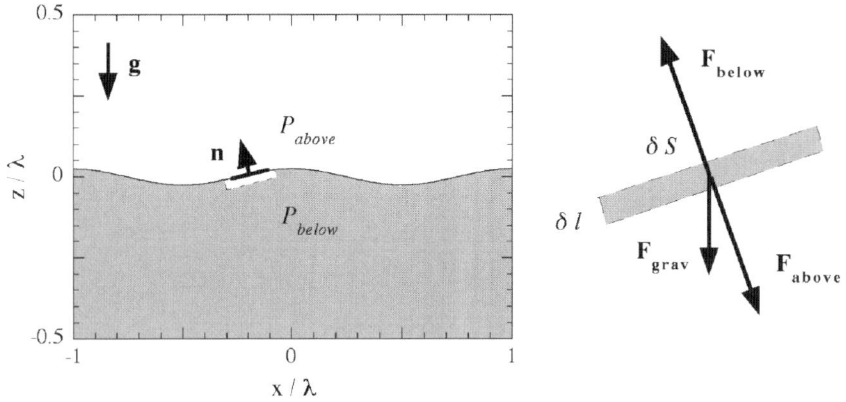

Fig. 6.10 The pressure balance at the free surface

of a small mass $\delta m = \rho\delta S\delta l$. These are the pressure acting on the upper surface $\delta F_{out} = P_0\delta S$, the force acting at the lower surface of the element $\delta F_{in} = P_{in}\delta S$ and the gravity, $g\delta m$ (Fig. 6.10). According to the Newton law we have

$$\rho\delta S\delta l\frac{d\mathbf{u}}{dt} = (P_0 - P_{in})\mathbf{n}\delta S + \mathbf{g}\rho\delta S\delta l.$$

For the infinitely small width of the element, $\delta l \to 0$, we obtain the condition of pressure balance, $P_0 = P_{in}$. Since pressure of the gas above the fluid interface is not perturbed the condition of the pressure balance is equivalent to the condition of zero perturbed pressure just below the fluid surface.

According to the continuity condition, the fluid particles at the surface moves together with the fluid, that is the normal velocity of the fluid close to the surface must be equal to the normal velocity of the surface element, $u_{surface} = (\mathbf{n}\cdot\mathbf{u})_{z=\zeta(x,y)}$, where the vector normal to the surface has the components

$$\mathbf{n} = \left(\frac{1}{\sqrt{1+(\nabla\zeta)^2}}; -\frac{\nabla\zeta}{\sqrt{1+(\nabla\zeta)^2}}\right).$$

In the simplest case of a planar fluid surface oscillating up and down the continuity boundary condition is simply $u_z = \partial z/\partial t$. In the case of small amplitude linear gravitational waves the continuity condition has the same form in spite of ripples on the water surface. Indeed, both terms in the left and the right sides in (6.11.6) are of the first order, and account of non-zero curvature of the fluid surface will bring nonlinear terms, which are of the second order in $a/\lambda << 1$. Indeed, for the normal component of velocity of a fluid particle close to the surface we have with accuracy of the first order terms

$$u_n = \mathbf{n}\cdot\mathbf{u}|_{z=\zeta}(n_z u_z + n_x u_x) = \frac{1}{\sqrt{1+(\partial\zeta/\partial x)^2}}\left(u_z|_{z=\zeta} - \frac{\partial\zeta}{\partial x}u_x|_{z=\zeta}\right) \approx u_z|_{z=\zeta}.$$

That is, since the amplitude of the wave oscillations is small, the displacement ζ is small, and we can suppose, to the same degree of approximation, that the vertical component of the velocity of points on the surface is simply the time derivative of ζ. Writing z-component of velocity with the help of velocity potential we come to the continuity condition

$$u_n = u_z|_{z=\zeta} = \frac{\partial\zeta}{\partial t} = \left(\frac{\partial\phi}{\partial z}\right)_{z=\zeta}. \tag{6.11.7}$$

Taking time derivative of (6.11.6) and substituting (6.11.7) we obtain

$$\left(\frac{\partial\phi}{\partial z} + \frac{1}{g}\frac{\partial^2\phi}{\partial t^2}\right)_{z=0} = 0. \tag{6.11.8}$$

Two Eqs. (6.11.4) and (6.11.8) completely specify the problem for gravitational waves of small amplitudes at the surface, if depth is large compared to the wavelength of the waves. If bottom is not too far, the equations must be complemented by boundary conditions at the bottom, which is condition that normal component of the velocity vanishes at the bottom, i.e. $u_z = \partial\phi/\partial z = 0$ at $z = -h$.

Let us consider waves on the surface of a liquid whose area is unlimited, propagating along the x-axis and uniform in the y-direction, and suppose the wavelength is small in comparison with the depth of the liquid, which means that we shall consider the liquid as infinitely deep. In such a wave, all quantities are independent of y. We shall seek a solution, which is a simple periodic function of time and of the coordinate x, i.e. the velocity potential is

$$\phi(z,x,t) = f(z)\psi(x,t) = f(z)\cos(kx - \omega t), \tag{6.11.9}$$

where $\omega = 2\pi/T$ is the circular frequency (or simply the frequency) of the wave, and $k = 2\pi/\lambda$ is the wave number with λ being the wavelength.

Then, substituting (6.11.9) in (6.11.4), we obtain

$$\nabla^2\phi = \Delta\phi = \left[\frac{d^2f}{dz^2} - k^2 f\right]\cos(kx - \omega t) = 0. \tag{6.11.10}$$

The general solution to Eq. (6.11.10) is

$$\phi(x,z,t) = [A_1\exp(kz) + A_2\exp(-kz)]\cos(kx - \omega t), \tag{6.11.11}$$

where A_1 and A_2 are constants depending on boundary conditions. The solution, which is limited as $z \to -\infty$, i.e. into interior of the liquid, is

$$\phi(x, z, t) = A_1 e^{kz} \cos(kx - \omega t). \tag{6.11.12}$$

Substituting this solution into the boundary condition on the free surface (6.11.8), we obtain equation, which gives the relation between the wave number and the frequency of a gravity wave – *dispersion relation:*

$$\left(\frac{\partial \phi}{\partial z} + \frac{1}{g} \frac{\partial^2 \phi}{\partial t^2} \right)_{z=0} = \left(k - \frac{\omega^2}{g} \right) A_1 \cos(kx - \omega t) = 0. \tag{6.11.13}$$

Thus, the *dispersion relation* for the gravitational waves is

$$\omega^2 = kg. \tag{6.11.14}$$

The problem can be generalized for the case of gravity waves separating two liquids, the upper liquid being unbounded from above, and the lower liquid being unbounded below, with the density of the lower liquid ρ and of the upper liquid ρ', with $\rho > \rho'$. In this case the *dispersion relation* is

$$\omega^2 = \frac{\rho - \rho'}{\rho + \rho'} kg, \tag{6.11.15}$$

which is obviously transformed in (6.11.14) when $\rho' \to 0$.

Let us consider also the velocity distribution in the moving liquid. Taking the space derivatives of ϕ, given by (6.11.12), we have

$$\begin{aligned} u_x &= \partial \phi / \partial x = -A_1 k e^{kz} \sin(kx - \omega t), \\ u_z &= \partial \phi / \partial z = A_1 k e^{kz} \cos(kx - \omega t). \end{aligned} \tag{6.11.16}$$

We see that at any given point in space (i.e. for given *x, z*) the velocity vector rotates uniformly in the x,z-plane, with its magnitude remaining constant, and the velocity diminishes exponentially as we go deeper into the liquid. The fluid particles move along circles with the radius, which diminishes exponentially with increasing depth.

6.12 The Rayleigh-Taylor Instability

In the previous section we considered the gravity waves at the interface between two fluids of different density, when the fluid of higher density is below, while the fluid of less density is above. Particular case of such configuration is the gravity waves at the surface of water, when the density of air is negligible compared with the density of water, $\rho' << \rho$, so that we can put $\rho' = 0$. What happens if the lower density fluid supports from below the heavy fluid in a gravitational field? This configuration corresponds to many physical problems, in particular, related to problem of energy cumulating. For example, this is the configuration of a matter accelerated by high temperature gases from strong explosion. Similar problem is the plasma compressed or confined by the pressure of magnetic field. In the last case the magnetic field can be viewed as massless fluid.

Obviously the interface between two fluids cannot be ideally planar; there are always infinitely small disturbances at the surface $z = \zeta(x, y, t)$ arising due to a noise. Let us consider development of small perturbations in the case of a semi-infinite fluid of density ρ supported by a low-density gas, which density is negligibly small. From the mathematical point of view the problem is almost identical to the problem of gravity waves with the only exception that the gravity acceleration now directed not from the light to the heavy fluid as on Fig. 6.9, but in the opposite direction. This means that we can use formulas derived in previous section, but must change the sign of gravity acceleration in all formulas. In particular, we obtain formula the dispersion relation

$$\omega^2 = -gk, \tag{6.12.1}$$

which means that frequency is pure imaginary: $\omega = \pm i\sqrt{gk}$, and the amplitude of small oscillations, $\zeta(x, y, t) = A\exp(ikx - i\omega t)$, growth exponentially fast the positive imaginary ω as $\exp(\sqrt{gk}t)$. This instability of the interface surface separating two liquids of different densities is called the *Rayleigh-Taylor instability*; its growth rate is

$$\sigma = i\omega = \sqrt{gk}. \tag{6.12.2}$$

The Rayleigh-Taylor instability always develops when a heavy liquid is supported in a gravitational field by a lower density liquid, or if a heavy liquid is accelerated by a liquid of smaller density. In general, the condition of the instability is

$$\mathbf{g} \cdot \nabla\rho < 0. \tag{6.12.3}$$

In usual life we can see manifestation of the Rayleigh-Taylor instability when we turn over the glass of water. The water flows out of the glass even though the atmospheric pressure is more than enough for keeping the water in the glass. This is manifestation of the Rayleigh-Taylor instability that makes the water surface unstable and very fast results in a water jet flowing from the glass. Of course, the final result – water jet flowing from the glass is the nonlinear stage of the instability. It is well know that if we put a sheet of paper attached to the glass of water, and then carefully turn over the glass, the paper remains attached to the glass and water will not flow out. The role of the paper is that it damps small initial perturbations on the water surface and stops the instability.

So far we considered perturbations of small amplitude. With the growth of the perturbation's amplitude, when perturbation amplitudes become larger and the condition of linearity breaks down, the problem becomes nonlinear. The nonlinear effects as well as other physical effects that are not taken into consideration in scope of ideal fluid approximation, limit the growth of the initial disturbances. Also there are different factors, which may limit from below the growth of short wavelength perturbations. For example, this is the surface tension; another factors are dissipations, surface curvature, etc. Interesting example of the Rayleigh-Taylor instability is the gas bubbles rising to the water surface. It is evident from what was said above, that the gas layer or a large bubble under the water surface is unstable. Manifestation of the instability in this case is a gas bubble rising to the water surface. Of course, the instability in that case is complicated by influence of the surface tension and the surface curvature. However, using dimensional considerations we can obtain a simple estimate for the velocity of the rising bubble. From the parameters of the problem: the gravity acceleration g, the density of the fluid ρ (we assume that the density of the gas is negligible compared to the fluid density and put it equal to zero) and the bubble radius, we can form only one combination of the velocity dimension: $U_{bubble} \propto \sqrt{gR}$. This means that as larger the bubble is, as the faster it flows up to the surface. Though, the obtained estimate for the bubble velocity agrees with experimental observations, its application is limited: bubbles of very large radius are broken into bubbles of smaller radius, while bubble of very small radius will collapse due to the surface tension faster than it flows up.

6.13 Sound Waves

We proceed now to the study oscillatory motion with small amplitude in a compressible fluid. Such a small amplitude oscillatory motion is called *sound waves*. When propagating in a gas or fluid, a sound wave causes alternate compression and rarefaction at each point in the fluid. Since the oscillations are supposed to be small, the characteristic velocity of the fluid particles is small also. The relative changes in the fluid density and pressure are supposed to be small also. We write the variables for pressure, P and density ρ in the form

$$P = P_0 + P', \qquad \rho = \rho_0 + \rho' \tag{6.13.1}$$

where P_0 and ρ_0 are the constant equilibrium pressure and density, and P' and ρ' are their small variations in the sound wave ($P' << P_0$, and $\rho' << \rho_0$).

Substituting $\rho = \rho_0 + \rho'$ in the equation of continuity and neglecting small quantities of the second order (we consider P', $\rho'$and **u** being small quantities of the first order; we shall verify from the obtained solution that **u** is indeed small quantity of the same order as P′, ρ′), we obtain

$$\frac{\partial \rho'}{\partial t} + \rho_0 \nabla \cdot \mathbf{u} = 0. \tag{6.13.2}$$

The term $(\mathbf{u} \cdot \nabla)\mathbf{u}$ in Euler equation may be neglected since it is of the second order. Substituting expressions for pressure and density in the form (6.13.1) and neglecting small quantities of the second order, we obtain in the same first order approximation

$$\frac{\partial \mathbf{u}}{\partial t} + \frac{1}{\rho_0} \nabla P' = 0. \tag{6.13.3}$$

The condition that the linearized Eqs. (6.13.2) and (6.13.3) should be applicable to the propagation of sound waves is that the velocity of the fluid particles in the wave should be small compared with the velocity of sound: $u \ll a_s$. Below we shall see that this condition can be obtained, for example, from the requirement that $\rho' \ll \rho_0$.

Equations (6.13.2) and (6.13.3) contain the unknown **u**, P′, ρ′. To eliminate one of them we notice that a sound wave in an ideal fluid is adiabatic, like any other motion in an ideal fluid. Hence the small change P′ in the pressure is related to the small change ρ′ in the density by

$$P' = \left(\frac{\partial P}{\partial \rho_0}\right)_s \rho' = a_s^2 \rho' \tag{6.13.4}$$

where $a_s = \sqrt{(\partial P/\partial \rho)_s}$ is the sound velocity.

Substituting for ρ′ according to this equation in the linearized Eq. (6.13.2), we find

$$\frac{\partial P'}{\partial t} + \rho_0 \left(\frac{\partial P}{\partial \rho_0}\right)_s \nabla \cdot \mathbf{u} = \frac{\partial P'}{\partial t} + \rho_0 a_s^2 \nabla \cdot \mathbf{u} = 0. \tag{6.13.5}$$

The two Eqs. (6.13.3) and (6.13.5) for contain **u** and P′ giving a complete description of the sound wave.

As it was discussed above, the flow with small oscillations is potential. In order to express all the unknowns in terms of one of them, it is convenient to introduce the velocity potential by putting $\mathbf{u} = \nabla \phi$. Then, Eq. (6.13.3) can be written as

$$\frac{\partial}{\partial t}(\nabla \phi) + \frac{1}{\rho_0} \nabla P' = 0, \quad \text{or} \quad \nabla\left(\frac{\partial \phi}{\partial t} + \frac{1}{\rho_0} P'\right) = 0. \tag{6.13.6}$$

From the last equation we have

$$P' = -\rho_0 \frac{\partial \phi}{\partial t}, \tag{6.13.7}$$

where we omit constant integrating Eq. (6.13.6) since the velocity potential can be taken with arbitrary additive function ϕ_0, such that $\nabla\phi_0 = 0$.

Substituting P' from (6.13.7) and $\mathbf{u} = \nabla\phi$ in the Eq. (6.13.5) we obtain

$$\frac{\partial^2 \phi}{\partial t^2} - a_s^2 \Delta\phi = 0. \tag{6.13.8}$$

Applying the gradient operator to (6.13.8), we find that each of the three components of the velocity **u** satisfies the equation having the same form. Also differentiating (6.13.8) with respect to time we see that the pressure P' and therefore ρ' also satisfies the equation having the same form, which is a *wave equation*.

For a plane wave propagating along only one coordinate (x) and when and the flow is completely homogeneous in the yz-plane, the wave Eq. (6.13.8) becomes

$$\frac{\partial^2 \phi}{\partial t^2} - a_s^2 \frac{\partial^2 \phi}{\partial x^2} = 0. \tag{6.13.9}$$

Solution of this equation can be obtained replacing independed variables x and t by the new variables $\xi = x - a_s t$ and $\eta = x + a_s t$. In these variables (6.13.9) becomes

$$\frac{\partial^2 \phi}{\partial \xi \partial \eta} = 0, \tag{6.13.10}$$

which is easily integrated

$$\phi = F_1(x - a_s t) + F_2(x + a_s t), \tag{6.13.11}$$

where F_1 and F_2 are arbitrary functions of their arguments.

To be definite, we shall take the density as $\rho' = F_1(x - a_s t) + F_2(x + a_s t)$. Let, for example, $F_2 = 0$, so that $\rho' = F_1(x - a_s t)$. The meaning of this solution is evident. In any plane x = constant the density varies with time, and at any given time it is different for different x, but it is the same for coordinates and times such that $x - a_s t = \text{constant}$, or for coordinates $x = \text{constant} + a_s t$. This means that, if at some instant and at some point the fluid density has a certain value, then after a time t the same value of the density is found at a distance $a_s t$ along the x-axis from the original point. The same is true of all the other quantities in the wave. Thus the pattern of motion is propagated through the medium in the x-direction with a velocity a_s, which is called the velocity of sound. The functional dependence of the type $F = F_1(x - a_s t)$ represents what is called a traveling plane wave propagating in the positive direction of the x-axis. It is evident that another function $F = F_2(x + a_s t)$ represents a traveling plane wave propagating in the opposite direction.

Of the three components of the velocity $\mathbf{u} = \nabla\phi$ in a plane wave, only $u_x = \partial\phi/\partial x$ is not zero. Thus the fluid velocity in a sound wave is in the direction of the wave propagation, therefore, sound waves in a fluid are longitudinal.

In a travelling plane wave, the velocity $u_x \equiv u$ is related to the pressure P' and the density ρ' in a simple manner. Putting $\phi = F(x - a_s t)$, we find $u = \partial\phi/\partial x = F'(x - a_s t)$ and $P' = -\rho_0 \partial\phi/\partial t = \rho_0 a_s F'(x - a_s t) = \rho_0 a_s u$, where F' means a derivative of function F. Thus, we find

$$u = P'/\rho a_s. \tag{6.13.12}$$

Substituting in (6.13.12) expression for P' from (6.13.6) $P' = a_s^2 \rho'$, we find the relation between the velocity and the density variation in the sound wave:

$$u = a_s \rho'/\rho, \tag{6.13.13}$$

which confirm the condition made at the beginning that P', $\rho'$and $\mathbf{u}$ are small quantities of the same first order.

Let us calculate the velocity of sound in an ideal gas. The equation of state for ideal gas is $PV = \frac{P}{\rho} = RT/\mu$, where R is the universal gas constant and μ is the molecular weight. Substituting expression for pressure from the equation of state in the formula for sound velocity, $a_s = \sqrt{(\partial P/\partial \rho)_s}$, and taking into account that the adiabatic compressibility is related to isothermal compressibility by the following thermodynamic formula

$$\left(\frac{\partial P}{\partial \rho}\right)_S = \frac{c_P}{c_V}\left(\frac{\partial P}{\partial \rho}\right)_T = \gamma\left(\frac{\partial P}{\partial \rho}\right)_T, \tag{6.13.14}$$

where $\gamma = c_P/c_V$ is adiabatic constant, we obtain for the velocity of sound in ideal gas

$$a_s = \sqrt{\gamma RT/\mu}. \tag{6.13.15}$$

Since adiabatic constant usually depends only slightly on the temperature, the velocity of sound in the gas may be supposed proportional to the square root of the temperature, and for a given temperature it does not depend on the pressure. It should be noticed that the velocity of sound in the gas is of the same order as the mean thermal velocity of molecules in the gas; the magnitude of the adiabatic constant is $\gamma = 5/3$ and $7/5$ for mono-atomic and two-atomic gases, respectively.

In an important case of monochromatic waves all quantities are harmonic functions of the time. Importance of the monochromatic waves is due to any wave can be represented as a sum of superposed monochromatic plane waves with various wave vectors and frequencies. This decomposition of a wave into monochromatic waves is simply an expansion as a Fourier series or integral, which is called also spectral resolution. It is convenient to write such functions as the real part of a complex quantity. For example, we put for the velocity potential

$$\phi = \mathrm{Re}\{\phi_0(x, y, z) \cdot e^{-i\omega t}\}, \tag{6.13.16}$$

where $\omega = 2\pi/T$ is the frequency of the wave, and here T is period.

Then the function $\phi_0(x, y, z)$ satisfies the equation

$$\Delta\phi_0 + \frac{\omega^2}{a_s^2}\phi_0 = 0. \tag{6.13.17}$$

Let us consider a monochromatic traveling plane wave, propagating in the positive direction of the x-axis, so that all quantities are functions of argument $(x - a_s t)$ only, and the potential of the form (6.13.16) obtained by solving the Eq. (6.13.17) is

$$\phi = \mathrm{Re}\{A \exp[-i\omega(t - x/a_s)]\}, \tag{6.13.18}$$

where A is the complex amplitude.

Writing this as $A = ae^{i\alpha}$ with real constants a and α, we obtain the solution with *a* being amplitude of the wave, and α being the phase:

$$\phi = a\cos(\omega x/a_s - \omega t + \alpha). \tag{6.13.19}$$

Let us denote by $\mathbf{n}$ a unit vector in the direction of wave propagation. The vector

$$\mathbf{k} = \frac{\omega}{a_s}\mathbf{n} = \frac{2\pi}{\lambda}\mathbf{n} \tag{6.13.20}$$

is called the wave vector, and its magnitude $k = |\mathbf{k}|$ the wave number; here λ is the wavelength. In terms of the wave number the expression (6.13.18) can be written as

$$\phi = \mathrm{Re}\{A \exp[-i(\omega t - \mathbf{kr})]\}. \tag{6.13.21}$$

6.14 One-Dimensional Traveling Waves

In the previous section discussing sound waves, we assumed that the amplitude of oscillations is small, so that the equations of motion were linear. A particular solution of these equations was a plane wave, that is any function of $(x \pm a_s t)$, corresponding to a traveling wave whose distribution of density, velocity, etc., along the direction of propagation, moves with velocity a_s, remaining unchanged. Since u, ρ and P in such a wave are functions of the same quantity $(x \pm a_s t)$, they can be expressed as functions of one another, in which the coordinates and time do not explicitly appear, $P = P(\rho)$, so on.

When the wave amplitude is not small, these simple relations do not hold. However, a general solution of the exact equations of motion can be obtained in the form of a traveling plane wave, which is a generalization of the solution $F(x \pm a_s t)$ of the approximate equations valid for small amplitudes. In the absence of shock waves the flow is adiabatic. If the gas is homogeneous at some initial instant, then its entropy is constant at all times. The pressure is thus a function of the density only $P = P(\rho)$, and we assume also that the velocity can be expressed as a function of the density $u = u(\rho)$.

In a plane wave, propagating in the x-direction, all quantities depend on x and t only, the velocity is $u_x = u$, $u_y = u_z = 0$. So that the continuity and the Euler equations are

$$\frac{\partial \rho}{\partial t} + \frac{\partial(\rho u)}{\partial x} = 0, \tag{6.14.1}$$

$$\frac{\partial u}{\partial t} + u\frac{\partial u}{\partial x} + \frac{1}{\rho}\frac{\partial P}{\partial x} = 0. \tag{6.14.2}$$

Using the fact that u is a function of ρ or P only, we can write these equations as

$$\frac{\partial \rho}{\partial t} + \frac{d(\rho u)}{d\rho}\frac{\partial \rho}{\partial x} = 0, \tag{6.14.3}$$

$$\frac{\partial u}{\partial t} + \left(u + \frac{1}{\rho}\frac{\partial P}{\partial u}\right)\frac{\partial u}{\partial x} = 0. \tag{6.14.4}$$

From the relation: $d\rho = (\partial\rho/\partial t)dt + (\partial\rho/\partial x)dx$ follows that

$$\frac{(\partial\rho/\partial t)}{(\partial\rho/\partial x)} = -\left(\frac{\partial x}{\partial t}\right)_\rho. \tag{6.14.5}$$

Then, using the relation (6.14.5), we obtain from (6.14.3)

$$\left(\frac{\partial x}{\partial t}\right)_\rho = -\frac{(\partial\rho/\partial t)}{(\partial\rho/\partial x)} = \frac{d(\rho u)}{d\rho} = u + \rho\frac{du}{d\rho}, \tag{6.14.6}$$

and from (6.14.4) follows

$$\left(\frac{\partial x}{\partial t}\right)_u = -\frac{(\partial u/\partial t)}{(\partial u/\partial x)} = u + \frac{1}{\rho}\frac{dP}{du}. \qquad (6.14.7)$$

Since the value of ρ uniquely determines that of u, the derivatives for constant ρ and constant u are the same, i.e. $(\partial x/\partial t)_u = (\partial x/\partial t)_\rho$, so that from (6.14.6) and (6.14.7) we obtain

$$\rho\frac{du}{d\rho} = \frac{1}{\rho}\frac{dP}{du} = \frac{1}{\rho}\frac{dP}{d\rho}\frac{d\rho}{du} = \frac{1}{\rho}a_s^2\frac{d\rho}{du}. \qquad (6.14.8)$$

From (6.14.8) follows

$$\frac{du}{d\rho} = \pm\frac{a_s}{\rho}, \qquad u = \pm\int\frac{a_s}{\rho}d\rho = \pm\int\frac{1}{a_s\rho}dP. \qquad (6.14.9)$$

The Eq. (6.14.9) gives the general relation between the velocity and the density or pressure in the wave. In a wave of small amplitude, when $\rho = \rho_0 + \rho'$, and $\rho' << \rho_0$, (6.14.9) gives in the first approximation $u = a_{s0}\rho'/\rho_0$, i.e. the usual formula for sound wave (Sect. 6.13). Combining (6.14.9) and (6.14.7) we obtain

$$\left(\frac{\partial x}{\partial t}\right)_u = u + \frac{1}{\rho}\frac{dP}{du} = u + \frac{1}{\rho}\frac{\partial P}{\partial \rho}\frac{\partial \rho}{\partial u} = u + \frac{1}{\rho}a_s^2\left(\pm\frac{\rho}{a_s}\right) = u \pm a_s(u),$$

which is integrated and gives

$$x = [u \pm a_s(u)]\cdot t + f(u), \qquad (6.14.10)$$

where f(u) is an arbitrary function of the velocity, and $a_s(u)$is given by (6.14.9).

This general solution was first obtained by B. Riemann (1880). It determines the velocity (and therefore all other quantities) as an implicit function of x and t, i.e. the wave profile at every instant. According to (6.14.10) the point where the velocity has a given value moves with constant velocity, which means that in this sense, the solution obtained is a traveling wave. The two signs in (6.14.10) correspond to waves propagating in the positive or negative x-directions relative to the gas. The flow described by the solution (6.14.9) and (6.14.10) is called a *simple wave*.

Let us write out the relations for a simple wave explicitly for an ideal gas. We recall that for adiabatic process, $\rho T^{1/(1-\gamma)} = \text{constant}$. Since sound velocity is proportional to $\sqrt{T}$, we have $P/\rho^\gamma = \text{const}$ and $\rho = \rho_0(a_s/a_{s0})^{2/(\gamma-1)}$. Substituting this in (6.14.9) we obtain

$$u = \pm \frac{2}{\gamma - 1} \int da_s = \pm \frac{2}{\gamma - 1}(a_s - a_{s0}). \tag{6.14.11}$$

Integration constant in (6.14.11) corresponds to $a_s = a_{s0}$ at the point in the wave for which $u = 0$. Correspondingly, using (6.14.11), we can write down

$$a_s = a_{s0} \pm \frac{\gamma - 1}{2} u, \tag{6.14.12}$$

$$\rho = \rho_0 \left(1 \pm \frac{\gamma - 1}{2} \frac{u}{a_{s0}}\right)^{2/(\gamma-1)}, \tag{6.14.13}$$

$$P = P_0 \left(1 \pm \frac{\gamma - 1}{2} \frac{u}{a_{s0}}\right)^{2\gamma/(\gamma-1)}. \tag{6.14.14}$$

Since $a_s > 0$, we conclude that $u \le \frac{2}{\gamma-1} a_{s0}$. It is convenient to present (6.14.10) in the form

$$u = F\left[x - \left(\frac{\gamma + 1}{2} u \pm a_{s0}\right) \cdot t\right], \tag{6.14.15}$$

where F is another arbitrary function.

It is seen from (6.14.12) that the velocity in a direction opposite to that of the propagation of the wave (relative to the gas itself) is of limited magnitude. For a wave propagating in the positive x-direction we have

$$-u \le \frac{2a_{s0}}{\gamma - 1}. \tag{6.14.16}$$

A simple wave described by formulae (6.14.12), (6.14.13), and (6.14.14) is essentially different from the limiting case of waves of small amplitude. The velocity of a point in the wave profile is

$$U = u \pm a_s. \tag{6.14.17}$$

It may be viewed as a superposition of the propagation of a disturbance relative to the gas with the velocity of sound and the movement of the gas itself with velocity u. The velocity u is now a function of the density, and therefore is different for different points in the profile. Thus, in the general case of a plane wave with arbitrary amplitude, there is no definite constant “wave velocity”. Since the velocities of different points in the wave profile are different, the profile changes its shape in the course of time.

Let us consider a wave propagating in the positive x-direction, for which $U = u + a_s$. The derivative of $u + a_s$ with respect to the density is

$$\frac{d(u + a_s)}{d\rho} = \frac{a_s}{\rho} + \frac{da_s}{d\rho} = \frac{1}{\rho}\frac{d(\rho a_s)}{d\rho} = \frac{a_s^2}{\rho}\frac{d(\rho a_s)}{dP} = \frac{\rho^2 a_s^5}{2}\left(\frac{\partial^2 V}{\partial P^2}\right)_s > 0, \qquad (6.14.18)$$

where we used formula (6.14.9): $du/d\rho = a_s/\rho$.

Let's take x-derivative of general solution (6.14.10) with sign plus, we obtain

$$t\frac{d(u + a_s)}{d\rho}\frac{\partial \rho}{\partial x} = 1. \qquad (6.14.19)$$

From (6.14.18) steams that for $t > 0$ also $\partial\rho/\partial x > 0$, and it is easy to see that also $\partial P/\partial x > 0$ and $\partial u/\partial x > 0$. However, for full time derivative we have, using continuous equation

$$\frac{d\rho}{dt} = \frac{\partial \rho}{\partial t} + u\frac{\partial \rho}{\partial x} = -\rho\frac{\partial u}{\partial x},$$

which means that $d\rho/dt < 0$ and $dP/dt < 0$, and also $du/dt < 0$.

This means that density and pressure are decreasing in every fluid elements of fluid in course of its move in space accompanying by rarefaction. Such motion is called *rarefaction wave*. In the form $x/t = u + a_s$ the solution (6.14.10) represents the rarefaction wave, which moves with velocity x/t relative to the laboratory system, and the head of the rarefaction wave moves with the local sound velocity $x/t - u = a_s$ relative to the gas.

Contrary to the rarefaction wave for a wave propagating in the negative x-direction the velocity of propagation of a given point in the wave profile increases with the density. If a_{s0} is the velocity of sound for the equilibrium density ρ_0, then in compressed region, $\rho > \rho_0$ and $a_s > a_{s0}$, while in rarefactions $\rho < \rho_0$ and $a_s < a_{s0}$.

The inequality of the velocity of different points in the wave profile causes its shape to change in the course of time: the points of compression move forward and those of rarefaction are left behind as it is shown in Fig. 6.11. Finally, the profile may become such that the function $\rho(x)$ for given t is no longer one-valued, when three different values of ρ correspond to the same value of x (Fig. 6.11c), i.e. the wave profile is turned-over. This is, of course, physically impossible. In reality, discontinuities are formed where $\rho(x)$ is not one-valued, and $\rho(x)$ is consequently one-valued everywhere except at the discontinuities themselves. The wave profile then has the form shown by the dashed line in Fig. 6.11c. When the discontinuity is formed, the wave ceases to be a simple wave. The cause of this can be briefly stated thus: when surfaces of discontinuity are present, the wave is reflected from them, and therefore ceases to be a wave traveling in one direction. The assumption on which the whole derivation is based, namely that there is a one-to-one relation between the various quantities, consequently ceases to be valid in general.

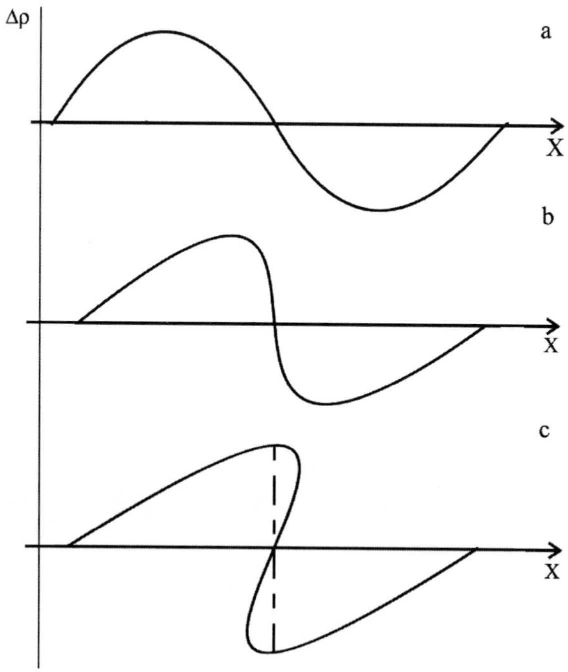

Fig. 6.11 Formation of discontinuity in a simple wave

The presence of discontinuities (shock waves) results in the dissipation of energy. The formation of discontinuities therefore leads to a marked damping of the wave. When the discontinuity is formed, the highest part of the wave profile is cut off. In the course of time, as the profile is bent over, its height becomes less, and the profile is smoothed to one with smaller amplitude, i.e. the wave is damped. It is clear from the above that discontinuities must ultimately be formed in every simple wave, which contains regions where the density decreases in the direction of propagation.

6.15 Flow in a Pipe Ahead of the Moving Piston

Although the wave is no longer a simple one when a discontinuity has been formed, the time and place of formation of the discontinuity can be determined analytically. We saw that the occurrence of discontinuities is mathematically due to the fact that, in a simple wave the quantities P, ρ and u become many-valued functions of x at times greater than a certain value t_0, whereas for $t < t_0$ they are one-valued functions. Therefore, the time t_0 is the time of formation of the discontinuity. It is evident from geometrical considerations that, at the instant t_0, the curve $u = u(x)$ as a function of x becomes vertical at some point $x = x_0$, which is the point where the function is subsequently many-valued. Analytically, this means that the derivative $(\partial u/\partial x)_t$, becomes infinite, and $(\partial x/\partial u)_t$ becomes zero. It is also clear that, at the instant t_0, the curve $u = u(x)$ must lie on both sides of the vertical tangent, since otherwise $u(x)$ would already be many-valued. In other words, the point $x = x_0$ must be, not an extremum of the function $x(u)$, but a point of inflexion, and therefore the second derivative $(\partial^2 x/\partial u^2)_t$ must also vanish. Thus the place and time of formation of the shock wave are determined by two equations

$$(\partial x/\partial u)_t = 0, \qquad (\partial^2 x/\partial u^2)_t = 0. \tag{6.15.1}$$

For an ideal gas, we found

$$x = \left[\frac{\gamma + 1}{2} u + a_{s0}\right] \cdot t + f(u). \tag{6.15.2}$$

Then, the Eqs. (6.15.1) give

$$t = -\frac{2}{\gamma + 1} f'(u), \qquad f''(u) = 0. \tag{6.15.3}$$

From this condition we can obtain explicit expressions for the time and place of formation of the discontinuity.

Let us use the above general theoretical consideration for a flow in a semi-infinite cylindrical pipe ($x > 0$) formed ahead of a moving piston. We assume that the piston is pushed into the pipe, and begins to move at time $t = 0$ with velocity $U_p = at$. In this case we find from (6.15.2) at the surface of the piston ($x = 0$)

$$f(u) = f(at) = -a_{s0}t - \frac{\gamma}{2} at^2, \tag{6.15.4}$$

and from (6.15.2) we obtain

$$x - (a_{s0} + \frac{\gamma + 1}{2} u)t = f(u) = -\frac{a_{s0}}{a} u - \frac{\gamma}{2a} u^2. \tag{6.15.5}$$

It is seen from (6.15.5) that the gas velocity monotonically decreases in a simple compression wave formed ahead of the piston and it vanishes ahead of the wave, at $x = a_{s0}t$. The time and location of the breakdown of the compression wave are defined by the two conditions (6.15.3). In fact, the shock wave is formed at the boundary where the simple wave adjoins a gas at rest. Here also the curve $u = u(x)$ must become vertical, i.e. the derivative $(\partial x/\partial u)_t$ must vanish at the time when the discontinuity occurs. However, in this case the second condition is simply that the velocity vanish at the boundary of the gas at rest, so that $(\partial x/\partial u)_t = 0$ for $u = 0$. From (6.15.5) we have $f'(u = 0) = -a_{s0}/a$, and substituting this in (6.15.3) we obtain time and location for the shock wave formation

$$t_{sh} = \frac{2}{\gamma + 1}\frac{a_{s0}}{a}, \qquad x_{sh} = a_{so} t_{sh} = \frac{2}{\gamma + 1}\frac{a_{s0}^2}{a}. \tag{6.15.6}$$

Note that discontinuity that must appear in a sound wave leads to a strong damping of the wave. It must be remarked, however, that this happens only for a sufficiently strong sound wave. Before nonlinear effects can develop, the usual effects of viscosity and thermal conduction will damp amplitude of a weak sound eliminating appearance of higher order effects in the amplitude. In the case of a strong sound wave and with account of damping due to viscosity, the time of the discontinuity formation is

$$t_{sh} = \frac{\lambda}{\pi U_0(\gamma + 1)} \exp(\mu t_{sh}), \tag{6.15.7}$$

where the sound wave is $u = U_0 \sin(\omega t)$, ω and λ are of frequency and wavelength, and $\mu \propto \nu/\lambda^2$. It is seen that solution to (6.15.7) exists only for large enough amplitudes of the sound wave, if $U_0 > \mu e\lambda/\pi(\gamma + 1)$.

Problems

6.1. Determine the shape of the surface of an incompressible fluid subject to a gravitational field, contained in a cylindrical vessel, which rotates about its vertical axis with a constant angular velocity Ω.

6.2 Calculate and draw streamlines of the flows with the velocity components:

(a) $u_x = Uy/L$, $u_y = Ux/L$; b) $u_x = U\frac{yL}{x^2 + y^2}$, $u_y = -U\frac{xL}{x^2 + y^2}$;

(b) $u_x = U\exp(-kx)\cos(ky)$, $u_y = U\exp(-kx)\sin(ky)$.

Are these flows incompressible? Irrotational? Find the velocity potential and the stream function.

6.3. Use the Bernoulli equation to find the time dependence of water level in the tank h(t) shown in the figure.

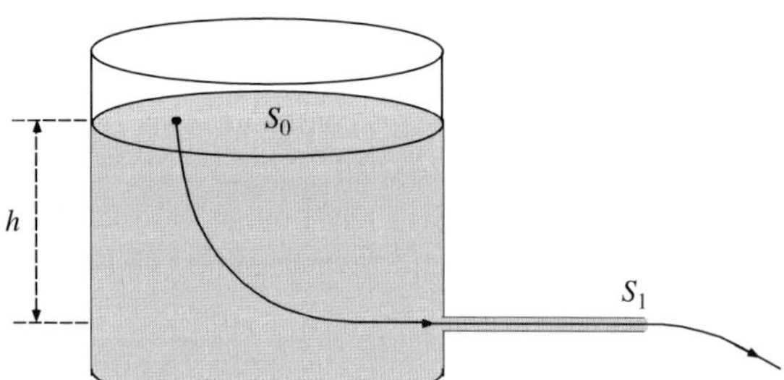

6.4. An infinitely light gas supports from below a layer of a heavy fluid. Thickness of the fluid layer is L, the upper surface of the fluid is free. Find the dispersion relation of the Rayleigh-Taylor instability at the low surface of the layer. Is the instability growth rate reduced in comparison with the maximal possible growth rate of the Rayleigh-Taylor instability $\sigma_{max} = \sqrt{gk}$?

6.5. Find position and time of the shock wave formed ahead of the piston, which starts moving with acceleration at $t = 0$.

Chapter 7
Energy Dissipation in Gases and Liquids

In Chap. 6 we considered dynamics of ideal gases and fluids assuming that there is no energy dissipation. In reality gases have non-zero viscosity and thermal conduction. In this chapter we shall consider the effect of energy dissipation, which is the result of the thermodynamic irreversibility of the motion. This irreversibility always occurs due to internal friction (viscosity) and irreversible energy diffusion – thermal conduction.

7.1 Viscous Fluids

In order to obtain the equations describing the motion of a viscous fluid, we have to include additional terms in the equation of motion. The equation of continuity, as it is seen from its derivation, is equally valid for any fluid, whether viscous or not. Euler's equation, on the other hand, requires modification. The equations of motion of a viscous fluid can be obtained by adding the expression which gives the irreversible "viscous" transfer of momentum in the fluid to the right-hand side of Euler's equation obtained for ideal fluid (we will not be considering gravity here)

$$\rho \frac{d\mathbf{u}}{dt} = \rho \left(\frac{\partial u_i}{\partial t} + u_k \frac{\partial u_i}{\partial x_k} \right) = -\nabla P + (\text{viscous force}) \tag{7.1.1}$$

Processes of internal friction occur in a fluid only when different fluid particles move with different velocities, so that there is a relative motion between various parts of the fluid. The viscous force is actually a friction force on the surface separating fluid layers. The physical reason for the viscous force is random thermal motion of fluid particles accompanied by particles exchange between two neighboring layers. Thus, the friction force between the layers must depend on the space derivatives of the velocity. Since the viscous force is a surface force, it is convenient to introduce the force per unit surface area called viscous stress, σ. If the friction force acting along the element surfaces in x-direction, so the x-component of the stress is

M.A. Liberman, *Introduction to Physics and Chemistry of Combustion*,
DOI: 10.1007/978-3-540-78759-4_7, © Springer-Verlag Berlin Heidelberg 2008

$$\sigma \equiv F_{\text{friction}}/S_{\text{layer}} = \eta \frac{\partial u_x}{\partial y}. \tag{7.1.2}$$

The viscous force acting on the fluid element is

$$(\text{viscous force}) = \left(\eta \frac{\partial u_x}{\partial y}\bigg|_{y+\delta y} - \eta \frac{\partial u_x}{\partial y}\bigg|_{y} \right) \delta S \approx \frac{\partial}{\partial y}\left(\eta \frac{\partial u_x}{\partial y} \right) \delta S \delta y$$
$$= \frac{\partial}{\partial y}\left(\eta \frac{\partial u_x}{\partial y} \right) \delta V, \tag{7.1.3}$$

and the equation of momentum transfer for this simplified case of motion along x-axis takes the form

$$\rho \delta V \frac{du_x}{dt} = -\frac{\partial P}{\partial x} \delta V + \frac{\partial}{\partial y}\left(\eta \frac{\partial u_x}{\partial y} \right) \delta V$$

or

$$\rho \frac{du_x}{dt} = -\frac{\partial P}{\partial x} + \frac{\partial}{\partial y}\left(\eta \frac{\partial u_x}{\partial y} \right).$$

Here η is the viscosity coefficient. The viscosity coefficient depends on a particular fluid or gas. Typically, the viscosity coefficients are very small for gases, for example, for air $\eta = 1.8 \cdot 10^{-4}$g/s cm, for hydrogen $\eta = 9.5 \cdot 10^{-5}$g/s cm, for oxygen $\eta = 2.1 \cdot 10^{-4}$g/s cm, for nitrogen $\eta = 1.84 \cdot 10^{-4}$g/s cm. For fluids the viscosity coefficients are larger: for water $\eta = 0.01$ g/s cm, for gasoline $\eta = 6.5 \cdot 10^{-3}$g/s cm, for glycerin $\eta = 8.5$ g/s cm.

In general case, the quantity $\sigma = \sigma_{ik}$ is tensor, which is called the stress tensor. Its form can be established in a very general form. If the velocity gradients are small, we may suppose that the momentum transfer due to viscosity represents a linear function of the first order space derivatives of the velocity. The stress tensor obviously must vanish when the whole fluid undergoes to a uniform rotation, since there is no internal friction for such motion, and there should not be terms independent on velocity derivatives, since the stress tensor must vanish for uniform motion with a constant velocity. Therefore the stress tensor may contain only symmetrical combinations of the derivatives $\partial u_i/\partial x_k$. It can be shown that the most general tensor of rank two satisfying the above conditions is

$$\sigma_{ik} = \eta \left(\frac{\partial u_i}{\partial x_k} + \frac{\partial u_k}{\partial x_i} - \frac{2}{3} \delta_{ik} \frac{\partial u_i}{\partial x_i} \right) + \zeta \delta_{ik} \frac{\partial u_i}{\partial x_i}, \tag{7.1.4}$$

with coefficients $\eta > 0$ and $\zeta > 0$ which are called the first and the second coefficients of viscosity, and δ_{ik} being the unit tensor:

$$\delta_{ik} = \begin{pmatrix} 1 & 0 & 0 \\ 0 & 1 & 0 \\ 0 & 0 & 1 \end{pmatrix}.$$

Thus, we came to the general form of the momentum equation with the viscous force

$$\rho \frac{du_i}{dt} = -\frac{\partial P}{\partial x_i} + \rho g_i + \frac{\partial}{\partial x_k} \eta \left(\frac{\partial u_i}{\partial x_k} + \frac{\partial u_k}{\partial x_i} - \frac{2}{3} \delta_{ik} \frac{\partial u_j}{\partial x_j} \right) + \frac{\partial}{\partial x_i} \left(\zeta \frac{\partial u_j}{\partial x_j} \right). \quad (7.1.5)$$

This is the most general form of the equations of motion of a viscous fluid. The quantities η and ζ are functions of pressure and temperature, and therefore are not constant throughout the fluid, so that η and ζ cannot be taken outside the gradient operator. However, in most cases, the viscosity coefficients do not change noticeably in the fluid, and they may be regarded as constant. Under these assumptions the Eq. (7.1.5) can be considerably simplified, written in vector form, known as *the Navier-Stokes equation*

$$\rho \frac{\partial \mathbf{u}}{\partial t} + \rho (\mathbf{u} \cdot \nabla) \mathbf{u} = -\nabla P + \rho \mathbf{g} + \eta \nabla^2 \mathbf{u} + (\zeta + \frac{1}{3} \eta) \nabla (\nabla \cdot \mathbf{u}). \quad (7.1.6)$$

The Navier-Stokes equation becomes considerably simpler if the fluid may be regarded as incompressible, so that $\nabla \cdot \mathbf{u} = 0$; the last term on the right side of (7.1.6) vanishes, and we obtain

$$\frac{\partial \mathbf{u}}{\partial t} + (\mathbf{u} \cdot \nabla) \mathbf{u} = -\frac{1}{\rho} \nabla P + \frac{\eta}{\rho} \nabla^2 \mathbf{u} + \mathbf{g}. \quad (7.1.7)$$

We see that the viscosity of an incompressible fluid is determined by only one coefficient, and the stress tensor takes the simple form

$$\sigma_{ik} = -P \delta_{ik} + \eta \left(\frac{\partial u_i}{\partial x_k} + \frac{\partial u_k}{\partial x_i} \right). \quad (7.1.8)$$

Frequently used the ratio $\nu = \eta/\rho$, which is called the *kinematic viscosity* (contrary to the dynamic viscosity coefficients). The kinematic viscosity coefficients at T = 20 C are: for air $\nu = 0.15\ cm^2/s$, for glycerin $\nu = 6.8\ cm^2/s$, for water $\nu = 0.01\ cm^2/s$.

7.2 Energy Dissipation in Viscous Fluids

The presence of viscosity results in the dissipation of energy, which is finally transformed into heat. For the considerably simplified case of an incompressible fluid, the total kinetic energy is

$$E_{kin} = \frac{1}{2}\rho \int u^2 dV. \qquad (7.2.1)$$

Taking the time derivative of (7.2.1), writing $\partial(\frac{1}{2}\rho u^2) = \rho u_i(\partial u_i/\partial t)$ and substituting for $\partial u_i/\partial t$ the expression for it from the Navier-Stokes equation:

$$\frac{\partial u_i}{\partial t} = -\frac{1}{\rho}\frac{\partial P}{\partial x_i} - u_k \frac{\partial u_i}{\partial x_k} + \frac{1}{\rho}\frac{\partial \sigma_{ik}}{\partial x_k}, \qquad (7.2.2)$$

we obtain for the incompressible fluid ($\nabla \cdot u = 0$):

$$\frac{\partial}{\partial t}\left(\frac{1}{2}\rho u^2\right) = -\nabla \cdot \left[\rho \mathbf{u}\left(\frac{1}{2}u^2 + \frac{P}{\rho}\right) - \mathbf{u}\cdot\hat{\sigma}\right] - \sigma_{ik}\frac{\partial u_i}{\partial x_k}. \qquad (7.2.3)$$

The expression in brackets in the right side of (7.2.3) is the energy flux density in the fluid, where the first term is the energy flux due to the actual transfer of fluid mass, and the second term is the energy flux due to processes of internal friction. The presence of viscosity results in a momentum flux. A transfer of momentum, however, always involves a transfer of energy. Integrating (7.2.3) over volume V, we obtain

$$\frac{\partial}{\partial t}\int \left(\frac{1}{2}\rho u^2\right) dV = -\oint \left[\rho \mathbf{u}\left(\frac{1}{2}u^2 + \frac{P}{\rho}\right) - \mathbf{u}\cdot\hat{\sigma}\right] \cdot d\mathbf{S} - \int \sigma_{ik}\frac{\partial u_i}{\partial x_k} dV. \quad (7.2.4)$$

The first term on the right gives the rate of change of the kinetic energy of the fluid in volume V owing to the energy flux through the surface bounding V. The integral in the second term is consequently the decrease per unit time in the kinetic energy owing to dissipation. If the integration is extended to the whole volume of the fluid, the surface integral vanishes since the velocity vanishes at infinity, and taking into consideration that σ_{ik} is symmetrical tensor, we find the energy dissipated per unit time in the whole fluid

$$\frac{\partial}{\partial t}E_{kin} = -\frac{1}{2}\int \sigma_{ik}\left(\frac{\partial u_i}{\partial x_k} + \frac{\partial u_k}{\partial x_i}\right) dV \qquad (7.2.5)$$

Substituting in (7.2.5) expression for the tensor σ_{ik} given by (7.1.8), we obtain for the energy dissipation in an incompressible fluid

$$\frac{\partial}{\partial t}E_{kin} = -\frac{1}{2}\eta \int \sigma_{ik}\left(\frac{\partial u_i}{\partial x_k} + \frac{\partial u_k}{\partial x_i}\right)^2 dV. \qquad (7.2.6)$$

The dissipation leads to a decrease in the mechanical energy, i.e. $\frac{\partial}{\partial t}E_{kin} < 0$. Since the integral in (7.2.6) is positive, we conclude that the viscosity coefficient η is always positive too.

Finally, some words should be said about *boundary conditions* for a viscous fluid, which are somewhat different from the boundary conditions for an ideal fluid. Let us consider boundary conditions at a rigid wall. A fluid cannot penetrate into a rigid wall, therefore the normal velocity component should vanish at a wall both for ideal and viscous fluids, $u_n = \mathbf{u} \cdot \mathbf{n} = 0$. However, unlike an ideal fluid, a viscous fluid flow cannot slip along a wall since the friction force stops fluid elements close to the wall. Therefore, in the case of a viscous fluid the tangential components of the velocity along a wall are also zero $\mathbf{u}_t = 0$. Combining both of these conditions we can formulate the boundary condition at a rigid wall: *velocity of a viscous fluid is zero at the wall,* $\mathbf{u}|_{\text{rigid wall}} = 0$.

7.3 Thermal Conduction

A complete system of equations of fluid dynamics must contain five equations. For a fluid in which processes of thermal conduction and internal friction occur, one of these equations is the equation of continuity, the Euler equations are replaced by the Navier-Stokes equations. While for an ideal fluid the fifth equation is the equation of conservation of entropy, in a viscous fluid this equation does not hold since irreversible processes of energy dissipation occur in it. So far we have considered flows in fluids and gases of constant density that are not affected by energy release and heat transfer. However, if some part of a gas is heated, then density of the gas drops in comparison with the surroundings and the buoyant force pushes the heated gas of the smaller density upwards, leading to the effect called convection. In this chapter we shall consider the processes of heat transfer and flows caused by density difference.

7.3.1 The Equation of Thermal Conduction

We now shall take into account the transport processes related to viscosity and thermal conduction:

$$\frac{\partial}{\partial t}\left(\rho\varepsilon + \frac{1}{2}\rho u^2\right) + \nabla \cdot \left[\rho \mathbf{u}\left(\varepsilon + \frac{P}{\rho} + \frac{1}{2}u^2\right)\right] = \\ = -\{(\text{thermal conduction}) + (\text{Viscous losses})\} \tag{7.3.1}$$

The first term on the left side of (7.3.1) is the rate of change of energy in unit volume of the fluid, and the second term is the energy flux due to the hydrodynamic flow and to work of pressure forces on the fluid. Two terms on the right are the energy flux due to random thermal motion of fluid particles. In a viscous fluid the law of conservation of energy holds: the change per unit time in the total energy of the fluid in any volume must be equal to the total flux of energy through the surface bounding that volume. The energy flux density, however, now has a different form. Besides the second term on the left of (7.3.1) – the flux $\rho\mathbf{u}(H + u^2/2)$ due to the simple transfer of mass by the motion of the fluid, there is also a flux due to processes of internal friction, which is given by the vector $\mathbf{u} \cdot \sigma$, with components $u_i\sigma_{ik}$. In addition, if the temperature of the fluid is not constant throughout its volume, there will be a transfer of heat, which is called thermal conduction. This signifies the direct molecular transfer of energy from points where the temperature is high to those where it is low. It does not involve macroscopic motion, and occurs even in a fluid at rest. We denote the heat flux density due to thermal conduction by $\mathbf{q}$:

$$\textit{Energy flux per unit surface} = q = -\kappa\nabla T. \qquad (7.3.2)$$

The coefficient κ is called *the coefficient of thermal conduction* (or the thermal conductivity). This coefficient characterizes how efficient the process of energy transfer is for a particular type of fluid, gas or solids. The minus sign in (7.3.2) appears since energy is transferred from hot to cold layers in the direction opposite to the direction of the temperature gradient. It is always positive, that is seen from the fact that the energy flux must be from points at a high temperature to those at a low temperature, i.e. $\mathbf{q}$ and ∇T must be in opposite directions. The coefficient κ is in general a function of temperature and pressure.

Accordingly, the general law of conservation of energy is given by the equation

$$\frac{\partial}{\partial t}\left(\rho\varepsilon + \frac{1}{2}\rho u^2\right) + \nabla \cdot \left[\rho\mathbf{u}\left(\varepsilon + \frac{P}{\rho} + \frac{1}{2}u^2\right)\right] = \nabla \cdot (\kappa\nabla T + \sigma \cdot \mathbf{u}). \qquad (7.3.3)$$

The Eq. (7.3.3) could be taken to complete the system of fluid-mechanical equations of a viscous fluid. By virtue of the continuity equation and the Navier-Stokes equation, and using the thermodynamic relation $d\varepsilon = TdS - PdV = TdS + (P/\rho^2)d\rho$, it is convenient to present (7.3.3) in another form

$$\rho T\left(\frac{\partial S}{\partial t} + \mathbf{u} \cdot \nabla S\right) = \sigma_{ik}\frac{\partial u_i}{\partial x_k} + \nabla \cdot (\kappa\nabla T). \qquad (7.3.4)$$

The expression on the left side of (7.3.4) is the total time derivative of the entropy, multiplied by ρT. The quantity dS/dt gives the rate of change of the entropy of a unit mass of fluid as it moves about in space, with $\rho TdS/dt$ being the quantity of heat gained per unit volume. In the presence of dissipations the production of the entropy is associated with viscosity and thermal conduction on the right side of the Eq. (7.3.4), where the first term is the energy dissipated into heat by viscosity, and the second is the heat conducted into the volume.

7.3.2 *Thermal Conduction in an Incompressible Fluid*

The equation of thermal conduction is considerably simplified if the fluid velocity is small compared to the sound speed, $u << a_s$. In this case the pressure variations occurring as a result of the motion, which can be evaluated as $\Delta P \propto \rho \cdot u^2$, are small, and the variation in the density and in the other thermodynamic quantities caused by them may be neglected. However, a non-uniformly heated fluid is not completely incompressible in the sense that the density varies with the temperature, so that this variation cannot in general be neglected. Therefore, even at small velocities, we can suppose the pressure constant, but not the density. Taking into account that the specific heat at constant pressure is $C_P = T(\partial S/\partial T)_P$, we can write for the terms in the left side of the Eq. (7.3.4)

$$T\frac{\partial S}{\partial t} = T\left(\frac{\partial S}{\partial T}\right)_P \frac{\partial T}{\partial t} = C_P\frac{\partial T}{\partial t}, \text{ and } T\nabla S = C_P\nabla T. \qquad (7.3.5)$$

Then, Eq. (7.3.4) takes the form

$$\rho C_P\left(\frac{\partial T}{\partial t} + \mathbf{u}\cdot\nabla T\right) = \sigma_{ik}\frac{\partial u_i}{\partial x_k} + \nabla\cdot(\kappa\nabla T), \qquad (7.3.6)$$

When the fluid may be supposed incompressible in the usual sense, and if the temperature differences in the fluid is small so that σ and κ does not change with the temperature, then we obtain the equation of heat transfer in an incompressible fluid in the form:

$$\frac{\partial T}{\partial t} + \mathbf{u}\cdot\nabla T = \nabla(\chi\nabla T) + \frac{\nu}{2C_P}\left(\frac{\partial u_i}{\partial x_k} + \frac{\partial u_k}{\partial x_i}\right)^2, \qquad (7.3.7)$$

where $\nu = \eta/\rho$ is the kinematic viscosity, and $\chi = \kappa/\rho C_P$ is the thermal conductivity.

Let us compare viscosity and thermal conduction terms on the right side of the Eq. (7.3.7) for the case of subsonic flow, which characteristic velocity is much less than the speed of sound, $U << a_s$, and typical length scale of the flow is L. Taking into consideration that the sound speed for an ideal gas is

$$a_s^2 = \gamma P/\rho = (\gamma - 1)C_P T,$$

we can evaluate viscous term as

$$\frac{\nu}{2C_P}\left(\frac{\partial u_i}{\partial x_k} + \frac{\partial u_k}{\partial x_i}\right)^2 \propto \frac{\nu}{C_P}\frac{U^2}{L^2},$$

and the thermal conduction term as

$$\nabla(\chi\nabla T) \propto \chi\frac{T}{L^2} \approx \chi\frac{a_s^2}{(\gamma-1)C_P L^2}.$$

Contribution of the viscous and thermal conduction can be characterized by the ratio

$$(\gamma-1)\frac{\nu}{\chi}\frac{U^2}{a_s^2}. \tag{7.3.8}$$

The dimensionless ratio of the coefficients of viscosity and thermal diffusion is called the *Prandtl number*

$$\mathrm{Pr} = \frac{\nu}{\chi} = \frac{\eta C_P}{\kappa}. \tag{7.3.9}$$

For gases the Prandtl number is of the order of unity, for air Pr = 0.733 at 20 C. For liquids the Prandtl number can be larger than unity, for example, for water Pr = 6.75. Therefore, we conclude that the equation of energy transfer for a subsonic flow can be taken in the form

$$\frac{\partial T}{\partial t} + \mathbf{u}\cdot\nabla T = \nabla(\chi\nabla T). \tag{7.3.10}$$

The equation of heat transfer is particularly simple for an incompressible fluid at rest, in which the transfer of energy takes place entirely by thermal conduction. Omitting the terms, which involve the velocity, we obtain

$$\frac{\partial T}{\partial t} = \chi\Delta T. \tag{7.3.11}$$

This equation is known as the *Fourier equation.*

It should be mentioned that the applicability of the thermal conduction equation in the form (7.3.11) to fluids is actually very limited. The reason is that, in fluids in a gravitational field, even a small temperature gradient usually results in considerable convection, parts of the fluid of lower density flow up. Hence we can actually have a fluid at rest with a non-uniform temperature distribution only if the direction of the temperature gradient is opposite to that of the gravitational force, or if the fluid is very viscous.

Let Q be the quantity of heat generated by external sources of heat, for example, a chemical reaction in unit volume of the fluid per unit time. In general, Q is a function of time and coordinates. Then the heat balance equation with the heat source, i.e. the equation of thermal conduction, is

$$\rho C_P\left(\frac{\partial T}{\partial t} + \mathbf{u}\cdot\nabla T\right) = \nabla\cdot(\kappa\nabla T) + Q. \tag{7.3.12}$$

The boundary conditions on the equation of thermal conduction hold at the boundary between two media: the temperatures of the two media must be equal at the boundary, $T_1 = T_2$. The heat flux out of one medium must be equal the heat flux into the other medium. Denoting $\partial T/\partial n$ the derivative of T along the normal to the surface, we obtain the boundary condition in the form

$$\kappa_1 \frac{\partial T_1}{\partial n} = \kappa_2 \frac{\partial T_2}{\partial n}. \tag{7.3.13}$$

If there are external sources of heat on the surface of separation which generate an amount of heat Q_{ext} on unit area in unit time, then (7.3.13) must be replaced by

$$\kappa_1 \frac{\partial T_1}{\partial n} - \kappa_2 \frac{\partial T_2}{\partial n} = Q_{ext}. \tag{7.3.14}$$

The boundary condition at a rigid wall is $(T)_{\text{at the wall}} = T_0$, and the condition of fixed thermal flux to the wall is $(dT/dx_n)_{\text{at the wall}} = q$.

7.3.3 Heat Propagation

Consider solution of the *Fourier equation* for some simple problems. Let at initial instant, $t = 0$, amount of heat Q was released at some small volume, near the origin of the coordinate system $r = 0$ (local explosion). Later on, at $t > 0$, heat propagates in space, and we want to find temperature variation $T = T(\mathbf{r}, t)$. The solution to the problem is given by the Fourier equation (7.3.11) together with condition

$$\int T(x, y, z) d^3x = Q, \tag{7.3.15}$$

where $Q = E/\rho c_P$ if this is process at constant pressure, and $Q = E/\rho c_V$ for the process at constant volume; E is the density of the energy released. Two Eqs. (7.3.11) and (7.3.15) are equivalent to the Eq. (7.3.12) with the sources of energy in the form of delta-function, $q = Q\delta(r)\delta(t)$.

To solve the problem, let us expand T(r,t) in the Fourier integral

$$T(r, t) = \int T_k(t)\, e^{i\mathbf{kr}} \frac{d^3k}{(2\pi)^3}, \; T_k(t) = \int T(r, t)\, e^{-i\mathbf{kr}} d^3x. \tag{7.3.16}$$

Substituting (7.3.16) in the Fourier equation (7.3.11) we obtain for each Fourier component $T_k(t)e^{i\mathbf{kr}}$ equation

$$\frac{dT_k}{dt} = k^2 \chi T_k, \tag{7.3.17}$$

so that

$$T_k(t) = T_{0k} e^{-k^2 \chi t}. \tag{7.3.18}$$

Let the initial temperature distribution is $T(t = 0, r) = T_0(r)$, then we can write for the coefficients T_{0k} in (7.3.18)

$$T_{0k} = \int T(\mathbf{r}') \, e^{-i\mathbf{k}\mathbf{r}'} d^3x'. \tag{7.3.19}$$

Substituting $T_k(t)$ in (7.3.16), we obtain

$$T(r, t) = \int T_0(\mathbf{r}') e^{-k^2 \chi t} \, e^{i\mathbf{k}(\mathbf{r}-\mathbf{r}')} d^3x' \frac{d^3k}{(2\pi)^3}. \tag{7.3.20}$$

Integral on d^3k in (7.3.20) can be presented as a product of three identical integrals

$$\int_{-\infty}^{+\infty} e^{-\alpha k_x^2} \cos(x - x') k_x dk_x = \sqrt{\frac{\pi}{\alpha}} e^{-\frac{(x-x')^2}{4\alpha}}, \text{ where } \alpha = \chi t. \tag{7.3.21}$$

Similar integrals but with $\sin(x - x')$ instead of $\cos(x - x')$ vanish because of sin is even function. The simplest way to calculate integral of the form (7.3.21) is the following. Consider integral

$$I(\beta) = \int_{-\infty}^{+\infty} e^{-\alpha x^2} \cos \beta x \, dx$$

as function of the parameter β, and calculate its derivative relative to β

$$\frac{dI(\beta)}{d\beta} = -\int_{-\infty}^{+\infty} e^{-\alpha x^2} x \sin \beta x \, dx = \frac{1}{2\alpha} \int_{-\infty}^{+\infty} d\left(e^{-\alpha x^2}\right) \sin \beta x =$$

$$\frac{1}{2\alpha} e^{-\alpha x^2} \sin \beta x \Big|_{-\infty}^{+\infty} - \frac{\beta}{2\alpha} \int_{-\infty}^{+\infty} e^{-\alpha x^2} \cos \beta x \, dx = -\frac{\beta}{2\alpha} I(\beta)$$

Thus we came to equation

$$\frac{dI(\beta)}{d\beta} = -\frac{\beta}{2\alpha} I(\beta)$$

Which can be elementary solved, giving $I(\beta) = Ce^{-\beta^2/4\alpha}$. For $\beta = 0$ we have

$$I(0) = \int_{-\infty}^{+\infty} e^{-\alpha x^2} dx = \sqrt{\pi/\alpha}.$$

Thus, we obtain $I(\beta) = \sqrt{\frac{\pi}{\alpha}} e^{-\beta^2/4\alpha}$.

Finally, we obtain solution for the temperature distribution in the following form

$$T(\mathbf{r}, t) = \frac{1}{8(\pi\chi t)^{3/2}} \int T_0(\mathbf{r}') \exp\left\{\frac{(\mathbf{r} - \mathbf{r}')^2}{4\chi t}\right\} d^3x' \tag{7.3.22}$$

If initial temperature distribution depends on x-coordinate only, then after integrating on $dy'dz'$ in (7.3.22) we obtain

$$T(x, t) = \frac{1}{2(\pi\chi t)^{1/2}} \int T_0(x') \exp\left\{\frac{(x - x')^2}{4\chi t}\right\} d^3x'. \tag{7.3.23}$$

Let us assume that all heat was released at the one point at $\mathbf{r} = 0$ at $t = 0$, which can be presented as

$$T_0(r) = \text{const} \cdot \delta(r), \int T(x, y, z) d^3x = Q. \tag{7.3.24}$$

Integrating (7.3.22) with this initial temperature distribution gives

$$T(\mathbf{r}, t) = \frac{Q}{8(\pi\chi t)^{3/2}} \exp\left(-\frac{r^2}{4\chi t}\right). \tag{7.3.25}$$

It is seen form the obtained solution that temperature decreases at the point $r = 0$ with time as $1/t^{3/2}$, and simultaneously temperature of the surrounding gas is gradually increasing. The heating and characteristic length of the heat penetration in the surrounding gas are defined mainly by the exponent in (7.3.25), so that width of the heating zone is

$$L \propto \sqrt{\chi} t. \tag{7.3.26}$$

If initially heat was released at the plane $x = 0$, then the temperature distribution at the latter instants is

$$T(x, t) = \frac{Q}{2(\pi\chi t)^{3/2}} \exp\left(-\frac{x^2}{4\chi t}\right) \tag{7.3.27}$$

Notice that the obtained temperature distribution and, in particular, instant character of the temperature increase up to infinity is due to the assumption that thermal conductivity coefficient does not depend on temperature, though temperature increase at large distances is infinitely small according to the Gauss character of temperature distribution (7.3.27). For more realistic situation, when χ is a decreasing function of temperature, the heat will propagate at a finite distance at every given instant of time.

In solving this problem we assumed that gas is motionless. In reality this is not true. When energy is released at some location, the compression wave or shock wave will start from the location of the energy release. The compression wave propagates with velocity of the order of sound velocity, a_s, while heat propagates due to thermal conduction with the velocity

$$\frac{dx}{dt} \propto \frac{d}{dt}\sqrt{\chi t} \propto \sqrt{\frac{\chi}{t}} \propto \frac{\chi}{x} \propto \frac{l}{x} a_s, \tag{7.3.28}$$

where we used that from kinetic theory $\chi = la_s$, with l being an average distance between molecule collision (mean free path). If total amount of energy released at $x=0$ is not very large, then we came to conclusion that the heat propagates with velocity much less that the sound speed, since the scales in the problem of question are much larger than the mean free path. In this case the gas velocity is small compared to the sound velocity, and the process of heat propagation proceeds with almost constant pressure. This is what we discussed considering thermal mechanism of the flame propagation. In the opposite case, when energy release is large, compression wave became a shock wave after it left the location of the energy release. This is the case of strong explosion. Now the process is pure gasdynamic one and thermal conductivity of the gas does not important.

7.4 Stationary Flow of Incompressible Viscous Fluid

7.4.1 Flow of Viscous Incompressible Fluid in a Duct

Let us consider the flow of a viscous fluid enclosed between two parallel planes, one of which (upper plane in Fig. 7.1) is moving with a constant velocity **U** relative to the other. We shall choose the planes as the xz -plane with the x-axis in the direction of **U**. A fluid layer of thickness L (distance between the plates) separates the plates. We choose the x-axis to be along the channel walls and the y-axis perpendicular to the walls.

Obviously, that all the quantities depend only on y, and the fluid velocity everywhere is in x-direction, which follows from the symmetry of the problem. Boundary conditions on the plates may be written as

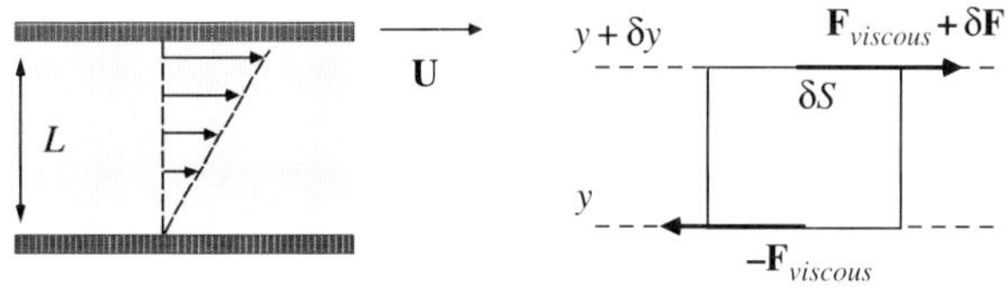

Fig. 7.1 The viscous flow between two plates and viscous forces acting on a fluid element

$$u_x(y = 0) = 0;\ u_x(y = L) = U.$$

We assume that there is no pressure gradient in the flow and that the gravitational force is negligible. Taking into account the geometry of the flow, it is natural to look for solution in the form $\mathbf{u} = (u_x; 0; 0)$. Since we consider a steady flow, then the flow velocity should be independent of time and the x-coordinate, that is $u_x = u_x(y)$. In that case the continuity equation is satisfied identity

$$\nabla \cdot \mathbf{u} = 0 \Rightarrow \frac{\partial u_x}{\partial x} + \frac{\partial u_y}{\partial y} = 0, \tag{7.4.1}$$

since $u_y = 0$ and $\partial u_x/\partial x = 0$.

In the case of a flow in the x-direction with zero pressure gradient and zero gravitational force the Navier-Stokes equation for the incompressible fluid can be written in the form

$$\frac{\partial \mathbf{u}}{\partial t} + (\mathbf{u} \cdot \nabla)\mathbf{u} = -\frac{1}{\rho}\nabla P + \nu\nabla^2\mathbf{u}, \tag{7.4.2}$$

where $\nu = \eta/\rho$ is the kinematic viscosity, and in our case (7.4.2) becomes

$$\rho\frac{\partial u_x}{\partial t} + \rho(\mathbf{u} \cdot \nabla)u_x = \eta\nabla^2 u_x.$$

The left-hand side of the equation is zero since velocity is independent of time, and the nonlinear part of substantive derivative is zero, since $(\mathbf{u} \cdot \nabla)u_x = u_x\frac{\partial u_x}{\partial x} = 0$. Then the Navier-Stokes equation is reduced to the equation

$$\eta\nabla^2 u_x = \eta\frac{\partial^2 u_x}{\partial y^2} = 0,$$

which has a solution

$$u_x = Ay + B$$

with the numerical coefficients A and B that have to be determined from the boundary conditions. Substituting the obtained solution into the boundary conditions we obtain $B = 0$, $AL = U$, so that the velocity distribution in the channel is: $u_x = Uy/L$.

7.4.2 The Poiseuille Flow

Now let us consider a viscous flow between two parallel planes separated by a fluid layer of a thickness 2R. We take the x-axis directed along the channel walls and the y-axis perpendicular to the walls with the origin $y = 0$ in the centre of the channel. It is clear that all quantities depend only on y, and that the fluid velocity is everywhere in the x-direction. Taking into account the geometry of the flow, it is natural to look for solution in the form $\mathbf{u} = (u_x; 0; 0)$ with the x-coordinate of the velocity $u_x = u_x(y)$. The continuity equation then satisfied identically

$$\nabla \cdot \mathbf{u} = 0 \Rightarrow \frac{\partial u_x}{\partial x} + \frac{\partial u_y}{\partial y} = 0 \tag{7.4.3}$$

From the Navier-Stokes equation (7.4.2) without gravitational force ($\mathbf{g}=0$) and for steady flow we have

$$0 = -\frac{\partial P}{\partial x} + \eta \frac{\partial^2 u_x}{\partial y^2}, \qquad 0 = -\frac{\partial P}{\partial y}. \tag{7.4.4}$$

The second of Eq. (7.4.4) means that the pressure is independent of y-co-ordinates, i.e. it is constant across the depth of the fluid between the planes, so that $P=P(x)$. Therefore the first term in the first Eq. (7.4.4) depends only on the variable x, while the second is a function of y only. This is possible only if both terms are constant, which implies the constant pressure gradient $dP/dx = \frac{P_1 - P_2}{L} \equiv \Pi = \text{const}$, which can be written as the following pressure distribution

$$P = P_1 - \frac{P_1 - P_2}{L} x. \tag{7.4.5}$$

The equation for the flow velocity becomes

$$\eta \frac{\partial^2 u_x}{\partial y^2} = -\Pi, \tag{7.4.6}$$

which solution is

$$u_x = -\frac{\Pi}{2\eta} y^2 + Ay + B. \tag{7.4.7}$$

Substituting the obtained solution into the boundary conditions at the walls: $u_x = 0$, at $y = \pm R$, we have

$$-\frac{\Pi}{2\eta} R^2 + AR + B = 0, \quad -\frac{\Pi}{2\eta} R^2 - AR + B = 0.$$

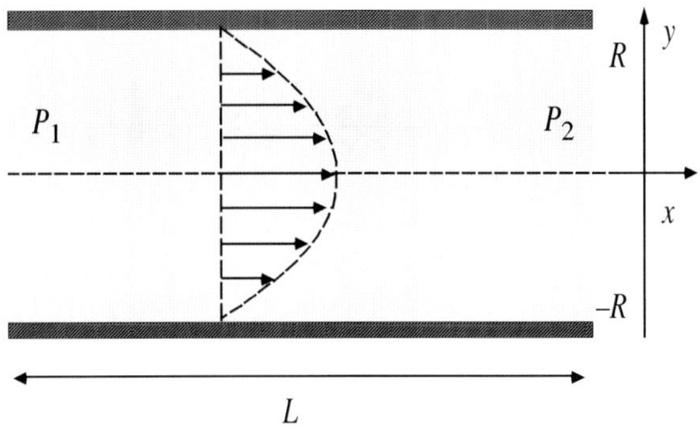

Fig. 7.2 The Poiseuille flow in a rectangular channel

From these equations the integration constants are: $A=0$, $B=\frac{\Pi}{2\eta}R^2$, and the velocity distribution is

$$u_x = \frac{\Pi}{2\eta}\left(R^2 - y^2\right) = \frac{P_1 - P_2}{2L\eta}\left(R^2 - y^2\right). \tag{7.4.8}$$

The velocity for the Poiseuille flow varies parabolically across the fluid (Fig. 7.2), reaching its maximum value in the middle of the channel axis, at $y=0$:

$$u_{max} = \frac{P_1 - P_2}{2\eta L}R^2. \tag{7.4.9}$$

For the velocity averaged over the channel width we obtain

$$\bar{u} = -\frac{R^2}{3\eta}\frac{dP}{dx} = -\frac{R^2}{3\eta}\frac{P_1 - P_2}{L}.$$

The tangential friction force acting on one of the fixed plane wall is

$$\sigma_{xy} = \eta\left(\frac{du}{dy}\right)_{y=0} = -R\frac{dP}{dx}.$$

Let us discuss some interesting conclusions, which are not obvious. It follows from (7.4.8) and (7.4.9) that the flow velocity increases if: (1) pressure gradient increases; (2) the channel becomes wider or (3) the fluid is less viscous (for smaller viscosity). It is also interesting to calculate the mass flux through the channel. Since the flow is two-dimensional, we calculate the mass flux per unit length in x-direction J/L_x

$$J/L_x = \int_{-R}^{R} \rho u_x dy = \frac{\rho\Pi}{2\eta}\int_{-R}^{R}\left(R^2 - y^2\right)dy = \frac{2}{3}\frac{P_1 - P_2}{\eta L}\rho R^3. \tag{7.4.10}$$

The mass flux increases linearly with pressure gradient, while the dependence of the mass flux on the channel width is much stronger $J \propto R^3$.

The expressions (7.4.8)–(7.4.10) give an exact stationary solution to the Navier-Stokes equation for a stationary flow of viscous fluid in a channel. If such flow can really exist it depends on dimensionless parameters of the problem, which are the Reynolds number and the ratio of the tube length to the characteristic tube width L/R. For an infinitely long tube $L/R >> 1$, the Reynolds number remains the only dimensionless parameter of the flow. Taking the maximal velocity of the flow as a characteristic velocity, the Reynolds number may be expressed through the pressure gradient in the tube

$$Re = \frac{\rho u_{max} R}{\eta} = \frac{P_1 - P_2}{2\eta^2 L} \rho R^3. \tag{7.4.11}$$

Experiments show that for small and moderate values of the Reynolds number the flow in a channel remains stationary and laminar. However, as soon as the Reynolds number exceeds some critical value $Re > Re_c$, then transition to a turbulent non-stationary flow occurs. The critical value of the Reynolds number measured experimentally is about 10^3.

Still the velocity distribution in a channel may be different from the Poiseuille formula even in the case of a laminar flow, if the channel length is finite. Obviously, close to the end of a channel the velocity distribution is determined by the boundary conditions for the incoming flow. The typical boundary condition at the entry is the condition of uniform velocity independent of the transverse coordinate y: $u_x = const = J/2\rho RL_x$ at $x = 0$ with the mass flux per unit length determined by the pressure gradient Eq. (7.4.5). In that case fluid passes certain distance from the channel entry before the stationary velocity profile Eq. (7.4.8) is established as shown in Fig. 7.3. The length X needed to establish the Poiseuille flow is called the entry length.

From dimensional considerations one may write the general dependence for the entry length $X/R = f(Re)$. Experiments show that the entry length increases linearly with the pressure gradient. The entry length depends strongly on the channel width $X \propto R^4$, therefore it is rather difficult to observe the Poiseuille flow in wide tubes, even if the flow is still laminar.

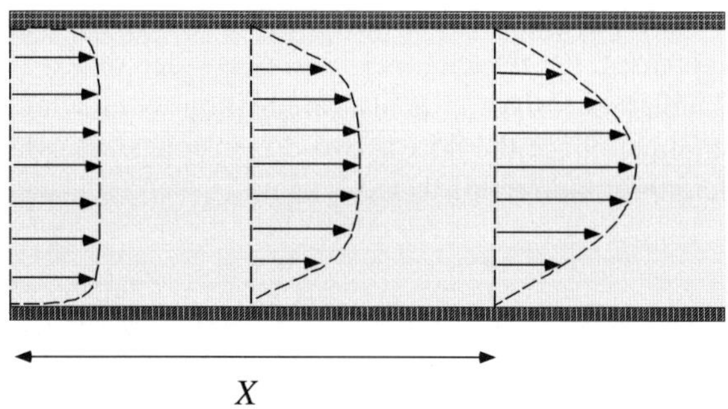

Fig. 7.3 The effect of an entry length in a channel

7.4.3 A Stationary Viscous Flow in a Cylindrical Tube

Let us consider now a viscous flow in a cylindrical tube of radius R. As before, we suppose that the length of the tube is L (along the z-axis) and pressure at the different ends of the tube is P_1 and P_2, $(P_2 < P_1)$. The fluid velocity is evidently directed along the z-axis at all points, and is a function of x and y only. The equation of continuity is satisfied identically, while the x and y components of the Navier-Stokes equation again give $\partial P/\partial x = \partial P/\partial y = 0$, i.e. the pressure is constant over the cross-section of the tube.

It is convenient to solve the problem in cylindrical coordinates with the z-axis directed along the tube axis (see Fig. 7.4). The boundary condition at the tube walls are: $u_z(r = R) = 0$. Due to the symmetry of the flow, the velocity and pressure distributions are stationary and independent of the angle coordinate θ. We look for the velocity distribution in the form $\mathbf{u} = (0; 0; u_z)$ with $u_z = u_z(r)$. The continuity equation is satisfied identically

$$\nabla \cdot \mathbf{u} = \frac{1}{r}\frac{\partial}{\partial r}(r u_r) + \frac{1}{r}\frac{\partial u_\theta}{\partial \theta} + \frac{\partial u_z}{\partial z} = 0. \tag{7.4.12}$$

The Navier-Stokes equation for z and r components gives

$$0 = -\frac{\partial P}{\partial z} + \eta \frac{1}{r}\frac{\partial}{\partial r}\left(r \frac{\partial u_z}{\partial r}\right), \tag{7.4.13}$$

$$0 = -\frac{\partial P}{\partial r}, \tag{7.4.14}$$

where we used the expression for the Laplacian in polar coordinates (see Appendix).

As before, we see that the pressure is constant over the cross-section of the tube and varies $dP/dz = -\Pi = -\frac{P_1 - P_2}{L}$ only along the tube axis, with the pressure gradient being constant. Then the Eq. (7.4.13) becomes

$$\eta \frac{1}{r}\frac{\partial}{\partial r}\left(r \frac{\partial u_z}{\partial r}\right) = -\Pi. \tag{7.4.15}$$

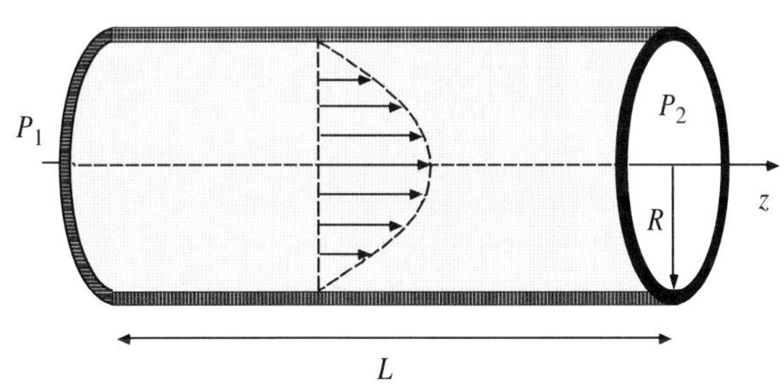

Fig. 7.4 Poiseuille flow in a tube

Integration gives

$$u_z = -\frac{\Pi}{4\eta} r^2 + A \ln r + B. \tag{7.4.16}$$

The second term diverges at the tube axis. Since the velocity must remain finite at the centre of the tube, we must take $A = 0$. The boundary condition at the tube wall determines the other integration constant from the requirement that $u_z(r = R) = 0$

$$B = \frac{\Pi}{4\eta} R^2.$$

We then find the velocity distribution

$$u_z = \frac{\Pi}{4\eta}\left(R^2 - r^2\right). \tag{7.4.17}$$

The velocity distribution is parabolic again similar to the two-dimensional flow (7.4.8) with the only difference in the numerical coefficient. The maximal velocity is achieved at the tube axis

$$u_{max} = \frac{P_1 - P_2}{4\eta L} R^2, \tag{7.4.18}$$

Let us calculate the mass Q of fluid passing per unit time through any cross-section of the tube. A mass $\rho u_z 2\pi r dr$ passes per unit time through an annular element $2\pi r dr$ of the cross-sectional area, and the total mass is

$$Q = 2\pi \int_0^R \rho u_z r dr = \frac{\pi}{8} \frac{P_1 - P_2}{\eta L} \rho R^4.$$

For the average velocity of the flow we obtain

$$\bar{u} = \frac{P_1 - P_2}{8\eta L} R^2.$$

The interesting feature of the Poiseuille flow in a cylindrical tube is very strong dependence of the mass flux on the tube radius $Q \propto R^4$. For example, if we increase the tube radius twice, then the mass flux increases 16 times. Similar to the two-dimensional case the laminar Poiseuille flow in a cylindrical tube holds only if the Reynolds number of the flow is small enough.

7.5 Dimensional Analysis: The Law of Similarity

We shall consider steady flow, for example, a flow past a solid body in which case the velocity of the main stream must be constant. We shall suppose the fluid incompressible. The kinematic viscosity $\nu = \eta/\rho$ in the Navier-Stokes equations is the only parameter, which characterizes the fluid. The unknown functions, which have to be determined by solving the equations, are the velocity $\mathbf{u}$ and the ratio of the pressure to the constant density P/ρ. Moreover, the flow depends, through the boundary conditions, on the shape and dimensions of the body moving through the fluid and on its velocity. Since the shape of the body is supposed given, its geometrical properties are determined by one linear dimension, which we denote by L. Let the velocity of the main stream is u. Then any flow is specified by three parameters, $\nu = \eta/\rho$, u, L. These quantities have the following dimensions:

$$\nu = [l]^2[t]^{-1} = \mathrm{cm}^2\mathrm{sec}^{-1}, \quad [u] = [l][t]^{-1} = \mathrm{cm}\cdot\mathrm{sec}^{-1} \quad [L] = [l] = \mathrm{cm}.$$

One can verify that only one dimensionless quantity can be formed from the above three parameters. This combination is called the *Reynolds number*:

$$\mathrm{Re} = \frac{uL}{\nu} = \frac{\rho uL}{\eta}. \tag{7.5.1}$$

It is convenient to introduce the scaled dimensionless variables

$$\mathbf{u}' = \mathbf{u}/U_0, \ \mathbf{x}' = \mathbf{x}/L_0, \qquad t' = tU_0/L_0, \qquad P' = P/\rho_0 U_0^2. \tag{7.5.2}$$

The hydrodynamic equations in the new variables may be rewritten in the form

$$\frac{U_0}{L_0}\nabla' \cdot \mathbf{u}' = 0, \text{ or } \nabla' \cdot \mathbf{u}' = 0, \tag{7.5.3}$$

$$\frac{U_0^2}{L_0}\frac{\partial \mathbf{u}'}{\partial t'} + \frac{U_0^2}{L_0}(\mathbf{u}' \cdot \nabla')\mathbf{u}' = -\frac{U_0^2}{L_0}\nabla' P' + \frac{\eta_0 U_0}{\rho_0 L_0^2}(\nabla')^2\mathbf{u}', \text{or}$$

$$\frac{\partial \mathbf{u}'}{\partial t'} + (\mathbf{u}' \cdot \nabla')\mathbf{u}' = -\nabla' P' + \frac{1}{\mathrm{Re}_0}(\nabla')^2\mathbf{u}', \tag{7.5.4}$$

which shows that $\mathrm{Re}_0 = \rho_0 U_0 L_0/\eta_0$ is the only dimensionless parameter of the problem. Any other dimensionless parameter can be written as a function of the *Reynolds number*. Since the only dimensionless parameter is the Reynolds number, it is evident that the velocity distribution obtained by solving the equations of incompressible flow is given by a function having the form $\mathbf{u} = U_0\mathbf{f}(\mathbf{r}/L; \mathrm{Re})$. It is evident that, in two different flows of the same type (for example, flow past spheres with different radii by fluids with different viscosities), the velocities $\mathbf{u}/U_0$ are the same functions of the ratio $\mathbf{r}/L$ if the

Reynolds number is the same for each flow. Flows, which can be obtained from one another by simply changing the unit of measurement of coordinates and velocities, are called similar. The law of similarity says that the flows of the same type with the same Reynolds number are similar.

Let us consider the flow caused by the body of characteristic L_1, which moves with velocity U_1 in a fluid with some density ρ_1 and viscosity η_1. In this case the scaled variables become

$$\mathbf{u}'' = \mathbf{u}/U_1,\ \mathbf{x}'' = \mathbf{x}/L_1,\ t'' = tU_1/L_1,\ P'' = P/\rho_1 U_1^2,$$

and we obtain the same Eqs. (7.5.3) and (7.5.4) for the new variables with some other value of the Reynolds number $Re_1 = \rho_1 U_1 L_1/\eta_1$. Since the geometrical shapes of the original body are the same, then the boundary conditions for the scaled equations of fluid dynamics are also the same, and from the mathematical point of view two problems are completely identical, if

$$Re_0 = Re_1 \qquad \Rightarrow \qquad \frac{\rho_0 U_0 L_0}{\eta_0} = \frac{\rho_1 U_1 L_1}{\eta_1}.$$

Suppose that we perform experiments with the model in the same fluid as for the original model ($\rho_0 = \rho_1$ and $\eta_0 = \eta_1$). If the new model is 100 times smaller $L_0/L_1 = 100$, then the model should move 100 times faster $U_1/U_0 = 100$.

Suppose that we have found experimentally the following flow for the one model

$$\mathbf{u}_{\text{model}} = \mathbf{U}(\mathbf{x}, t).$$

Then the scaled flow of dimensional variables is described by the function

$$\mathbf{u}' = \frac{1}{U_1}\mathbf{U}(\mathbf{x}/L_1; tU_1/L_1)$$

and the flow for another model is given by the expression

$$\mathbf{u} = U_0\mathbf{u}' = \frac{U_0}{U_1}\mathbf{U}\frac{L_0}{L_1}\mathbf{x}; \frac{U_0 L_1}{U_1 L_0}t.$$

If we know the drag force acting on the model $F = F_{\text{model}}$, then the scaled force is given by the expression $F' = \frac{F_{\text{model}}}{\rho_1 U_1^2 L_1^2}$, because the dimension of a force is equal to the dimension of pressure $\rho_1 U_1^2$ multiplied by the dimension of a surface area L_1^2. The real drag on the body is calculated as

$$F = \frac{\rho_0 U_0^2 L_0^2}{\rho_1 U_1^2 L_1^2} F_{\text{model}}.$$

If we take into account the gravitational force with the gravity acceleration $\mathbf{g} = g\hat{\mathbf{e}}_z$, and perform similar scaling, then the scaled Navier-Stokes equation takes the form

$$\frac{\partial \mathbf{u}'}{\partial t'} + (\mathbf{u}' \cdot \nabla')\mathbf{u}' = -\nabla' P' + \frac{1}{Re}(\nabla')^2\mathbf{u}' + \frac{1}{Fr}\mathbf{e}'_z, \qquad (7.5.5)$$

where $Fr = U^2/gL$ is the *Froude number, which* shows the relative strength of the inertial force compared to the gravity force. When gravity is taken into consideration, then two flows may be mathematically identical only if both dimensionless numbers of the flow are equal: $Re_1 = Re_2$, $Fr_1 = Fr_2$ and the boundary conditions (the geometrical shapes of the objects) are the same. Obviously, this condition is much more restrictive. Particularly, we cannot decrease the size of the experimental model in comparison with the real object, unless we take another fluid to perform experiments with another density and another viscosity coefficient.

Another important dimensionless hydrodynamic number is the *Mach number,* which is the ratio of typical velocity of a flow to the sound speed, $M = U/a_s$. This number characterizes importance of compressibility of a flow. For example, the Mach number defined as the shock velocity divided by the sound speed in front of the shock measures strength of a shock wave. For shock waves it is always $M > 1$. On the contrary, the condition of an incompressible flow is $M << 1$.

For the reference purpose the dimensions of different physical quantities expressed through the basic values are given below.

Velocity: $[u] = [L][t]^{-1}$, acceleration: $[g] = [du/dt] = [L][t]^{-2}$,
Frequency: $[\sigma] = [\Omega] = [t]^{-1}$,
surface area: $[S] = [L]^2$, volume: $[V] = [L]^3$, density: $[\rho] = [m][L]^{-3}$,
force: $[F] = [mdu/dt] = [m][L][t]^{-2}$, pressure: $[P] = [F/L^2] = [m][L]^{-}1[t]^{-2}$,
momentum: $[mu] = [m][L][t]^{-1}$, energy: $[mu^2] = [m][L]^2[t]^{-2}$,
momentum density: $[\rho u] = [m][L]^{-2}[t]^{-1}$, energy density:
$[\rho u^2] = [m][L]^{-1}[t]^{-2}$.

Dimension of other quantities can be found from hydrodynamic equations. For example, consider Navier-Stokes equation and compare dimensions of different terms in this equation: $[\rho\partial\mathbf{u}/\partial t] = [\mu\nabla^2\mathbf{u}]$. From here we can see that

$$[\mu] = \frac{[\rho\partial\mathbf{u}/\partial t]}{[\nabla^2\mathbf{u}]} = [m][L]^{-1}[t]^{-1}.$$

The situation becomes especially interesting when there are no free dimensionless parameters in a problem under consideration. As an example, we consider the gravitational waves in deep water, $kL >> 1$. In this case propagation of the waves is independent of the depth L of the fluid, and the problem is characterized by three-dimensional parameters, which are the fluid density, the gravitational acceleration and the wave number of perturbations. It is impossible to form any dimensionless number from the quantities characterizing of the problem: $[\rho] = [m][L]^{-3}$, $[g] = [L][t]^{-2}$, and $[k] = [L]^{-1}$. However, we can form a combination of frequency dimension as

$$[\Omega] = [t]^{-1} = \sqrt{[g][k]}.$$

Then we may conclude that $\Omega = C\sqrt{gk}$ in the dispersion relation. Exact solution to the problem gives us the value $C = 1$. If the water layer has a finite thickness L, then we can create one dimensionless number kL. In this case the general expression for the dispersion relation becomes

$$\Omega = \sqrt{gk}\ f(kL).$$

Suppose that we have found experimentally, that the phase velocity Ω/k is independent of the wave-number k for long wavelength perturbations with $kL << 1$, then the dimensional analysis gives

$$\Omega/k = \sqrt{g/k}\ f(kL) = C_1\sqrt{g/k}(kL)^{1/2}.$$

So that the dispersion relation for the shallow water waves takes the form $\Omega = C_1 k\sqrt{gL}$.

Finally, let us compare terms of the thermal conductivity transfer with mechanical energy transfer in the equation of thermal conduction. Their ratio is dimensionless *Peclet number*

$$Pe = \frac{\rho C_P UL}{\kappa} = \frac{UL}{\chi} \propto \frac{UL}{\nu} \propto Re. \qquad (7.5.6)$$

The Pecle number in gases is of the order of Reynolds number since the coefficient of thermal conductivity and kinematic viscosity are very close for gaseous mixtures.

7.6 Flow with Small Reynolds Numbers

The Navier-Stokes equation is considerably simplified in the case of flow with small Reynolds numbers. The Navier-Stokes equation for a steady flow of an incompressible fluid is

$$(\mathbf{u}\cdot\nabla)\mathbf{u} = -\frac{1}{\rho}\nabla P + \frac{\eta}{\rho}(\nabla)^2\mathbf{u} \qquad (7.6.1)$$

The first term in (7.6.1) is of the order of u^2/L, and the viscosity term is of the order of $\eta u/\rho L^2$. The ratio of these two terms is the Reynolds number, therefore the term in the left-hand part of (7.6.1) may be neglected if the Reynolds number is small, and the Navier-Stokes equation for small Reynolds numbers reduces to a linear equation

$$\nabla P - \eta \Delta \mathbf{u} = 0. \tag{7.6.2}$$

Equation (7.6.2) together with the equation of continuity, which in the case of incompressible fluid is $\nabla \mathbf{u} = 0$ determines the flow in the case of small Reynolds numbers. It is can be also useful to note that taking the curl of Eq. (7.7.2), we come to the system of equations for velocity only

$$\Delta (\nabla \times \mathbf{u}) = 0.$$

Classical example of application of these equations is uniform motion of a sphere in a viscous fluid. Solution of this problem gives well-known Stocks formula for the drag force on a sphere of radius R slowly moving in a fluid

$$F = 6\pi \eta R u.$$

This expression can be obtained with accuracy of the numerical factor from dimensional consideration, if we notice that density does not appear in the Eq. (7.6.2), and therefore the force must be expressed only in terms of η, u and R. The only combination with the dimension of force, which can be formed from these quantities, is $\eta R u$.

7.7 Turbulence: Stability of Steady Viscous Flow

In principle, for any problem of viscous flow under given steady conditions, and formally for an arbitrary Reynolds numbers, there must exist an exact steady solution of the equations of fluid dynamics. However, the existence of the exact solution is not enough condition for realizing a real flow in Nature. The solution must also be stable against infinitesimal small disturbances, which inevitably present in a flow. Stability of the solution means that any small perturbations, which arise must decrease in the course of time. On the contrary if the small perturbations, which always may occur in the flow, tend to grow with time, the flow is unstable and cannot exist actually as a steady flow.

Let us suppose that the steady solution of the Navier-Stokes equation for the incompressible fluid is known, $\mathbf{u} = \mathbf{u}_0(\mathbf{r})$, and we superpose on this solution a non-steady small perturbation, $\mathbf{u}_1(\mathbf{r}, t)$, such that the resulting velocity $\mathbf{u} = \mathbf{u}_0 + \mathbf{u}_1$ satisfies the equation of motion. Substituting $\mathbf{u} = \mathbf{u}_0 + \mathbf{u}_1$ in the equation of motion, we have for the incompressible fluid

$$\frac{\partial \mathbf{u}}{\partial t} + (\mathbf{u} \cdot \nabla)\mathbf{u} = -\frac{1}{\rho}\nabla P + \nu\nabla^2\mathbf{u} \tag{7.7.1}$$

$$\nabla \cdot \mathbf{u} = 0, \tag{7.7.2}$$

where $\mathbf{u} = \mathbf{u}_0 + \mathbf{u}_1$, $P = P_0 + P_1$, and the known functions $\mathbf{u}_0$ and P_0 satisfy the unperturbed equations

$$(\mathbf{u}_0 \cdot \nabla)\mathbf{u}_0 = -\frac{1}{\rho}\nabla P_0 + \nu\nabla^2\mathbf{u}_0, \tag{7.7.3}$$

$$\nabla \cdot \mathbf{u}_0 = 0. \tag{7.7.4}$$

Subtracting (7.7.3–4) from (7.7.1–2) and omitting terms above the first order in $\mathbf{u}_1$, we obtain linear equations for the perturbations

$$\frac{\partial \mathbf{u}_1}{\partial t} + (\mathbf{u}_0 \cdot \nabla)\mathbf{u}_1 + (\mathbf{u}_1 \cdot \nabla)\mathbf{u}_0 = -\frac{1}{\rho}\nabla P_1 + \nu\nabla^2\mathbf{u}_1, \tag{7.7.5}$$

$$\nabla \cdot \mathbf{u}_1 = 0, \tag{7.7.6}$$

with the boundary conditions that the perturbations vanish either at infinity or on the solid surfaces bounded the flow.

The general solution of the system of homogenous linear equations (7.7.5–6) with coefficients that are functions of the coordinates only, and not of the time, can be represented as a sum of particular solutions, in which the time dependence is given by $\exp(-i\omega t)$, where the frequencies ω are determined by the solution of the Eqs. (7.7.5–6) with the appropriate boundary conditions. If the imaginary parts of frequencies are positive, the perturbations will unlimited increase with time, and once occur they will indefinitely increase; the flow is unstable with respect to these perturbations. The flow is stable if imaginary parts of frequencies are negative.

Such a mathematical program of stability investigation is extremely complicated in general case. The experimental data indicates that, in general, steady flow is stable for sufficiently small Reynolds numbers, and when Reynolds number increases and reaches some critical value $Re_{cr} \cong 10/100$, the flow becomes unstable with respect to infinitely small perturbations. For example, transition from laminar to turbulent motion in a wake behind a moving cylinder was observed in experiments with $Re_{cr} \approx 30$. As a matter of fact, close to stability limits with the Reynolds number being close to the critical value $Re \approx Re_c$ only one perturbation mode is usually unstable. The wavelength of the unstable linear mode determines the length scale of the fluid motion at the nonlinear stage of the instability. Far from the stability limit, for the flow with larger values of the Reynolds number $Re >> Re_c$, more perturbation modes become unstable. Nonlinear interaction of modes leads to some complex and even chaotic fluid motion, which can be viewed as the well-developed turbulence. Well-developed turbulence is characterized by irregular pulsations of fluid velocity in time and space involving many modes of different frequencies and wavelengths.

7.7.1 Instability of Steady Flow at Large Reynolds Numbers

Though there is no complete theory of turbulence yet, some useful qualitative conclusions can be obtained form simple consideration. We consider the instability of the non-steady flow at large Reynolds numbers, slightly larger than Re_{cr} (L.D. Landau 1944). For $Re < Re_{cr}$ the imaginary part of the complex frequency, $\omega = \omega_1 + i\gamma_1$ is negative for all possible small perturbations, $\gamma_1 < 0$. For $Re = Re_{cr}$ there is one frequency whose imaginary part is zero, and for $Re > Re_{cr}$ this imaginary part becomes positive but still it is small compared to the real part ω_1. The function $\mathbf{u}_1$ corresponding to this frequency can be written as

$$\mathbf{u}_1 = A\, e^{\gamma_1 t}\, e^{-i\omega_1 t}\, \mathbf{f}_1(x, y, z), \tag{7.7.7}$$

where A is constant, and $\mathbf{f}_1(x, y, z)$ is some complex function of coordinates.

The obtained expression $A(t) = A\, e^{\gamma_1 t}\, e^{-i\omega_1 t}$ for amplitude of the function (7.7.7) can be valid only during short interval of time. Indeed, $e^{\gamma_1 t}$ increases rapidly with time, and the method used for obtaining solution of Eqs. (7.7.5–6) is applicable only when perturbations are small. In reality of course the modulus of the amplitude does not increase without limit, but tends to some finite value, which can be determined for the Reynolds numbers slightly larger than Re_{cr}. Let us find the time derivative of $|A(t)|$. For very small time interval when the expression (7.7.7) is still valid, we have

$$\frac{d|A(t)|^2}{dt} = 2\gamma_1 |A(t)|^2. \tag{7.7.8}$$

In fact this is just the first term in series of power expansion of A and A^+. As the modulus $|A(t)|$ increases, still remaining small, subsequent higher order terms in this expansion of the third order in A, forth order, etc., must be taken into account. However, since we are not interested in the exact value of the derivative $d|A(t)|^2/dt$, but in its time average taken over large interval of time compared to the period $2\pi/\omega_1$, all the odd-order terms, which contain the periodic factor vanish on averaging. Notice, that since $\omega_1 >> \gamma_1$, the period $2\pi/\omega_1$ is small compared to the time $1/\gamma_1$, so that terms with the factor $e^{-i\omega_1 t}$ are fast oscillating periodic terms and they vanish over averaging. The fourth order terms, which is proportional to $A^2 A^{*2} = |A|^4$, does not vanish on averaging, so that we obtain

$$\left\langle \frac{d|A|^2}{dt} \right\rangle = 2\gamma_1 |A|^2 - \alpha |A|^4, \tag{7.7.9}$$

where symbol $\langle\rangle$ means time averaging, and the numerical coefficient α can be either positive or negative.

We are interested in the case where an infinitesimal perturbation becomes unstable for $Re > Re_{cr}$, which corresponds to $\alpha > 0$. Omitted for simplicity symbol $\langle\rangle$, the corresponding solution to the Eq. (7.7.9) can be written as

$$\frac{1}{|A|^2} = \frac{\alpha}{2\gamma_1} + \text{const} \cdot e^{-2\gamma_1 t}, \tag{7.7.10}$$

It is clear form (7.7.10) that for large time, $|A|^2$ tends asymptotically to a finite limit:

$$|A|^2 = \frac{2\gamma_1}{\alpha}.(7.7.11) \tag{7.7.11}$$

The quantity γ_1 itself is some function of the Reynolds number, which can be expanded as a power-series of $(\text{Re} - \text{Re}_{cr})$. Since, $\gamma_1(\text{Re}_{cr}) = 0$ by the definition of the critical Reynolds number, the first non-zero term of such an expansion is

$$\gamma_1 = \text{const} \cdot (\text{Re} - \text{Re}_{cr}). \tag{7.7.12}$$

Substituting (7.7.12) in (7.7.11) we find that $|A|$ is proportional to the square root of $(\text{Re} - \text{Re}_{cr})$

$$|A| \propto \sqrt{(\text{Re} - \text{Re}_{cr})}. \tag{7.7.13}$$

In the case of $\alpha < 0$, the two terms in the expansion are not sufficient to determine the limiting amplitude and one must include a negative term of higher (sixth) order, $\beta|A|^6$ with $\beta > 0$. The corresponding result will be

$$|A|^2_{max} = \frac{\alpha}{2\beta} \pm \sqrt{\frac{\alpha^2}{4\beta^2} + \gamma_1 \frac{2|\alpha|}{\beta}}. \tag{7.7.14}$$

The dependence of $|A|^2_{max}$ for positive and negative α is shown in Figs. 7.5 and 7.6, respectively.

For the both cases there can no be steady flow, when $\text{Re} > \text{Re}_{cr}$. However, in the second case the unperturbed flow is metastable in the range $\text{Re}'_{cr} < \text{Re} < \text{Re}_{cr}$.

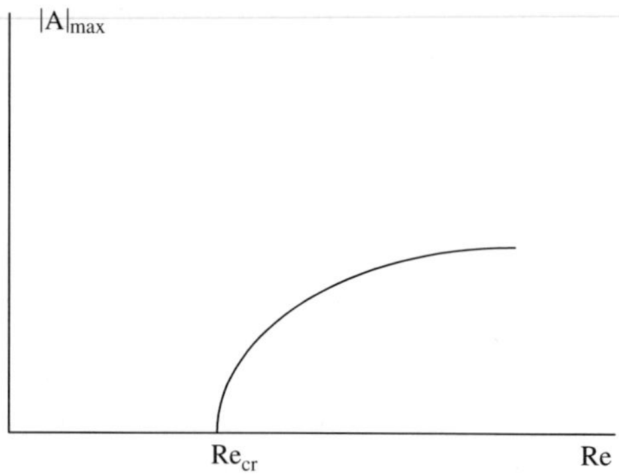

Fig. 7.5 $|A|_{max}$ versus Re for positive α, Eq. (7.7.13)

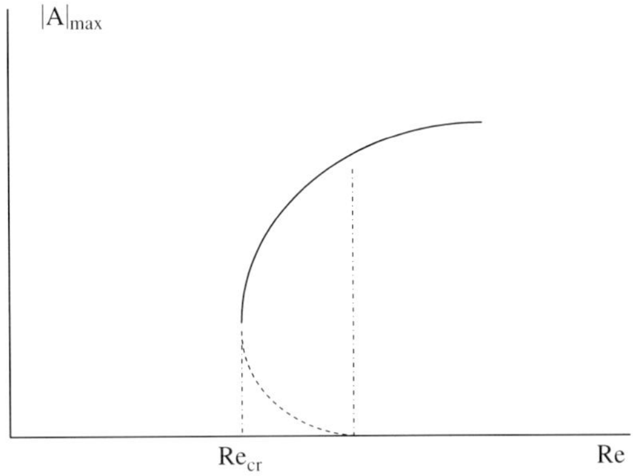

Fig. 7.6 $|A|_{max}$ versus Re for negative α, Eq. (7.7.14)

It is stable with respect to infinitesimal perturbations, but unstable with respect to perturbations of finite amplitude.

Thus, with the increase of the Reynolds number the non-steady flow can be represented as a superposition of the steady flow, $\mathbf{u}_0(\mathbf{r})$ and a periodic flow $\mathbf{u}_1(\mathbf{r}, t)$ with a small but finite amplitude in the form (7.7.13). As the Reynolds number increases further, the separation of the velocity into simple parts $\mathbf{u}_0(\mathbf{r})$ and $\mathbf{u}_1(\mathbf{r}, t)$ is no longer meaningful. In this case the velocity of the flow is no longer a simple trigonometric function, but it will become periodic pulsating flow.

The steady flow of a viscous liquid in a tube losses its stability in an unusual manner. Since the flow is uniform in x-direction along the tube, the unperturbed velocity distribution $\mathbf{u}_0(\mathbf{r})$ is independent of x. Therefore we can look for solutions of the Eqs. (7.7.5–6) in the form

$$\mathbf{u}_1 = e^{i(kx - \omega t)}\mathbf{f}(y, z) \tag{7.7.15}$$

At some value of $Re = Re_{cr}$, $\gamma = Im\{\omega\}$ first becomes zero for some value of k, however the real part of $\omega(k)$ is not zero. As we discussed before, for values of the Reynolds numbers only nearly exceeding Re_{cr}, the range of values of k for which $\gamma(k) > 0$ is small and lies near the point for which $\gamma(k)$ is a maximum, i.e. $d\gamma(k)/dk = 0$. Let us imagine that a slight perturbation occurs in some part of the flow. It represents a wave packet obtained by superposing a series of components of the form (7.7.15). In the course of time, the component for which $\gamma(k) > 0$ will be amplified, while remaining will be damped. This amplified wave packet thus formed also will be carried downstream with the velocity equal to the group velocity $d\omega(k)/dk$of the packet. Since we are considering now waves whose wave numbers lie in a small range near the point where $d\gamma(k)/dk = 0$, the quantity $d\omega(k)/dk \cong d(\text{Real part}\{\omega\})/dk$ is real, and therefore it is actual velocity of the propagation of the packet. The fact that $Im\{\omega\}$ is positive means now merely amplification of the perturbations, which move downstream. This means that two different scenarios are possible. It is possible that the perturbation increases without limit in the course of time at any point fixed in space, and this kind of instability with respect to any infinitesimal perturbations is called absolute instability. On the other hand, the packet may be shifted faster than the perturbation growths, so that in any point fixed in space the perturbation tends to zero as time goes to infinity. This kind of instability is called convective instability. It is thought that this second kind of instability takes place for the Poiseuille flow, though it was not rigorously proved yet. The complete analytical investigation of the flow stability was not performed even for such simple geometry as 2D tubes – Poiseuille flow between two parallel planes.

7.7.2 Fully Developed Turbulence: Kolmogorov Theory

The phenomenological approach to turbulence may help in understanding nature of some flows, though it has all shortcomings of a phenomenological approach. At very large Reynolds numbers turbulent flow is characterized by the presence of strongly irregular disordered time variation of the velocity at each point. This is called fully developed turbulence. In order to be able to describe turbulent pulsations of different length scales from large length scales comparable in size to the scale of the average flow L to the smallest length scale λ, at which turbulence is suppressed by viscosity, we introduce the concept of the mean velocity, denoted $\mathbf{u}$. The difference $\mathbf{u}' = \mathbf{v} - \mathbf{u}$ between the true velocity $\mathbf{v}$ and the mean velocity varies very irregularly and we shall call it pulsating or fluctuating part of the velocity. Such a pulsating strongly irregular motion, which is superposed on the mean flow, can be viewed as the superposition of turbulent eddies of different sizes. When we mention the size of an eddy, we mean the order of magnitude of the distance over which the velocity varies considerably. When the Reynolds number increases, the first appears the largest eddies. As larger the Reynolds number as smaller eddies should be, and they appear later. For very large Reynolds numbers, eddies of all sizes are present in the flow: from the largest one to the smallest size. The largest eddy, whose size is of the order of the region in which the flow takes place, has the largest amplitude and they called *fundamental* or *external* scale of turbulence, L. The frequencies corresponding to the large eddies are of the order of u/L. The small eddies correspond to large frequencies and they may be regarded as a fine structure superposed on the fundamental eddies. Only a small part of the turbulent kinetic energy resides in the small eddies. The Reynolds number introduced as $Re = uL/\nu$ determines the properties of a given flow. Besides, we can also introduce the Reynolds number for turbulent eddies, $Re_\lambda = u_\lambda \lambda/\nu$, where λ is the order of magnitude of a given eddy, and u_λ is its velocity. Evidently, for large $Re = uL/\nu$ the Reynolds numbers $Re_\lambda = u_\lambda \lambda/\nu$ are also large. Note that large Reynolds numbers are equivalent to small viscosity. We therefore can conclude that for large-scale motion and for large scale eddies, which are the basis of any turbulent flow, the viscosity is not important, i.e. there is no appreciable energy dissipation in large eddies. The viscosity becomes important only for the smallest eddies, whose Reynolds number, $Re_\lambda = u_\lambda \lambda / \nu \cong 1$, and which are of no importance for general character of the turbulent flow. We thus come to the conclusion that the dissipation of energy may be regarded as a continuous energy flow from large turbulent eddies to the smallest eddies of size λ_0, where the kinetic energy is finally transformed into heat due to viscosity. It is of course necessary that an external source of energy to be present, which should continually supply energy to the large eddies, to maintain a steady state. Since the viscosity of the fluid is important only for the smallest eddies, we may say that all the quantities pertaining to eddies of sizes $\lambda >> \lambda_0$ do not depend on viscosity. Under these circumstances the qualitative conclusion about well-developed turbulence can be made from similarity arguments.

Let us consider the case of a well-developed turbulence with a broad spectrum of turbulent modes in dynamical equilibrium. The dynamical equilibrium implies that the energy transfer between different turbulent modes can be viewed as a steady flow of energy from the eddy of large scale to the small eddies of scale, $\lambda = L_\nu$ with the characteristic Reynolds number about unity $Re(L_\nu) \approx 1$, where energy is dissipated by viscosity. Let us define ε to be an average amount of energy density (the energy per unit mass); the average amount of the energy density dissipated per unit time is

$$q = \frac{d\varepsilon}{dt}. \tag{7.7.16}$$

The dimension of the energy dissipation rate is

$$[q] = \left[\frac{d\varepsilon}{dt}\right] = \left[u^2\right]\frac{[u]}{[\lambda]}.$$

The dimensional analysis gives us the dependence of the turbulent pulsation velocity on the scale of the order of λ as

$$u_\lambda \propto (q\lambda)^{1/3}. \tag{7.7.17}$$

This relation determines the velocity spectrum of a well-developed turbulence is called the *Kolmogorov spectrum*. As one can see, characteristic velocity of turbulent pulsations increases for modes of larger scales (larger wavelengths).

In a sense we can introduce meaning of turbulent viscosity, combining the magnitude of kinematic viscosity from the dimension variables: ρ, u_λ, λ. The only combination is

$$\nu_{turb} = u_\lambda \lambda. \tag{7.7.18}$$

From the expression (7.7.18) steams that $\nu_{turb}/\nu = Re_\lambda$. The energy dissipation can be expressed using $\nu_{turb} = u_\lambda \lambda$ as

$$q \cong \nu_{turb}\left(\frac{u_\lambda}{\lambda}\right)^2 \propto \frac{u_\lambda^3}{\lambda}. \tag{7.7.19}$$

Important is that though energy dissipation occurs at the smallest scale due to viscosity, the same scale of energy (q) defines also motion at the large scale. At the same time the local properties of turbulence – the properties of motion at the intermediate scales, which are large compared to λ, but small compared to L, do not depend on L or u'. The order of magnitude of the pulsating velocity is given by the expression (7.7.17), which says that the order of magnitude of the velocity variation on a small scale is proportional to this scale in 1/3.

Let us consider some fluid element. It moves during time τ the way of the order of $u\tau$. Therefore, the order of magnitude of the velocity variation during time τ, which is smaller than the characteristic time of the whole flow, $T \cong L/u$, can be obtained substituting $u\tau$ instead of λ in (7.7.17)

$$u_\tau \cong (qu\tau)^{1/3}. \tag{7.7.20}$$

The velocity variation within the given fluid element, u'_τ depends on the local properties of turbulence, and therefore can depend on q only. From dimension consideration, constructing the velocity dimension from q and τ, we obtain for the order of magnitude

$$u'_\tau \cong \sqrt{q\tau}. \tag{7.7.21}$$

Let us consider the local Reynolds number Re (λ), where energy is dissipated by viscosity. Taking into account that the rate of the energy density dissipation on the large scale is defined by the velocity variation of the flow, Δu on scale of the order of L, which is $q \cong (\Delta u)^3/L$, we can write

$$Re(\lambda) \cong \frac{u_\lambda \lambda}{\nu} \cong \frac{(q\lambda)^{1/3}\lambda}{\nu} \cong \frac{\Delta u \lambda^{4/3}}{\nu L^{1/3}} \cong Re(L)\left(\frac{\lambda}{L}\right)^{4/3}, \tag{7.7.22}$$

where $Re(L) = \Delta u L/\nu$ is the *global Reynolds number* of the turbulent flow.

As it was said the characteristic Reynolds number, which specifies the energy dissipation, is defined by the condition $Re(\lambda) \approx 1$, thus with account (7.7.22)

$$\lambda_0 \cong L/(Re)^{3/4}. \tag{7.7.23}$$

The dissipation length scale is also called the *Kolmogorov length scale*. We can say that scale of the order of L is the region of scale of kinetic energy, while scale of the order of λ_0 is the region of scales of energy dissipation. Well-developed turbulence is possible when the maximal length scale L and the dissipation length scale λ_0 differ by several orders of magnitude $L >> \lambda_0$. According to (7.7.23) the last condition requires large values of the Reynolds number of a turbulent flow

$$L >> \lambda_0 \cong L\mathrm{Re}^{-3/4} \qquad \Rightarrow \qquad \mathrm{Re} >> 1.$$

If we are interested in energy dq of one particular "monochromatic" harmonic with a wave number in the interval $[k;\ k + dk]$, then from (7.7.19) it follows

$$dq = \frac{dq}{dk} dk \propto q^{2/3} k^{-5/3} dk,$$

so that the spectrum characteristic of turbulent pulsation is

$$E(k) \propto q^{2/3} k^{-5/3}.$$

7.7.3 The Velocity Correlation Function

For any problem of viscous flow under given steady conditions the regular flow can be separated from pulsations by use of velocity averaging in time

$$\mathbf{u} = \mathbf{U} + \mathbf{u}', \tag{7.7.24}$$

$$\mathbf{U} \equiv \langle u \rangle = \frac{1}{\tau} \int_t^{t+\tau} \mathbf{u} dt, \tag{7.7.25}$$

where τ is the averaging interval, which has to be taken sufficiently large. The time interval must be at least as large as the maximal time period of turbulent pulsations. By definition, average value of the pulsation velocity is zero

$$\langle u' \rangle = \frac{1}{\tau} \int_t^{t+\tau} \mathbf{u}' dt = \frac{1}{\tau} \int_t^{t+\tau} (\mathbf{u} - \mathbf{U}) dt = \mathbf{U} - \mathbf{U} = 0.$$

In a sense irregular pulsations of a turbulent flow are similar to thermal motion of fluid particles in a laminar flow. While averaging over the thermal velocity distribution leads to equations of fluid dynamics, then we can expect that averaging of turbulent pulsations will give us hydrodynamic equations for the main flow.

Substituting the representation (7.7.24) into the continuity equation we have

$$\nabla \cdot (\mathbf{U} + \mathbf{u}') = 0,$$

and the time averaging gives the continuity equation for the main flow

$$\nabla \cdot \langle \mathbf{U} + \mathbf{u}' \rangle = 0 \qquad \Rightarrow \qquad \nabla \cdot \mathbf{U} = 0. \tag{7.7.26}$$

Averaging of the Navier-Stokes equation is not that simple, therefore we will consider one particular component of the equation (say, the x-component) in the case of a two-dimensional flow

$$\frac{\partial u_x}{\partial t} + (\mathbf{u} \cdot \nabla) u_x = -\frac{1}{\rho}\frac{\partial P}{\partial x} + \nu \nabla^2 u_x.$$

With the help of the continuity equation written as

$$u_x \nabla \cdot \mathbf{u} = 0$$

we can present the Navier-Stokes equation in the form

$$\frac{\partial u_x}{\partial t} + \frac{\partial}{\partial x} u_x^2 + \frac{\partial}{\partial y}\left(u_x u_y\right) = -\frac{1}{\rho}\frac{\partial P}{\partial x} + \nu \nabla^2 u_x.$$

Taking time average of the above equation we find

$$\frac{\partial U_x}{\partial t} + \frac{\partial}{\partial x}\langle u_x^2 \rangle + \frac{\partial}{\partial y}\langle u_x u_y \rangle = -\frac{1}{\rho}\frac{\partial \wp}{\partial x} + \nu \nabla^2 U_x,$$

where the average pressure is denoted by $\wp = \langle P \rangle$. As one can see, turbulent pulsations vanished after time averaging in all terms of the Navier-Stokes equation except those describing nonlinear inertial acceleration. Indeed averaging of the second term in the right-hand side gives us

$$\langle u_x^2 \rangle = \left\langle (U_x + u_x')^2 \right\rangle = \left\langle U_x^2 + 2U_x u_x' + (u_x')^2 \right\rangle = U_x^2 + \langle u_x' u_x' \rangle.$$

At this point one should remember, that square of pulsation velocity does not vanish with averaging, though average of the pulsation velocity is zero. Following the similarity between thermal motion of particles and turbulent pulsations one can say that the value $\langle u_x' u_x' \rangle$ plays the role of scaled kinetic energy of turbulent motion. In a similar way we obtain the expression for the third term

$$\langle u_x u_y \rangle = U_x U_y + \left\langle u_x' u_y' \right\rangle.$$

Then the Navier-Stokes equation of a turbulent flow may be reduced to

$$\frac{\partial U_x}{\partial t} + (\mathbf{U} \cdot \nabla) U_x = -\frac{1}{\rho}\frac{\partial \wp}{\partial x} + \nu \nabla^2 U_x - \frac{\partial}{\partial x}\langle u_x' u_x' \rangle - \frac{\partial}{\partial y}\left\langle u_x' u_y' \right\rangle,$$

or

$$\frac{\partial U_x}{\partial t} + (\mathbf{U} \cdot \nabla) U_x = -\frac{1}{\rho}\frac{\partial \wp}{\partial x} + \frac{\partial}{\partial x}\zeta'_{xx} - \frac{\partial}{\partial y}\zeta'_{yx}. \tag{7.7.27}$$

The Navier-Stokes equation for the average velocity of a turbulent flow takes the form of a usual Navier-Stokes equation with a new expression for the stress tensor

$$\zeta'_{xx} = \nu \frac{\partial U_x}{\partial x} - \langle u'_x u'_x \rangle, \zeta'_{yx} = \nu \frac{\partial U_x}{\partial y} - \left\langle u'_x u'_y \right\rangle. \tag{7.7.28}$$

The turbulence stress tensor contains the usual viscous stress and some additions due to the irregular pulsations called the *Reynolds stress*. We can also define an effective turbulent viscosity. For example, taking the yx-component of the stress tensor, we can present the second term in the same form as the first one related to the laminar viscosity with some new "viscosity" coefficient ν'

$$-\left\langle u'_x u'_y \right\rangle \equiv \nu' \frac{\partial U_x}{\partial y}. \tag{7.7.29}$$

Equation (7.7.29) is the definition of eddy viscosity, which is typically much larger than the normal laminar viscosity. Indeed, if the flow is characterized by the length scale L and the velocity scale U (which are usually comparable for the average flow and for the turbulent pulsations of a large scale), then the laminar and turbulent terms in the expression for stress may be evaluated as

$$\nu \frac{\partial U_x}{\partial y} \propto \nu \frac{U}{L}, \left\langle u'_x u'_y \right\rangle \propto U^2$$

and their effective ratio gives the Reynolds number of the flow

$$\left\langle u'_x u'_y \right\rangle / \nu \frac{\partial U_x}{\partial y} = \frac{UL}{\nu} = \mathrm{Re}.$$

Since turbulent flows take place at large values of the Reynolds number $\mathrm{Re} >> 1$, then the Reynolds stress is always much larger than the usual viscous stress and the eddy viscosity is much stronger than the laminar viscosity.

However, the eddy viscosity is not a parameter of a fluid that can be measured and tabulated. Turbulent viscosity depends on a particular flow and we have to know energy distribution of turbulent pulsations in order to calculate turbulent viscosity.

Early theories of turbulence (e.g. the Prandtl theory) employed phenomenological approach to eddy viscosity and similarity between turbulent pulsations and thermal motion of molecules. As follows from the kinetic theory, laminar viscosity is determined by particle free path $\lambda_{freepath}$, which is the average distance between two collisions of a particle that change particle momentum. Following the analogy between turbulent flows and thermal motion Prandtl introduced a *mixing length* λ_m as the distance passed by a fluid element before it changes velocity in turbulent pulsations. Assuming the length scale of pulsations to be comparable to the length scale of the average flow we come to the following estimate for the xy-component of the turbulent stress

$$u'_x \propto u'_y \propto \lambda_m \frac{\partial U_x}{\partial y}.$$

In this estimate we have assumed also isotropic turbulence with $u'_x \propto u'_y$. Then we can relate the expression for eddy viscosity to the average flow

$$\nu' \frac{\partial U_x}{\partial y} \equiv -\left\langle u'_x u'_y \right\rangle \propto \lambda_m^2 \left(\frac{\partial U_x}{\partial y} \right)^2$$

or

$$\nu' \propto \lambda_m^2 \frac{\partial U_x}{\partial y}. \tag{7.7.30}$$

The relation (7.7.30) is determined with the accuracy of a numerical coefficient. However, it has been obtained experimentally, that the following evaluation holds for many flows

$$\lambda_m \approx K \left(\frac{dU_x}{dy} \right) \Big/ \left(\frac{d^2 U_x}{dy^2} \right),$$

where K is a "universal" constant called the Von Karman coefficient and equal approximately $K \approx 0.4$ independent of boundary conditions and the Reynolds number.

7.8 Boundary Layer

Let us now consider a flow, which is characterized by large Reynolds number. As it was seen before, a very large Reynolds number $Re = \rho UL/\eta = UL/\nu$ mimics flow with very small viscosity, so that the fluid can be viewed as ideal one. However, such an approximation is not true near the wall. The boundary conditions at the wall require vanishing velocity at the wall. Thus, for a large Reynolds number velocity must drop down to zero within a thin layer near the wall, which is called a boundary layer and is characterized by large velocity gradients within the layer. The boundary of such a layer is not sharp one, which means that transition form the main flow to the flow in the boundary layer is more or less smooth, the flow inside the layer can be laminar or turbulent. Viscous forces smooth the velocity profile in a boundary layer from zero velocity at the wall to the outer velocity of the ideal flow far from the wall. Velocity drops in the boundary layer due to the viscosity of the fluid, which is not negligible in spite of the large Reynolds number. Formally, this means that the velocity gradients are large in the boundary layer, so that viscous terms in the equation of motion are not negligible even for small viscosity, $\nu = \eta/\rho$.

Let us consider as an example, a stationary flow in two-dimensional channel. Let xz is the plane of the wall, and x-axis is directed along the flow velocity. For such 2D problem, there is no z-component of velocity and velocity does not depend on co-ordinate z. Components of the Navier-Stokes equations and continuity equation for the incompressible fluid become

$$u_x \frac{\partial u_x}{\partial x} + u_y \frac{\partial u_x}{\partial y} = -\frac{1}{\rho}\frac{\partial P}{\partial x} + \nu\left(\frac{\partial^2 u_x}{\partial x^2} + \frac{\partial^2 u_x}{\partial y^2}\right), \tag{7.8.1}$$

$$u_x \frac{\partial u_y}{\partial x} + u_y \frac{\partial u_y}{\partial y} = -\frac{1}{\rho}\frac{\partial P}{\partial y} + \nu\left(\frac{\partial^2 u_y}{\partial x^2} + \frac{\partial^2 u_y}{\partial y^2}\right), \tag{7.8.2}$$

$$\frac{\partial u_x}{\partial x} + \frac{\partial u_y}{\partial y} = 0. \tag{7.8.3}$$

The motion of the fluid within the boundary layer is mainly along x-axis, which means that u_y is small compared to u_x.

Consider as an example, the boundary layer, which is formed in 2D the flow of fluid moving along a wide plate of negligible thickness. The inner structure of the boundary layer is defined by the continuity equation and the Navier-Stokes equation for an incompressible stationary viscous flow (7.8.1–3).

Because of a small thickness of the boundary layer, the motion in the layer is mainly along the surface of the plane – along x-axis. The scale of the velocity change in the x-direction is slow and characterized by the length scale L of the flow outside the boundary layer. Along y-axis the scale of the change of the velocity is of the order of the boundary layer thickness δ. This means that the velocity derivation along y-axis is large compared to its derivation along x-axis. From continuity equation (7.8.3) we can estimate

$$\frac{U_x}{L} \propto \frac{U_y}{\delta} \quad \Rightarrow \quad U_y \propto U_x \frac{\delta}{L}. \tag{7.8.4}$$

Let us consider different terms in the x-component of the Navier-Stokes equation (7.8.1) taking into account (7.8.4). The first and the second terms are $\frac{U_x U_y}{\delta} \propto \frac{U_x^2}{L}$. Viscous terms in the right hand side can be evaluated as $\frac{\nu U_x}{L^2}$, and $\frac{\nu U_x}{\delta^2}$, which means that we can neglect by $\frac{\partial^2 u_x}{\partial x^2}$ compared to $\frac{\partial^2 u_x}{\partial y^2}$. Evaluating terms from the y-component of the Navier-Stokes equation (7.8.2), we find that the first two terms in the left can be estimated as $\frac{\delta U_x^2}{L^2}$. Viscous terms in the right hand side are $\frac{\nu\delta U_x}{L^3}$, and $\frac{\nu U_x}{\delta L}$. Pressure derivatives are estimated from Eqs. (7.8.1) and (7.8.2) as $\frac{\partial P}{\partial x} \propto \rho \frac{U_x^2}{L}$ and $\frac{\partial P}{\partial y} \propto \rho \frac{\delta}{L} \frac{U_x^2}{L}$. We can see that $\frac{\partial P}{\partial y}$ is small compared to $\frac{\partial P}{\partial x}$ as $\frac{U_y}{U_x} \propto \frac{\delta}{L}$. Since pressure variations across the boundary layer (along the y-axis) are negligible in comparison with variations along the layer (along the x-axis), then we can assume that pressure in the boundary layer depends on the x-coordinate only $P = P(x)$ and in this approximation we can take $\frac{\partial P}{\partial y} = 0$. The last fact means that pressure in the boundary layer coincides with pressure in the ideal flow "along the wall" outside the boundary layer, so that we can take $P = P_0(x)$. Outside the boundary layer, the flow is potential and the pressure can be expressed using the Bernoulli equation

$$P + \frac{1}{2}\rho U^2 = \text{const.} \tag{7.8.5}$$

Then, for pressure derivative we obtain

$$\frac{1}{\rho}\frac{dP}{dx} = -U\frac{dU}{dx}. \tag{7.8.6}$$

In this way we came to the system of equations in the laminar boundary layer, which consists of the continuity equation and y-component of the Navier-Stokes equation and known as the *Prandtl equations*

$$\frac{\partial u_x}{\partial x} + \frac{\partial u_y}{\partial y} = 0. \tag{7.8.7}$$

$$u_x\frac{\partial u_x}{\partial x} + u_y\frac{\partial u_x}{\partial y} - \nu\frac{\partial^2 u_x}{\partial y^2} = -\frac{1}{\rho}\frac{dP}{dx} = U\frac{dU}{dx}. \tag{7.8.8}$$

(The x-derivative in the viscous term in (7.8.8) was neglected since it is small as δ^2/L^2). The boundary condition for the Eqs. (7.8.7–8) require that velocities vanish at the rigid wall

$$u_x(y = 0) = u_y(y = 0) = 0. \tag{7.8.9}$$

Outside of the boundary layer, far away from the wall, the velocity must approach asymptotically the velocity of the main flow

$$u_x = U(x) \text{ at } y \to \infty. \tag{7.8.10}$$

Let U_0 be a characteristic velocity of the flow (e.g. velocity of the flow at infinity). We introduce dimensionless variables as

$$x = Lx', \; y = Ly'/\sqrt{Re}, \; u_x = U_0 u'_x, \; u'_y = U_0 u_y/\sqrt{Re}. \qquad (7.8.11)$$

The Equations (7.8.7)–(7.8.8) being written in these variables become

$$\frac{\partial u'_x}{\partial x'} + \frac{\partial u'_y}{\partial y'} = 0 \qquad (7.8.12)$$

$$u'_x \frac{\partial u'_x}{\partial x'} + u'_y \frac{\partial u'_x}{\partial y'} - \frac{\partial^2 u'_x}{\partial y'^2} = U' \frac{dU'}{dx'}, \qquad (7.8.13)$$

where $U = U_0 U'$, $Re = U_0 L/\nu$.

Notice, that viscosity does not enter either Eqs. (7.8.12)–(7.8.13) or boundary conditions for these equations. Thus, we can conclude that solutions of (7.8.12)–(7.8.13) do not depend on the Reynolds number, which means that the motions in the boundary layer are similar, with distances and velocities along x-axis being unchanged, and distances and velocities along y-axis being proportional to $1/\sqrt{Re}$. Since dimensionless u'_x and u'_y do not depend on Re, they must be of the order of unity, and therefore, from (7.8.11) follows that

$$u_y \propto U_0/\sqrt{Re}, \text{ and } \delta \propto L/\sqrt{Re}. \qquad (7.8.14)$$

Consider solution for boundary layer for the case of zero pressure gradients and zero velocity gradients in the outside flow. Let velocity of the main flow is $U = \text{const}$. The boundary layer is described by the Eqs. (7.8.7)–(7.8.8) with zero velocity gradients

$$\frac{\partial u_x}{\partial x} + \frac{\partial u_y}{\partial y} = 0, \; u_x \frac{\partial u_x}{\partial x} + u_y \frac{\partial u_x}{\partial y} = \nu \frac{\partial^2 u_x}{\partial y^2}. \qquad (7.8.15)$$

As we saw, the variables in the solution of the Prandtl equations, u_x/U and $u_y\sqrt{L/\nu U}$ can be functions of only $x' = x/L$ and $y' = y\sqrt{U/\nu L}$. However, there is no parameter with dimension of length for the problem of flow around of semi-infinite plate. Therefore, u_x/U must be function of only

$$\frac{y'}{\sqrt{x'}} = y\sqrt{\frac{U}{\nu x}} \equiv \frac{y}{\delta(x)}, \text{ where } \delta(x) = \frac{x}{\sqrt{Re(x)}} = \frac{x}{\sqrt{\frac{Ux}{\nu}}} = \sqrt{\frac{\nu x}{U}},$$

which is combination of x' and y' where dimension of length does not enter, but x plays a role of variable length.

Thus, we shall introduce a dimensionless combination of the coordinates

$$\eta = \frac{y}{\delta(x)} \qquad (7.8.16)$$

and look for a solution in the form $u_x = u_x(\eta)$, $u_y = u_y(\eta)$.

It is also convenient to introduce the stream function ψ

$$u_x = \frac{\partial \psi}{\partial y}, u_y = -\frac{\partial \psi}{\partial x}, \tag{7.8.17}$$

so that the continuity equation is identically satisfied.

The stream function corresponding to the above conditions can be taken in the form

$$\psi = U\delta(x)f(\eta),$$

where $f(\eta)$ is dimensionless function.

Straightforward calculations give

$$u_x = U\frac{df(\eta)}{d\eta}, \quad u_y = -\frac{\partial \psi}{\partial x} = -Uf\frac{d\delta}{dx} - U\delta\frac{df}{d\eta}\frac{\partial \eta}{\partial x} = U\frac{d\delta}{dx}\left(\eta\frac{df}{d\eta} - f\right), \tag{7.8.18}$$

$$\frac{\partial u_x}{\partial x} = -\frac{U}{\delta}\frac{d\delta}{dx}\eta\frac{d^2f}{d\eta^2}, \frac{\partial u_x}{\partial y} = -\frac{U}{\delta}\frac{d^2f}{d\eta^2}, \frac{\partial^2 u_x}{\partial y^2} = -\frac{U}{\delta^2}\frac{d^3f}{d\eta^3}.$$

Substituting these expressions into the Navier-Stokes equation (7.8.15) we obtain equation for $f(\eta)$

$$\delta\frac{d\delta}{dx}f\frac{d^2f}{d\eta^2} + \frac{\nu}{U}\frac{d^3f}{d\eta^3} = 0 \tag{7.8.19}$$

Taking into account the expression for the boundary layer thickness,

$$\delta\frac{d\delta}{dx} = \frac{d}{dx}\left(\frac{\delta^2}{2}\right) = \frac{d}{dx}\left(\frac{1}{2}\frac{\nu x}{U}\right) = \frac{1}{2}\frac{\nu}{U},$$

Eq. (7.8.19) can be reduced to the form

$$f\frac{d^2f}{d\eta^2} + 2\frac{d^3f}{d\eta^3} = 0. \tag{7.8.20}$$

The boundary conditions with respect to the function *f* become

$$f(0) = \frac{df}{d\eta}(0) = 0 \text{ , and } \frac{df}{d\eta} = 1 \text{ for } \eta \to \infty. \tag{7.8.21}$$

Of interest is the derivative $df/d\eta$ not the function f, because $df/d\eta$ determines the scaled velocity distribution in the boundary layer.

The numerical solution of Eq. (7.8.20) is shown in Fig. 7.7. The respective velocity profile known as the Blasius velocity profile (*H. Blasius*, 1908) for the laminar boundary layer is shown in Fig. 7.8.

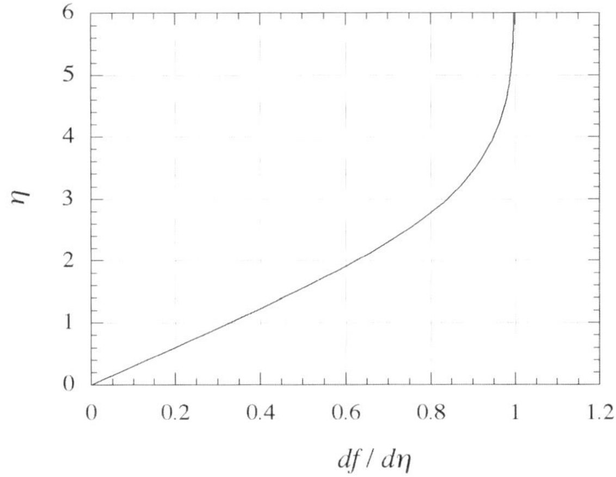

Fig. 7.7 The Blasius velocity profile in a boundary layer (*right*)

Using this numerical solution we can obtain more accurate formula for the thickness of the boundary layer. If we assume that the outer boarder of the boundary layer is the distance where velocity differs from the outside velocity U by less than 1%, then the numerical value of $df/d\eta = 0.99$ is at the value $\eta \approx 5$, and the layer thickness is

$$\delta = 5\sqrt{\nu x/U}. \tag{7.8.22}$$

Since velocity along the wall in a boundary layer is much larger than the transverse velocity $u_x >> u_y \propto \frac{\delta}{L} u_x$, then a boundary layer is similar to a shear flow with $\mathbf{u} \approx (u_x; 0; 0)$. It is known that a shear flow is unstable against the Kelvin-Helmholtz instability at sufficiently large Reynolds numbers and transition to turbulence takes place. Stability analysis of a stationary laminar boundary

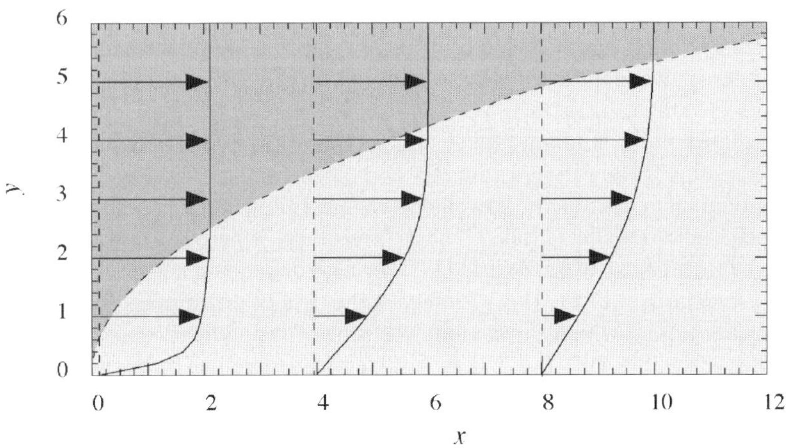

Fig. 7.8 Velocity distribution in a boundary layer (*left*)

layer shows that the boundary layer becomes unstable for the Reynolds number $Re_\delta > Re_c \approx 1.5 \cdot 10^3$, when flow in the boundary layer becomes turbulent.

The boundary layers are strongly influenced by the main outer flows. If the fluid accelerates in the outer flow $dU/dx > 0$ $(dP/dx < 0)$, then it pushes the boundary layer to the wall and makes it thin. On the contrary, deceleration of the fluid in the outer flow $dU/dx < 0$ makes the boundary layer wider.

Let us consider how the boundary layer is formed when liquid enter the tube. We assume that the velocity distribution is uniform at the entrance of the cross section of the tube, and the fluid velocity drops only within the boundary layer. The liquid flow entering the tube starts decelerating: parts of the flow adjoined outer edge of the boundary layer are decelerating as larger as it is far away from the entrance. The total flux of the liquid is constant, which means that simultaneously with shrinking of the outer part of the flow its central part near the axis is accelerating. This presses continues until the Poiseuille flow will be established asymptotically at a large distance far from the end of the tube. Thus, the Poiseuille flow established at the distance form the tube entrance where thickness of the boundary layer is of the order of the tube diameter. We can estimate the distance L, from the end of the tube where the Poiseuille flow is formed, using expression for the thickness of the boundary layer $\delta(x) = \sqrt{\frac{\nu x}{U}}$, where we put $\delta(x) \propto R$ – the tube radius. We then obtain

$$L \propto R^2 U/ = R \cdot Re, \tag{7.8.23}$$

which means that the transitional length is proportional to the Reynolds number and the tube radius: as larger velocity as far from the end the Poiseuille flow is formed.

Problems

7.1. Calculate pressure and velocity of a flow for a layer of viscous fluid of thickness h, which is flowing due to gravity from the inclined plane with the angle α to the horizontal.

7.2. Calculate velocity and amount of viscous fluid flowing between tubes with outer radius R_2 and inner radius R_1.

7.3. A model of a turbulent flow involves the following x- component of velocity field $u_x = \sum U_i(t) k_i^{-5/6} \cos(k_i x) \sin(k_i y)$, where k_i are wave numbers of turbulent pulsations and $U_i(t)$ are time dependent amplitudes. Find the y-component of the flow velocity in the case of incompressible fluid.

7.4. Consider liquid, which flows along the tube of radius R. Let temperature of the tube wall is $T = T_0(1 + kz)$. Find temperature distribution assuming the Poiseuille flow.

Chapter 8
Detonation and Shock Waves

8.1 Surfaces of Discontinuity

In the preceding chapters we have considered only flows such that all quantities (velocity, pressure, density, etc.) vary continuously. Flows are also possible, for which discontinuities in the distribution of these quantities occur. As we saw in Sect. 6.14, the physically meaningless ambiguities in the solution for simple compression waves indicate that the basic assumptions for obtaining continuous solutions become invalid. We must therefore expand the class of solutions of the initial equations by allowing for discontinuity solutions. Allowing for the ordinary discontinuity – discontinuity of the first kind, can eliminate the ambiguity. The expansion of the class of solutions necessitates revising the initial assumptions used for obtaining solutions. At the point of discontinuity the gradients are large, and consequently, the flow at the point of discontinuity cannot be considered as isentropic. As a matter of fact, the real flow is fairly well approximated by the introduction of a jump, satisfying certain conditions at the discontinuity surface together with the initial equations for the continuous flow. In this approach the shock waves are represented by a discontinuous surface of zero thickness. The boundary conditions on the surfaces of discontinuity must be chosen so as to characterize the flow on either side of the jump. From the formal mathematical point of view, the differential equations, obtained for continuous flow, now become invalid, however their integral prototypes – the conservation laws remain valid, and they determine the boundary conditions of the surface of discontinuity.

To formulate these conditions, we consider an element of the surface and use a coordinate system co-moving with this element, with the x-axis along the normal to the surface. Firstly, the mass flux must be continuous: the mass of gas coming from one side must equal the mass leaving the other side. The mass flux through the surface element considered per unit area is ρu_x. Thus we have

$$\rho_1 u_{1x} - \rho_2 u_{2x} \equiv [\rho u_x] = 0, \tag{8.1.1}$$

where the suffixes 1 and 2 refer to the two sides of the surface of discontinuity, and the difference between the values of any quantity on the two sides of the surface will be denoted by enclosing it in square brackets.

M.A. Liberman, *Introduction to Physics and Chemistry of Combustion*,
DOI: 10.1007/978-3-540-78759-4_8, © Springer-Verlag Berlin Heidelberg 2008

Next, the momentum flux must be continuous, i.e. the forces exerted by the gases on the two sides of the surface of discontinuity must be equal. The momentum flux per unit area is: $Pn_i + \rho u_i u_k n_k$, where $\mathbf{n}$ is the normal vector along the x-axis. The continuity of the x-component of the momentum flux gives

$$(P_1 + \rho_1 u_{1x}^2) - (P_2 + \rho_1 u_{2x}^2) \equiv \left[P + \rho u_x^2\right] = 0, \tag{8.1.2}$$

and for y, z-components we have

$$\left[\rho u_x u_y\right] = 0, \ [\rho u_x u_z] = 0. \tag{8.1.3}$$

Finally, from the continuity condition for the energy flux we have

$$\left[\rho u_x \left(\frac{1}{2}\mathbf{u}^2 + H\right)\right] = 0. \tag{8.1.4}$$

Notice, that there is no such equation for the continuity of the entropy flux. Moreover, the entropy production at the discontinuity depends on the amplitude of the jump.

Equations (8.1.2)–(8.1.4) form a complete system of boundary conditions at a surface of discontinuity. From them we can deduce the possibility of two types of surface of discontinuity. The first type, which exhibits zero mass flux through the surface of discontinuity, is called a tangential discontinuity. For a tangential discontinuity

$$\rho_1 u_{1x} = \rho_2 u_{2x} = 0. \tag{8.1.5}$$

Hence, the normal velocity component and the gas pressure are continuous at the surface of discontinuity:

$$u_{1x} = u_{2x} = 0, \ [P] = 0, \tag{8.1.6}$$

while the tangential velocities and the density as well as the other thermodynamic quantities except the pressure may be discontinuous.

In the second type of discontinuity, which is called *shock wave*, the mass flux is not zero, and correspondingly u_{1x} and u_{2x} are also not zero. Since in the shock wave $\rho u_x \neq 0$ we then have from (8.1.3)

$$[u_y] = 0, \ [u_z] = 0, \tag{8.1.7}$$

i.e. the tangential velocity is continuous at the surface of discontinuity for shock waves. The normal velocity, pressure, density and other thermodynamic quantities are discontinuous. Since the tangential velocities are continuous at the surface of discontinuity for shock waves, we can present the boundary conditions in the following form:

$$[\rho u_x] = 0, \tag{8.1.8}$$

$$[P + \rho u_x^2] = 0, \tag{8.1.9}$$

$$\left[\frac{1}{2}\mathbf{u}^2 + H\right] = 0, \tag{8.1.10}$$

8.2 The Shock Adiabatic

Let us now examine more closely the boundary conditions for shock waves. We have seen that in this type of discontinuity the tangential component of the gas velocity is continuous. We can therefore take a coordinate system in which the surface element considered to be at rest and the tangential component of the gas velocity is zero on both sides of discontinuity surface. Then we can write (8.1.8)–(8.1.10) expressing the enthalpy via the internal energy of the gas, $H = \varepsilon + P/\rho$

$$\rho_1 u_1 = \rho_2 u_2 = j, \tag{8.2.1}$$

$$\rho_1 u_1^2 + P_1 = \rho_2 u_2^2 + P_2, \tag{8.2.2}$$

$$\varepsilon_1 + P_1/\rho_1 + \frac{1}{2}u_1^2 = \varepsilon_2 + P_2/\rho_2 + \frac{1}{2}u_2^2, \tag{8.2.3}$$

where j denotes the mass flux density at the surface of discontinuity, and we shall use u to denote the normal component of velocity.

The first two equations can be used to find the velocity of the downstream flow relative to the velocity of upstream flow. Using relation, $u_1 = jV_1$, $V_1 = 1/\rho_1$, $V_2 = 1/\rho_2$ and excluding u_2 from (8.2.2) we obtain

$$u_1^2 = \frac{P_2 - P_1}{V_1 - V_2}V_1^2, \text{ and } u_2^2 = \frac{P_2 - P_1}{V_1 - V_2}V_2^2,$$

so that

$$u_1 - u_2 = \sqrt{\frac{(P_2 - P_1)(\rho_2 - \rho_1)}{\rho_1\rho_2}} = \sqrt{(P_2 - P_1)(V_1 - V_2)}, \tag{8.2.4}$$

We call gas 1 the one into which the shock wave moves, and gas 2, which remains behind the shock. Next we shall derive the relation, which steams from the conditions (8.2.1)–(8.2.4)

$$j^2 = (P_2 - P_1)/(V_1 - V_2). \tag{8.2.5}$$

$$\begin{aligned} \varepsilon_2(P_2, \rho_2) - \varepsilon_1(P_1, \rho_1) &= \frac{1}{2}(P_2 + P_1)(\rho_2 - \rho_1)/\rho_1\rho_2 \\ &= \frac{1}{2}(P_2 + P_1)(V_2 - V_1). \end{aligned} \tag{8.2.6}$$

The last equation gives the relation between P_2 and V_2 for given initial values of P_1 and V_1 and therefore, similar to the adiabatic equation it is called the *shock adiabatic* or the *Hugoniot adiabatic*.

Consider the shock adiabatic for an ideal gas, assuming constant specific heats and adiabatic exponent $\gamma = C_P/C_V$. Substituting the expression for the internal energy, $\varepsilon = \frac{1}{\gamma - 1}\frac{P}{\rho}$, we obtain

$$\frac{P_2}{P_1} = \frac{(\gamma + 1)\rho_2 - (\gamma - 1)\rho_1}{(\gamma + 1)\rho_1 - (\gamma - 1)\rho_2}. \tag{8.2.7}$$

This expression shows that: (1) the shock adiabatic is passing through the given point (P_1, V_1) corresponding to the state of gas 1 in front of the shock wave, which we shall call the initial point; (2) the compression downstream of the front, ρ_2/ρ_1 exhibits a monotonic increase as the strength of the shock increases, and tends to the limit

$$\frac{\rho_2}{\rho_1} = \frac{(\gamma + 1)}{(\gamma - 1)}, \tag{8.2.8}$$

as the strength of the shock tends to infinity, $P_2/P_1 \to \infty$. The limiting compression is $\rho_2/\rho_1 = 4$ for a monatomic gas ($\gamma = 5/3$) and $\rho_2/\rho_1 = 6$ for a diatomic gas ($\gamma = 7/5$).

We can also use the gas equation of state to express the ratio of the temperature across the shock front

$$\frac{T_2}{T_1} = \frac{(\gamma + 1)P_1 + (\gamma - 1)P_2}{(\gamma - 1)P_1 + (\gamma + 1)P_2}\frac{P_2}{P_1}. \tag{8.2.9}$$

Introducing the Mach number, $M_1 = u_1/a_1$, which characterize intensity of the shock as a ratio of the shock wave velocity relative to the sound velocity, we find

$$\frac{\rho_2}{\rho_1} = \frac{M_1^2(\gamma + 1)}{2 + (\gamma - 1)M_1^2} \to \frac{\gamma + 1}{\gamma - 1}, \tag{8.2.10}$$

$$\frac{T_2}{T_1} = \frac{[2 + (\gamma - 1)M_1^2][2\gamma M_1^2 - (\gamma - 1)]}{M_1^2(\gamma + 1)^2} \to \frac{2\gamma(\gamma - 1)}{(\gamma + 1)^2}M_1^2, \tag{8.2.11}$$

$$\frac{P_2}{P_1} = \frac{2\gamma M_1^2 - (\gamma - 1)}{\gamma + 1} \to \frac{2\gamma}{\gamma + 1}M_1^2. \tag{8.2.12}$$

It must be emphasized that the shock adiabatic is not identical to the Poisson adiabatic. In particular, the equation of the shock adiabatic cannot be written in the form $f(P, V) = \text{const}$. The Poisson adiabatic for a given gas forms a one-parameter family of curves, but the shock adiabatic is determined by two parameters, the initial values P_1 and V_1. This has also the following important result: if two (or more) successive shock waves transform a gas first from state 1 to state 2 and then to state 3, the transition from state 1 to state 3 cannot in general be effected by the passage of any one shock wave.

Like the other thermodynamic quantities, the entropy is discontinuous at a shock wave. By the law of increase of entropy, the entropy of a gas can only increase during its motion. Hence the entropy S_2 of the gas, which has passed through the shock wave, must exceed its initial entropy S_1. The following fact should be emphasized. The presence of shock waves results in an increase in entropy in those flows, which can be regarded as motions of an ideal fluid in all space, the viscosity and thermal conductivity being zero. The increase in entropy signifies that the motion is irreversible, i.e. energy is dissipated. Thus the discontinuities are a means by which energy can be dissipated in the motion of an ideal fluid. The increase of the entropy for a gas passing the shock front can be expressed as

$$S_2 - S_1 = C_V \ln\left\{\frac{P_2}{P_1}\left[\frac{(\gamma-1)P_2/P_1 + (\gamma+1)}{(\gamma+1)P_2/P_1 + (\gamma-1)}\right]^{\gamma}\right\}. \qquad (8.2.13)$$

The increase in entropy in a shock wave has another important effect on the motion: even if we have potential flow in front of the shock wave, the flow behind it is, in general, rotational.

It can be proved that for shock waves of any intensity in a matter with normal compressibility, i.e. for $\left(\partial^2 V/\partial P^2\right)_S > 0$, the condition of the increase of the entropy for a gas passing the shock front means also that in a shock wave hold the inequalities

$$P_2 > P_1, \quad V_1 > V_2, \quad u_1 > a_{s1}, \quad u_2 < a_{s2}, \quad u_1 > u_2. \qquad (8.2.14)$$

The inequalities (8.2.14) mean that, when the gas passes through the shock wave, it is compressed, the pressure and density increased, and the shock wave moves supersonically relative to the gas ahead of it. Therefore, no perturbations starting from the shock wave can penetrate into the gas ahead the shock front. In other words, the presence of the shock has no effect on the state of gas in front of it. The condition $\left(\partial^2 V/\partial P^2\right)_S > 0$ can be violated for a medium which has a phase transition and for which the adiabatic therefore has a kink, for instance, near a gas-liquid critical point.

Finally, let us show that on the *shock adiabatic* (*Hugoniot adiabatic*) which is the curve of P_2 as a function of V_2 for the given quantities of (P_1, V_1) there is no a point such that the cord passing from the point (P_1, V_1) to any point (P_2, V_2) would be tangential to the shock adiabatic at this point. At such a point the slop of the chord passing from the point (P_1, V_1) to the point (P_2, V_2) must be minimum. From (8.2.1) and (8.2.2) follows that tangent of the angle of the cord between points (P_2, V_2) and (P_1, V_1) is

$$j^2 = \frac{P_2 - P_1}{V_1 - V_2}. \qquad (8.2.15)$$

Then condition for the minimum of the slop is $dj^2/dP_2 = 0$, which corresponds to the equality $u_2 = a_{s2}$ which contradicts to inequalities (8.2.14).

8.3 Detonation Adiabatic

In the type of combustion wave (slow combustion regime – deflagration) which we considered before, the chemical reaction propagated through the gaseous mixture of fuel and oxidants due to the heating (or diffusion, in general), which results from the direct transfer of heat from the burning gas to that which is still unburned. As we discussed already, another entirely different mechanism of propagation of combustion, involving shock waves is also possible. The shock wave heats and compresses the gas as it passes through the shock front so that the temperature of the combustible gas mixture behind the shock increases. If the shock wave is sufficiently strong, the rise in temperature, which it causes, may be sufficient to initiate combustion. The shock wave will then "ignite" the gas mixture as it moves through the mixture; i.e. the combustion will be propagated with the velocity of the shock, which is much faster than velocity of ordinary combustion wave, which we call flame, or slow combustion regime – deflagration. This mechanism of propagation of combustion is called *detonation*. The theory of detonation has been developed by Ya. B. Zel'dovich (1940), and independently by J. von Neumann (1942) and W. Döring (1943).

When the shock wave passes some point in the gas, the reaction begins at that point, and continues until all the gas there is burned, i.e. for a time τ_r, which characterizes the kinetics of the reaction. This time is, however, itself dependent on the strength of the shock, and decreases as the shock becomes stronger, on account of the increase in the reaction rate with rising temperature. It is therefore clear that a layer of chemical reaction zone moving with it in which combustion is occurring will follow the shock wave and the width of this layer is equal to the speed of propagation of the shock multiplied by the time τ_r. It is important that the width of detonation wave (width of the shock wave together with reaction zone) does not depend on the dimensions of any bodies that are present. When the characteristic dimensions of the problem are sufficiently large, we can regard the shock wave and the combustion zone following it as a single surface of discontinuity, which separates the burned and unburned gases. We shall call such a surface a *detonation wave*. Typical structure of a detonation, know as *Zel'dovich-von Neumann-Döring* model, consisting of a leading shock wave and following reaction zone, is presented in Fig. 8.1.

At a detonation wave the flux densities of mass, energy and momentum must be continuous. So that the relations derived previously for a shock adiabatic, which follow from continuity conditions alone, remain valid. These are

$$j \equiv \rho_1 u_1 = \rho_2 u_2 \tag{8.3.1}$$

$$P_1 + \rho_1 u_1^2 = P_2 + \rho_2 u_2^2, \tag{8.3.2}$$

$$H_1 + \frac{u_1^2}{2} = H_2 + \frac{u_2^2}{2}. \tag{8.3.3}$$

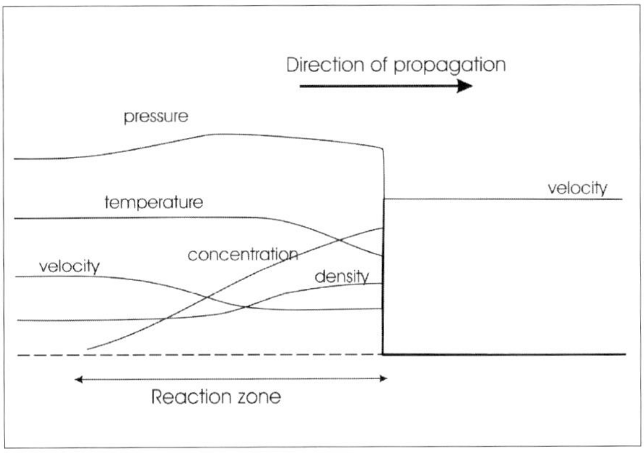

Fig. 8.1 Schematic structure of a detonation wave, Zel'dovich-von Neumann-Döring model

In particular, the equation

$$H_1 - H_2 + \frac{1}{2}(V_1 - V_2)(P_2 + P_1) = 0 \tag{8.3.4}$$

holds, which steams from continuity conditions (8.3.1–3).

Remember that we introduced specific volume as $V = 1/\rho$, and using the suffix *1* that always pertains to the unburned gas in the upstream flow ahead the shock front, and the suffix *2* pertains to the combustion products in the downstream flow behind detonation front. In particular, from Eqs. (8.3.1) and (8.3.2) we obtain for tangent of the angle of the cord between points (P_1, V_1) and (P_2, V_2).

$$j^2 = \frac{P_2 - P_1}{V_1 - V_2}, \tag{8.3.5}$$

The curve of P_2 as a function of V_2 given by the Eq. (8.3.4) is called the *detonation adiabatic*. Unlike the shock adiabatic that we considered earlier, this curve does not pass through the given initial point (P_1, V_1). The fact that the shock adiabatic passes through this point was due to the fact that the enthalpy H_1 and H_2 were the same functions of P_1, V_1 and P_2, V_2 respectively, whereas this does not now hold in detonation because of the chemical difference between the unburned and burned gases.

In Fig. 8.2 the continuous line shows the detonation adiabatic and the ordinary shock adiabatic for the unburned gas mixture is shown by dashed line for comparison purpose. The last one passes through the point (P_1, V_1) as it must be for the shock adiabatic for the unburned gas mixture. The detonation adiabatic always lies above the shock adiabatic, because a high temperature is reached in combustion, and the gas pressure is therefore greater than it would be in the unburned gas for the same specific volume.

It is obvious that formula for the mass flux density (8.3.5), which is consequence of the equations of continuity conditions for flux densities of mass and momentum only, holds for both shock wave and detonation. In the Fig. 8.2 of detonation adiabatic j^2 graphically is the slope of the chord passing from the

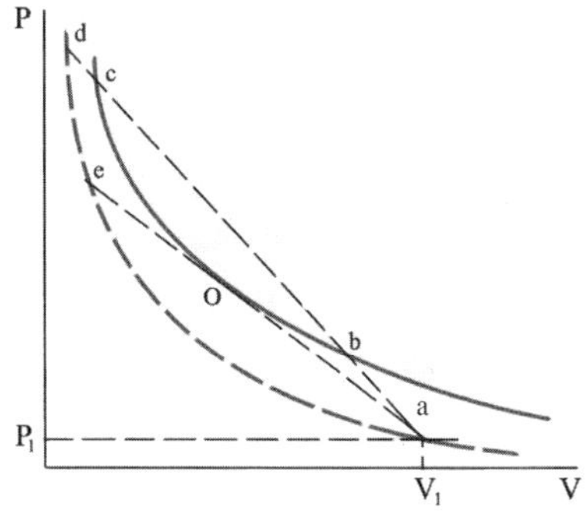

Fig. 8.2 Detonation adiabatic

point (P_1, V_1) to any point (P_2, V_2) on the detonation adiabatic. In Fig. 8.2 such chord is shown by the dashed line a-b-c-d. The immediate consequence, which is seen from Fig. 8.2, is that j^2 cannot be less than the slope of the tangent a – O to a curve of the detonation adiabatic. It is clear that the flux j^2 is the mass of the gas mixture, which is ignited per unit time per unit area of the surface of the detonation wave. We came to the conclusion that, in a detonation, this quantity cannot be less than a certain limiting value j_{min}, which depends on the initial state of the unburned gas mixture.

Since formula (8.3.5) is a consequence only of the conditions of continuity of the fluxes of mass and momentum, it therefore holds not only for the final state of the combustion products for a given initial state of the gas, but also for all intermediate states, in which only part of the reaction has been evolved. In other words, the pressure P and specific volume V of the gas in any state obey the linear relation

$$P = P_1 + j^2(V_1 - V), \tag{8.3.6}$$

which corresponds to the chord (a-b-c-d) in Fig. 8.2.

Below we shall follow a procedure suggested by Ya. B. Zel'dovich to investigate the variation of the state of the gas through the layer of finite width, which is actually a detonation wave. The leading front of the detonation wave is a true shock wave in the unburned gas (with index 1). In this region, the gas is compressed and heated to a state represented by the point d in Fig. 8.2 on the shock adiabatic of gas *1*. The chemical reaction begins in the compressed gas, and as the reaction proceed the state of the gas is represented by a point which moves down the chord (d-a) since the heat is evolved, the gas expands, and its pressure decreases. This continues until combustion is complete and the whole heat of the reaction has been evolved. The corresponding point is c, which lies on the detonation adiabatic representing the final state of the combustion products. The lower point *b* at which the chord (a-d) intersects the detonation adiabatic cannot be reached for a gas in which combustion is caused by compression and heating in a shock wave. Thus we conclude that the detonation is represented, not by the whole of the detonation adiabatic, but only by the upper part, lying above the point O where this adiabatic touches the straight line a-O drawn from the initial point a.

It has been shown previously that, at the point where $dj^2/dP_2 = 0$, i.e. where the shock adiabatic touches the chord a-O-e, the velocity u_2 is equal to the corresponding velocity of sound a_{s2}. This result can be obtained only from the conservation laws for the surface of discontinuity, and is therefore entirely applicable to the detonation wave also. On the ordinary shock adiabatic for a single gas there are no such points with $dj^2/dP_2 = 0$, as has been noticed in Sect. 8.2. On the detonation adiabatic, however, there is such a point, namely the point O. At the same time as $u_2 = a_{s2}$, we have at such a point the inequality $d(u_2/a_{s2})/dP_2 < 0$, and therefore $u_2 < a_{s2}$ at higher values of P_2, above O. Since detonation corresponds to the upper part only of the adiabatic, above the point O, we conclude that

$$u_2 \leq a_{s2}, \tag{8.3.7}$$

which means that a detonation wave moves relative to the gas just behind it with a velocity equal to or less than that of sound velocity. The equality $u_2 = a_{s2}$ holds for a detonation corresponding to the point O, which is called the *Chapman-Jouguet* point.

The velocity of the detonation wave relative to unburned gas *1*, it is always supersonic, including for the point O, i.e:

$$u_1 > a_{s1}. \tag{8.3.8}$$

Validity of inequality (8.3.8) is seen directly from Fig. 8.2. The velocity of sound a_{s1} in the figure is given by the slope of the tangent to the shock adiabatic for gas 1 by the dashed curve at the point a. The slope of the chord a-c gives velocity u_1, on the other hand, since as we seen in Sect. 8.2, $u_1 = jV_1$. Since all the chords concerned are steeper than the tangent, we always have $u_1 > a_{s1}$. Moving with supersonic velocity, the detonation wave, like a shock wave, does not affect the state of the gas in front of it. The velocity u_1 with which the detonation wave moves relative to the unburned gas at rest is the velocity of propagation of the detonation in the fresh unburned gas.

Since $u_1/V_1 = u_2/V_2 \equiv j$, and $V_1 > V_2$ (recall that $V_1 = 1/\rho_1$, and $V_2 = 1/\rho_2$), it follows that $u_1 > u_2$. The difference $u_1 - u_2$ is evidently the velocity of the combustion products relative to the unburned gas. This difference is positive, i.e. the combustion products move in the direction of propagation of the detonation wave.

It should be also mentioned the following important property of *Chapman-Jouguet* detonation wave. Since $ds_2/dj^2 > 0$, where *s* is entropy (Sect. 8.2), then at the point where j^2 has a minimum, entropy s_2 therefore also has a minimum. Obviously, this point is point O, and we conclude that this point corresponds to the minimal value of the entropy s_2 on the detonation adiabatic. The entropy s_2 also has a maximum at point O if we consider the change in state along the line a-O-e since the slopes of the curve and the tangent at O are the same. A displacement from point e to point O corresponds to the change of state as the combustion reaction occurs in the compressed gas, and this is accompanied by the evolution of heat and an increase in entropy. A passage from point O to point a, however, would correspond to the endothermic conversion of the combustion products into the original gases, with a decrease in entropy. If the detonation is caused by a shock wave, which is produced by some external source and is then incident on the gas, any point on the upper part of the detonation adiabatic may correspond to the detonation. A detonation, which is due to the combustion process itself, is of the most important case. Such detonation corresponds to the *Chapman-Jouguet* point, so that the velocity of the detonation wave relative to the combustion products behind it is equal to the sound velocity, while the velocity u_1 relative to the unburned gas has its minimal possible value.

8.4 Velocity of Detonation Wave

We consider now an ideal polytropic gas and derive the relations between the various quantities in a detonation wave. Substituting in the Eq. (8.3.4) the heat function for ideal gas in the form

$$H_{1,2} = H_{01,2} + c_p T_{1,2} = H_{01,2} + \frac{\gamma P_{1,2} V_{1,2}}{\gamma - 1}, \tag{8.4.1}$$

(recall that for ideal gas $P = (\gamma - 1)c_V \rho T$, and adiabatic constant is $\gamma = c_p/c_V$) we obtain

$$\frac{\gamma_2 + 1}{\gamma_2 - 1} P_2 V_2 - \frac{\gamma_1 + 1}{\gamma_1 - 1} P_1 V_1 - P_2 V_1 + P_1 V_2 = 2Q. \tag{8.4.2}$$

Here we introduced $Q = H_{01} - H_{02}$ as the heat of the reaction, reduced to the absolute zero of temperature.

The curve $P_2 = P_2(V_2)$ given by the Eq. (8.4.2) is a rectangular hyperbola that is detonation adiabatic curve. For a strong detonation when $P_2/P_1 \to \infty$ $(P_2/P_1 >> 1)$, the ratio of densities tends to a finite limit

$$\frac{\rho_2}{\rho_1} = \frac{V_1}{V_2} = \frac{\gamma_2 + 1}{\gamma_2 - 1}. \tag{8.4.3}$$

This is again the highest compression that can be achieved in a detonation wave similar to that one achieved in a shock wave.

All the formulae are simplified in the case of strong detonation waves, which are obtained when the heat evolved in the reaction is large compared with the internal heat energy of the original gas, i.e. when $Q >> c_V T_1$. In this case we can neglect the terms containing P_1 in the Eq. (8.4.2) and we obtain

$$P_2 \left(\frac{\gamma_2 + 1}{\gamma_2 - 1} V_2 - V_1 \right) = 2Q. \tag{8.4.4}$$

8.5 The Chapman-Jouguet Detonation

Since the Chapman-Jouguet detonation is of particular and practical interest, we shall consider it in more details. As it was shown above, at the Chapman-Jouguet point

$$j^2 = \frac{P_2 - P_1}{V_1 - V_2} = \frac{a_{s2}^2}{V_2^2} = \frac{\gamma_2 P_2}{V_2}. \tag{8.5.1}$$

From this equation together with the Eq. (8.3.6) we can express P_2 and V_2 in the form

$$P_2 = \frac{P_1 + j^2 V_1}{\gamma_2 + 1}, \text{ and } V_2 = \frac{\gamma_2 (P_1 + j^2 V_1)}{j^2(\gamma_2 + 1)} \tag{8.5.2}$$

Substituting these expressions in Eq. (8.4.2), replacing the flux j using expression $u_1 = jV_1$, and using again the expression $PV = (\gamma - 1)c_V T$, we shall obtain the equation for the velocity u_1:

$$u_1^4 - 2u_1^2[(\gamma_2^2 - 1)Q + (\gamma_2^2 - \gamma_1)c_{V1}T_1] + \gamma_2^2(\gamma_1 - 1)^2 c_{V1}^2 T_1^2 = 0. \tag{8.5.3}$$

From four roots of this equation, we are interested in two positive. From these two that differ by sign plus and minus in front of the second root square, should be taken root with plus which corresponds to the tangent that passes upwards from the point a in Fig. 8.2. The two signs in this case correspond to the fact that two tangents can be drawn from the point a to the detonation adiabatic: one upwards, as shown in Fig. 8.2, and the other downwards. The upward tangent, in which we are interested, has the steeper slope, and we accordingly take the plus sign

$$u_1 = \left\{\frac{\gamma_2 - 1}{2}[(\gamma_2 + 1)Q + (\gamma_1 + \gamma_2)c_{v1}T_1]\right\}^{1/2} + \\ + \left\{\frac{\gamma_2 + 1}{2}[(\gamma_2 - 1)Q + (\gamma_2 - \gamma_1)c_{v1}T_1]\right\}^{1/2} \tag{8.5.4}$$

Formula (8.5.4) determines the velocity of propagation of the detonation in terms of the temperature T_1 of the original gas mixture.

We can use Eqs. (8.5.1–2) to express P_2 and V_2 in terms of u_1 and T_1. We obtain after simple transformations

$$\frac{P_2}{P_1} = \frac{u_1^2 + (\gamma_1 - 1)c_{v1}T_1}{(\gamma_2 + 1)(\gamma_1 - 1)c_{v1}T_1}, \tag{8.5.5}$$

$$\frac{V_2}{V_1} = \frac{\gamma_2[u_1^2 + (\gamma_1 - 1)c_{v1}T_1]}{(\gamma_2 + 1)u_1^2}. \tag{8.5.6}$$

Equations (8.5.5) and (8.5.6) together with Eq. (8.5.4) determine the ratios of pressure and density between the combustion products and the unburned gas for the detonation wave propagating in the gas mixture of given pressure, density at the given temperature T_1.

Using formulae (8.5.4) and (8.5.5) and (8.5.6) and taking into account that velocity of the flow downstream (behind the detonation front) is $u_2 = u_1 V_2/V_1$, we can calculate u_2, which is

$$u_2 = \left\{\frac{\gamma_2 - 1}{2}[(\gamma_2 + 1)Q + (\gamma_1 + \gamma_2)c_{v1}T_1]\right\}^{1/2} + \\ + \frac{\gamma_2 - 1}{\gamma_2 + 1}\left\{\frac{\gamma_2 + 1}{2}[(\gamma_2 - 1)Q + (\gamma_2 - \gamma_1)c_{v1}T_1]\right\}^{1/2} \tag{8.5.7}$$

The velocity of the combustion products relative to the unburned gas mixture, i.e. the difference $u_1 - u_2$ can be than written down as

$$u_1 - u_2 = \left\{\frac{2}{\gamma_2 + 1}[(\gamma_2 - 1)Q + (\gamma_2 - \gamma_1)c_{v1}T_1]\right\}^{1/2}. \tag{8.5.8}$$

Finally, for the temperature of the combustion products we obtain the expression

$$T_2 = \frac{u_2^2}{\gamma_2(\gamma_2 - 1)c_{V2}}, \tag{8.5.9}$$

where $u_2 = a_2$ in case of the Chapman-Jouguet detonation.

All the formulae are simplified for strong detonation waves, when we can consider that $Q \gg c_V T_1$. In this case instead of (8.5.4)–(8.5.7) we obtain simple formulae

$$u_1 = \sqrt{2(\gamma_2^2 - 1)Q}, \tag{8.5.10}$$

$$u_1 - u_2 = \frac{1}{(\gamma_2 + 1)} u_1, \tag{8.5.11}$$

$$\frac{V_2}{V_1} = \frac{\gamma_2}{\gamma_2 + 1}, \tag{8.5.12}$$

$$\frac{P_2}{P_1} = \frac{2(\gamma_2 - 1)}{(\gamma_1 - 1)} \frac{Q}{c_{V1}T_1}, \tag{8.5.13}$$

$$T_2 = \frac{2\gamma_2}{(\gamma_2 + 1)} \frac{Q}{c_{V2}}. \tag{8.5.14}$$

For the ratio of the temperatures of the combustion products in detonation wave and in slow combustion (flame) calculated in Chap. 5 we obtain:

$$\frac{T_2^{det}}{T_2^{flame}} = \frac{2\gamma_2^2}{\gamma_2 + 1}. \qquad (8.5.15)$$

Since $\gamma_2 > 1$, the right hand side of (8.5.15) is always larger than one. For example, $2\gamma_2^2/(\gamma_2 + 1) = 1.309$ for $\gamma_2 = 1.2$, and $2\gamma_2^2/(\gamma_2 + 1) = 1.633$ for $\gamma_2 = 1.4$, which means that the temperature of the combustion products is always higher behind detonation wave compared to the temperature of combustion products behind a flame.

8.6 The Propagation of a Detonation Wave

Let us show that for the most important practical cases of the propagation of detonation waves in a gas mixture initially at rest, the regime of detonation is the Chapman-Jouguet detonation. Let us consider first the case of detonation propagating from the closed end of the tube. The boundary conditions in this case are that the gas velocity is zero in front of the detonation wave, which does not affect the state of the gas in front of it since velocity of the detonation wave is supersonic velocity. The boundary conditions also require that the gas velocity is zero at the closed end of the tube. Since the gas acquires a non-zero velocity when the detonation wave passes, the velocity must diminish in the region between the detonation wave and the closed end of the tube. In order to determine the resulting flow pattern, we notice that in this case there is no length parameter, which might characterize the conditions of flow along the tube. Therefore, the gas velocity can change either in a shock wave separating two regions where the velocity is constant or in a rarefaction wave.

Let us first assume that the detonation wave does not correspond to the Chapman-Jouguet detonation. Then its velocity of propagation relative to the gas behind is $u_2 < a_{s2}$. However, in this case it is impossible to satisfy the boundary conditions. Indeed, in this case, neither a shock wave nor a weak discontinuity, which is the forward front of a rarefaction wave, can follow the detonation wave. Indeed, the shock wave would propagate relative to the gas in front of it with a velocity exceeding a_{s2} and therefore overtake the detonation wave. Also is true for the weak discontinuity, which is the forward front of a rarefaction wave, which propagates with a velocity equal to a_{s2}, and also would overtake the detonation wave. Thus, with the above assumption, the velocity of the gas moving behind the detonation wave cannot decrease, i.e. the boundary condition at the closed end of the tube, at $x = 0$, cannot be satisfied. This condition can be satisfied only for a detonation wave corresponding to the Chapman-Jouguet detonation. In the latter case, $u_2 = a_{s2}$, and a rarefaction wave can follow the detonation wave. The rarefaction wave is formed at $x = 0$ when the detonation begins, and its forward front coincides with the detonation wave.

Thus, we came to the important result that a detonation wave ignited at the closed end of the tube must correspond to the Chapman-Jouguet point. It moves relative to the gas behind it with a velocity equal to the local velocity of sound. The detonation wave adjoins a rarefaction wave, in which the gas velocity (relative to the tube) falls monotonically to zero. The point where the velocity becomes zero is a weak discontinuity. Behind this discontinuity the gas is at rest.

Let us show that a detonation wave propagating from the open end of a tube is also the Chapman-Jouguet detonation. The pressure of the gas in front of the detonation wave must be equal to the original pressure, which equals the external pressure. It is evident that, in this case also, the velocity must decrease somewhere behind the detonation wave and even more – the velocity must change the sign. If the gas velocity would be constant between the detonation wave and the end of the tube, than the gas was being sucked into the open end of the tube from outside. However this is not possible, since the gas pressure in the tube is obviously greater than the external pressure because the pressure increases in a detonation wave. For the same reasons as in the previous case, the detonation wave must correspond to the Chapman-Jouguet detonation. The resulting flow behind the detonation should be as following. Immediately behind the detonation wave there is a rarefaction wave, in which the velocity decreases monotonically towards the end of the tube and it changes sign at some point. This means that, in the end section of the tube, the gas moves towards the open end and flows out of it. The velocity with which gas products leave the tube equals the local velocity of sound, and its pressure exceeds the external pressure.

8.7 Strong Explosion in Homogenous Atmosphere

When energy release is large, compression wave became a shock wave after it left the location of the energy release. In this case thermal conductivity of the gas does not important and the problem is pure gas-dynamic one. The idealized problem of strong explosion in a homogeneous atmosphere represents a typical example of this kind of problem.

Let us consider an ideal gas of density ρ_0 with constant specific heat, in which a large amount of energy E is liberated in a small volume during a short time interval. As a result a shock wave will be formed around location where energy was released, and it propagates in form of a strong spherical expanding shock wave from the point were the energy released. We shall consider the process at the stage when the shock wave moved away from the point of the explosion on the distance which is much larger compared to the size of the region were energy was originally released, but when shock wave is still very strong. This means that we can neglect by the initial pressure of the gas in comparison with the pressure behind the shock wave, which is equivalent to neglecting the initial internal energy of the gas, which was involved in motion in comparison to the explosion energy E. For such formulated problem, to a high degree of accuracy, the energy release can be considered to be instantaneous and occurring at a point. Assumption of a large intensity of the shock wave,

$$\frac{P_2}{P_0} \gg 1$$

means also that ratio of the densities in the shock wave is equal to its limiting value $\rho_2/\rho_0 = (\gamma + 1)/(\gamma - 1)$.

For thus formulating problem the motion of the gas is defined by only two parameters; initial density of the atmosphere, ρ_0 and by the amount of energy, E released in the explosion. The initial pressure and sound speed do not influence the process.

These two parameters, E and ρ_0 cannot be combined to yield scales with dimension of either length or time. Such motion is called *self-similar*. From ρ_0 and E and two independent variables r and t we can combine only one dimensionless parameter

$$\xi = r(\rho_0/Et^2)^{1/5}, \tag{8.7.2}$$

which is self-similar co-ordinate characterizing the motion of the gas.

We can state that the location of the shock wave at every moment of time must correspond to certain constant value of the self-similar co-ordinate ξ. Thus, motion and location $R = R(t)$ of the shock wave is defined by the value of ξ.

$$R(t) = \xi_0 \left(\frac{Et^2}{\rho_0}\right)^{1/5}, \tag{8.7.3}$$

where ξ_0 is some numerical constant, which can be found by solving equation of motion.

The propagation velocity of the shock wave in the laboratory system is

$$D = \frac{dR}{dt} = \frac{2R}{5t} = \xi_0 \frac{2}{5}\left(\frac{E}{\rho_0 t^3}\right)^{1/5} = \frac{2}{5}\xi_0^{5/2}\left(\frac{E}{\rho_0}\right)^{1/2} R^{-3/2}. \tag{8.7.4}$$

The density, pressure and velocity of the gas behind the shock wave can be expressed using formulae for a strong shock

$$\rho_2 = \rho_0 \frac{\gamma + 1}{\gamma - 1}, \; P_2 = \frac{2}{\gamma + 1}\rho_0 D^2, \; u_2 = \frac{2}{\gamma + 1} D. \tag{8.7.5}$$

As it is seen from (8.7.5) the pressure decreases with time as

$$P_2 \propto \rho_0 D^2 \propto \rho_0 \left(\frac{E}{\rho_0}\right)^{2/5} \propto \frac{E}{R^3}. \tag{8.7.6}$$

The fact that we neglected by the initial pressure of the gas means that we considered initial energy of the gas being negligible compared to the energy released in the explosion. It is therefore obvious that in this approximation the total energy in the region bounded inside the sphere of shock wave remains constant and equal to the energy released in the explosion.

$$E = \int_0^R 4\pi r^2 \rho(\varepsilon + u^2/2)dr = \int_0^R 4\pi r^2 dr\rho\left(\frac{a_s^2}{\gamma(\gamma-1)} + \frac{u^2}{2}\right)dr. \quad (8.7.7)$$

This is the condition of conservation of energy, which is used to determine unknown constant ξ_0 for exact solving the motion equations.

Instead of exact solution, which will require numerical solution of the ordinary differential equation, we consider a simple model, which gives the solution with quite high accuracy. Let us assume that the entire mass of gas involved in the motion behind the shock front is concentrated in a thin layer behind the shock front surface, and that the density of the layer is constant and equal to $\rho_2/\rho_0 = (\gamma+1)/(\gamma-1)$. The thickness of the layer is determined by conservation of mass

$$4\pi R^2 \Delta r \rho_2 = \frac{4\pi R^3}{3}\rho_0, \text{ or } \Delta r = \frac{R}{3}\frac{\rho_0}{\rho_2} = \frac{R}{3}\frac{\gamma-1}{\gamma+1}. \quad (8.7.8)$$

For example, $\Delta r/R = 0.055$ for $\gamma = 1.4$, and $\Delta r/R = 0.0435$ for $\gamma = 1.3$.

Let us denote the pressure at the inner side of the layer P_c and assume that this is the fraction α of the pressure P_2 behind the shock wave, i.e. $P_c = aP_2$. We shall consider the layer as being infinitely thin with all the mass concentrated within the layer. Then, the equation of motion is the Newton law

$$\frac{d}{dt}(Mu_2) = 4\pi R^2 P_c = 4\pi R^2 \alpha P_2, \quad (8.7.9)$$

where $M = \frac{4}{3}\pi R^3 \rho_0$ is the total mass in the layer, which is function of time.

The only one external force acts on the layer from inside, $4\pi R^2 P_c$, with P_c being the force per unit are, since the external pressure is assumed negligible. Expressing u_2 and P_2 in terms of the shock velocity $D = dR/dt$ and using (8.7.5), we obtain

$$\frac{1}{3}\frac{d}{dt}(R^3 D) = \alpha D^2 R^2. \quad (8.7.10)$$

Using new independent variable, by noting

$$\frac{d}{dt} = \frac{dR}{dt}\frac{d}{dR} = D\frac{d}{dR},$$

we obtain the equation

$$\frac{1}{3}D\frac{d}{dR}(R^3 D) = \alpha D^2 R^2, \quad (8.7.11)$$

which being integrated, gives

$$D = CR^{-3(1-\alpha)}. \quad (8.7.12)$$

The constants C and α are determined from the condition of energy conservation. Notice that that the pressure inside the sphere of radius R is $P_c = \alpha P_2$, which means that although we assumed that almost all the mass is concentrated in the layer, the space inside the sphere contains some small amount of the gas with the internal energy

$$E_{in} = \frac{1}{\gamma - 1}\frac{4}{3}\pi R^3 P_c.$$

The kinetic energy of the gas is $E_{kin} = Mu_2^2/2$. Thus, for the total energy, using $P_c = \alpha P_2$, and expressing u_2 in terms of D, and substituting for D its expression from (8.7.12) we obtain

$$E = \frac{4\pi}{3}\rho_0 C^2 \left[\frac{2\alpha}{\gamma^2 - 1} + \frac{2}{(\gamma + 1)^2}\right] \cdot R^{3-6(1-\alpha)}. \tag{8.7.13}$$

Since the energy released in the explosion is constant, the exponent of the variable R must be zero. This gives $\alpha = 1/2$. Then, the Eq. (8.7.13) defines the constant C:

$$C = \left[\frac{3}{4\pi}\frac{(\gamma - 1)(\gamma + 1)^2}{3\gamma - 1}\frac{E}{\rho_0}\right]^{1/2}. \tag{8.7.14}$$

Thus, form the Eq. (8.7.12) we have the following relations

$$D \propto R^{-3/2},\ P_2 \propto R^{-3},\ u_2 \propto R^{-3/2},\ R \propto t^{2/5}. \tag{8.7.15}$$

The coefficients in these relations can be expressed using (8.7.14), for example,

$$R = \left[\frac{75}{16\pi}\frac{(\gamma - 1)(\gamma + 1)^2}{3\gamma - 1}\frac{E}{\rho_0}\right]^{1/5} t^{2/5}. \tag{8.7.16}$$

The approximation used here is quite close to the exact solution of the problem. For example, for $\gamma = 1.4$ the exact solution gives $P_c = 0.35\,P_2$ and for $\gamma = 1.2$ it is $P_c = 0.41\,P_2$, while approximate solution gives $P_c = 0.5\,P_2$. The numerical coefficients in the brackets in (8.7.16) is 1.033 and 0.89 for $\gamma = 1.4$ and $\gamma = 1.2$, for exact solution, and 1.014 and 0.89 for the approximate solution, correspondingly.

Problems

8.1. Calculate the Chapman-Jouguet detonation velocity for a gaseous mixture of $H_2 + O_2$. (Assume no dissociation of the product gases and $Q = 57$ kcal/mol, initial temperature is 298 K).

8.2. Calculate the temperature and velocity of combustion products in Chapman-Jouguet detonation for a gaseous mixture of $H_2 + O_2$. (Assume no dissociation of the product gases, $Q = 57$ kcal/mol). Compare the temperature of detonation products with the temperature of the flame products.

8.3. Derive expressions for the pressure, velocity and density ratio in Chapman-Jouguet detonation.

8.4. Dynamics of a spherical blast wave resulting from a strong explosion is determined by the total energy of the explosion E and by the density of the surrounding gas ρ. Using the dimensional analysis, find how the radius of the blast wave increases in time. (The Sedov-Taylor blast wave solution).

8.5. Consider approximate solution for a spherical blast wave assuming that almost all mass of the air after explosion is accumulated within a thin layer behind the spherical shock wave expanding form the point of the explosion.

Chapter 9
Hydrodynamics of Propagating Flame

The simplest configuration of a slow combustion wave is a planar stationary flame. To large extend the modern flame theory was erected by Russian theorists Ya. B. Zel'dovich and D. A. Frank-Kamenetsky, who obtained the analytical formulas for the velocity and structure of a planar flame with chemical kinetics in the form of one-step Arrhenius law. Though the obtained formulas could not be used to calculate the velocity of a real flame with a complicated many steps reaction, but their work was the fundamental methodology that made up the basic platform for the whole flame theory. The next important step in the theory was due to the Darrieus-Landau (DL) solution for the problem of stability of propagating flames (Darrieus, 1939; Landau, 1944), where it has been shown that a planar flame front is absolutely unstable against small perturbations bending the flame front. The main conclusion of the DL theory was that a flame cannot propagate as a planar stationary front, but instead it becomes curved, convoluted, its velocity increases, and the flow in the flame can become turbulent.

9.1 Complete System of Equations for Propagating Flame

Dynamics of a flame is described by the hydrodynamic equations of mass, momentum and energy conservation with the account of reaction kinetics and transport processes of thermal conduction, fuel diffusion and viscosity. For the sake of simplicity a single irreversible reaction will be admitted, so that the governing equations are the following:

$$\frac{\partial}{\partial t}\rho + \frac{\partial}{\partial x_i}(\rho v_i) = 0, \tag{9.1.1}$$

$$\frac{\partial}{\partial t}(\rho v_i) + \frac{\partial}{\partial x_j}\left(\rho v_i v_j + \delta_{ij} P - \tau_{ij}\right) = \rho g_i, \tag{9.1.2}$$

$$\frac{\partial}{\partial t}\left(\rho e + \frac{1}{2}\rho v_i v_i\right) + \frac{\partial}{\partial x_i}\left(\rho v_i h + \frac{1}{2}\rho v_i v_j v_j + q_i - v_j \tau_{ij}\right) = \rho g_i v_i, \tag{9.1.3}$$

M.A. Liberman, *Introduction to Physics and Chemistry of Combustion*,
DOI: 10.1007/978-3-540-78759-4_9, © Springer-Verlag Berlin Heidelberg 2008

$$\frac{\partial}{\partial t}(\rho Y)+\frac{\partial}{\partial x_i}\left(\rho v_i Y-\kappa\frac{Le}{C_P}\frac{\partial Y}{\partial x_i}\right)=-\frac{\rho^n Y^n}{\rho_R^{n-1}\tau_R}\exp\left(-\frac{E}{T}\right),\tag{9.1.4}$$

where Y is the fuel mass fraction, $e = QY + C_V T$ is the internal energy, $h = QY + C_P T$ is the enthalpy, C_P and C_V are specific heats per unit mass at constant pressure and volume respectively, their ratio determines the adiabatic exponent, **g** is a gravity acceleration. In the development of the reaction the fuel fraction Y changes from 1 to 0. We consider a reaction of the order n with the energy release Q; the Arrhenius law gives the temperature dependence of the reaction rate with the activation energy E (taken in temperature units) and with constants of time dimension τ_R and density dimension ρ_R. Usually the reaction order is of no importance for hydrodynamic properties of a flame with the exception of the problem of flame dynamics in a closed burning chamber, where flames with the first, second and third order reactions behave in a different way. The stress tensor and the energy diffusion vector are given by the formulas

$$\tau_{ij}=\kappa\frac{Pr}{C_P}\left(\frac{\partial v_i}{\partial x_j}+\frac{\partial v_j}{\partial x_i}-\frac{2}{3}\frac{\partial v_k}{\partial x_k}\delta_{ij}\right),\tag{9.1.5}$$

$$qi=-\kappa\left(\frac{\partial T}{\partial x_i}-Le\frac{Q}{C_P}\frac{\partial Y}{\partial x_i}\right),\tag{9.1.6}$$

where Pr is the Prandtl number characterizing the relative strength of viscosity and thermal conduction and Le is the Lewis number that shows the relative role of fuel diffusion and thermal conduction. We assume that combustible mixture can be considered as a perfect gas with the equation of state

$$P=\frac{\gamma-1}{\gamma}C_P\rho T.\tag{9.1.7}$$

The set of Eqs. (9.1.1)–(9.1.7) is typically used in direct numerical simulations of flame dynamics. For a perfect gas the transport coefficients of thermal conduction, fuel diffusion and viscosity are proportional to each other. For example, the Prandtl number of a perfect gas depends on the adiabatic exponent $Pr = 4\gamma/(9\gamma-5)$ and changes within the limits $2/3 < Pr < 1$. For simplicity we assume also that adiabatic exponent and the specific heats do not change in the reaction.

9.2 Isobaric Approximation

The theoretical approach to flame dynamics is based on isobaric approximation. It was pointed out earlier, that flames propagate with the velocities much smaller than the sound speed. The Mach number, which is the ratio of the velocity of a plane stationary flame U_f and the sound speed in the fuel a_s, characterized the compressibility effects. Even for the fastest flames the Mach number is very small, $M = U_f/a_s \cong 0.01 << 1$. Since change of pressure is $\Delta P \propto \rho_f U_f^2$, so the relative change of pressure in a flame-generated flow is $\Delta P/P \propto \rho_f U_f^2/P_f \propto M^2 << 1$. At the same time the changes of temperature and density of a burning matter in a flame are about the temperature and the density themselves. As a result we can simplify the equation of state of the burning matter (9.1.7) neglecting the pressure changes, so that writing

$$\rho T = \rho_u T_u. \tag{9.2.1}$$

The equation of energy transfer can be also simplified in the isobaric approximation. Using (9.1.1), (9.1.2) and the equation of state (9.2.1) we can rewrite (9.1.3) in the form

$$\begin{aligned} &\rho\left(\frac{\partial}{\partial t} + \mathbf{v}\cdot\nabla\right)(C_P T + QY) - \frac{\partial P}{\partial t} - \mathbf{v}\cdot\nabla P - \frac{\partial}{\partial x_i}(v_j \tau_{ij}) \\ &= \nabla\cdot\left[\kappa\left(\nabla T + \mathrm{Le}\frac{Q}{C_P}\nabla Y\right)\right]. \end{aligned} \tag{9.2.2}$$

In the isobaric approximation the terms related to the pressure derivatives in (9.2.2) and to the viscous stress are small as $M^2 << 1$ and may be neglected when compared to the terms including changes of temperature and the fuel fraction. Therefore, the equation of energy is reduced to the equation of thermal conduction

$$\rho\left(\frac{\partial}{\partial t} + \mathbf{v}\cdot\nabla\right)(C_P T + QY) = \nabla\cdot\left[\kappa\left(\nabla T + \mathrm{Le}\frac{Q}{C_P}\nabla Y\right)\right]. \tag{9.2.3}$$

Thus, in the isobaric approximation flame dynamics is described by the hydrodynamic equations (9.1.1), (9.1.2), by the equation of thermal conduction (9.2.3), by the equation of chemical kinetics (9.1.4) and by the equation of state in the form (9.2.1).

In the case of equally strong thermal conduction and fuel diffusion (when the Lewis number is $\mathrm{Le} = 1$) further simplification is possible. In this case the internal enthalpy is conserved, $h = C_P T + QY = C_P T_u + Q$, and the fuel fraction is related linearly to the temperature of the burning matter

$$Y = 1 - C_P\frac{T - T_u}{Q}. \tag{9.2.4}$$

In the fuel (unburned gaseous mixture) one has $Y = 1$, while in the burnt products $Y = 0$ and the final temperature of the burned products can be calculated as $T_b = T_u + Q/C_P$. Taking into account the equation of state in the form (9.2.1), we obtain that in the isobaric approximation the ratio of the temperature of the burned matter and the temperature of the fuel determines the expansion coefficient, $\Theta \equiv \rho_u/\rho_b = T_b/T_u$. Therefore Θ can be expressed, using relation $T_b = T_u + Q/C_P$, through the energy release in the reaction

$$\Theta = 1 + Q/C_P T_u. \tag{9.2.5}$$

It should be pointed out that the relation between the expansion coefficient Θ and the energy release Q given by (9.2.5) is valid for any isobaric flame independent of the value of the Lewis number. Due to the enthalpy conservation the equation of thermal conduction (9.2.3) and the equation of chemical kinetics (9.1.4) may be replaced by a single equation of thermal conduction with the account of the energy release in the reaction expressed in terms of temperature only

$$\rho \frac{\partial T}{\partial t} + \rho \mathbf{v} \cdot \nabla T = \nabla \cdot \left(\frac{\kappa}{C_P} \nabla T \right) + Q_R(T) \tag{9.2.6}$$

with the reaction rate given by the formula

$$Q_R(T) = \frac{\rho_f^n T_u}{(\Theta - 1)^{n-1} \rho_R^{n-1} \tau_R} \left(\frac{T_b - T}{T} \right)^n \exp(-E/T). \tag{9.2.7}$$

Besides, it was found that in majority of practically interesting cases viscosity is of no importance for flame dynamics. Particularly, viscous effects do not important either at the linear or nonlinear stages of the DL instability. By this reason in many problems of flame dynamics one can use simplified equations of hydrodynamics for an ideal gas

$$\frac{\partial \rho}{\partial t} + \nabla \cdot (\rho \mathbf{v}) = 0, \tag{9.2.8}$$

$$\rho \frac{\partial \mathbf{v}}{\partial t} + \rho (\mathbf{v} \cdot \nabla) \mathbf{v} = -\nabla P + \rho \mathbf{g}. \tag{9.2.9}$$

Quite often instead of the complete set of hydrodynamic equations (9.1.1)–(9.1.7) analytical investigation of flame dynamics starts from the simplified set of Eqs. (9.2.6)–(9.2.9) and the isobaric equation of state (9.2.1).

In the case of slow combustion (deflagration), the isobaric approximation means that conservation of the momentum flux and energy flux is continuity of the pressure and enthalpy,

$$P_1 = P_2, \;\; H_1 = H_2.$$

Thus, we can write

$$H_1 = H_{01} + C_{P1} T_1 = H_2 = H_{02} + C_{P2} T_2,$$

where $H_{01} - H_{02} = Q$ is the heat released in the reaction.

Then we obtain the following relations between thermodynamic variables of the unburned and burned flow on both sides of the propagating flame:

$$P_1 = P_2,\; T_2 = (Q + C_{P1}T_1)/C_{P2},\; \frac{\rho_1}{\rho_2} = \frac{u_2}{u_1} = \frac{\gamma_1(\gamma_2 - 1)}{\gamma_2(\gamma_1 - 1)}\left(1 + \frac{Q}{C_{P1}T_1}\right).$$

9.3 Theory of Planar Flames

Let us consider a planar stationary flame propagating with the velocity U_f in the negative direction along the z-axis. In the reference frame co-moving with the flame the flow consists of a uniform flow of the fresh fuel mixture moving with the velocity $u_z = U_f$ towards the flame front, of the transition region where heating and reaction are in progress (which is the flame front itself) and of a uniform flow of the burned matter drifted away from the flame front. Our purpose is to find the structure of the heating and reaction region and the velocity of the planar flame front as a function of the thermal and chemical fuel parameters.

In the co-moving with the planar stationary flame co-ordinate system the hydrodynamic equations (9.2.8), (9.2.9) take the form

$$\frac{d}{dz}(\rho u_z) = 0, \tag{9.3.1}$$

$$\frac{d}{dz}\left(P + \rho u_z^2\right) = 0, \tag{9.3.2}$$

The corresponding integrals across the front are

$$\rho u_z = \rho_u U_f, \tag{9.3.3}$$

$$P + \rho u_z^2 = P_u + \rho_u U_f^2. \tag{9.3.4}$$

With the account of the mass conservation (9.3.3) the equation of thermal conduction takes the form

$$\rho_u U_f \frac{dT}{dz} = \frac{d}{dz}\left(\frac{\kappa}{C_P}\frac{dT}{dz}\right) + \frac{\rho_u T_u}{\tau_R}\frac{T_b - T}{T}\exp(-E/T) \tag{9.3.5}$$

with the boundary conditions in the fuel and in the burnt matter

$$T(z = -\infty) = T_u;\; T(z = \infty) = T_b,\; (\text{notice, that } u_u \equiv U_f). \tag{9.3.6}$$

The temperature of the burned matter depends only on the fuel temperature and the energy release in the reaction and may be expressed through the expansion coefficient as $T_b = \Theta T_u$. Equation (9.3.5) together with the boundary conditions (9.3.6) constitutes an eigenvalue problem, where the flame velocity U_f is the eigenvalue and the flame internal structure $T = T(z)$ is the eigenfunction.

The Zel'dovich-Frank-Kamenetsky theory of planar stationary flames takes into account that typically the activation energy of the reaction is very large, $E/T > E/T_b >> 1$. Because of the strong Arrhenius dependence, the last term in (9.3.5) differs from zero only in a thin reaction zone (that is as it was assumed by Mallard and LeChatelier, see Chap. 5), where temperature is nearly equal to the final value, $(T_b - T)/T_b \propto T_b/E$. Outside the reaction zone the equation of thermal conduction becomes

$$\rho_u U_f \frac{dT}{dz} = \frac{d}{dz}\left(\frac{\kappa}{C_P}\frac{dT}{dz}\right) \tag{9.3.7}$$

and after straightforward integration it gives the temperature distribution in the flame outside the reaction zone

$$\frac{T(z)}{T_u} = \begin{cases} 1 + (\Theta - 1)\exp(z/L_f), & z < 0 \\ \Theta, & z > 0 \end{cases} \tag{9.3.8}$$

with other hydrodynamic values depending on the temperature as $\rho = \rho_u T_u/T$, $u_z = U_f T/T_u$, $P = P_u + \rho_u U_f^2(1 - T/T_u)$. The solution (9.3.8) for the temperature distribution inside a planar flame is shown in Fig. 9.1 by the dashed curve. Solid curve in Fig. 9.1 depicts the numerical solution of the eigenvalue problem (9.3.5)–(9.3.6) for the case of the expansion coefficient $\Theta = 8$ and the activation energy of the reaction $E/T_b = 8.75$. One can see that the analytical solution is in a good agreement with the numerical one everywhere except the thin zone of active reaction, where the discontinuity of the temperature derivative of (9.3.8) is smoothed. The distributions of density, velocity and concentration for the same flame parameters are presented in Fig. 9.2.

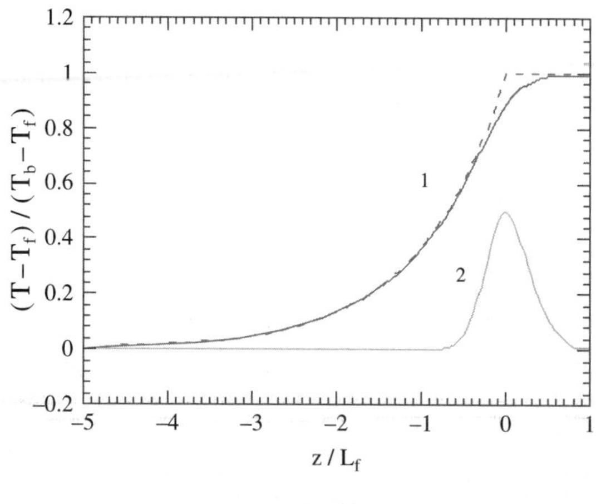

Fig. 9.1 Distribution of the scaled temperature (*curve 1*) and reaction rate (*curve 2*) scaled by the double maximum value in a planar flame with the expansion coefficient $\Theta = 8$ and the activation energy $E/RT_b = 8.75$. The *dashed line* corresponds to the analytical solution (9.3.5)

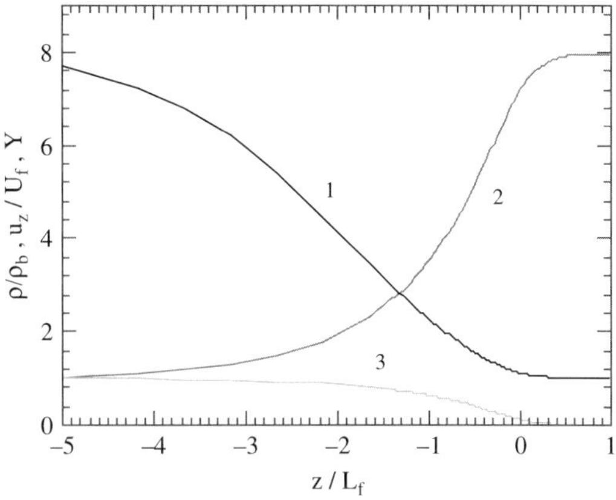

Fig. 9.2 Distribution of the scaled density (*curve 1*) velocity (*curve 2*) and the fuel fraction in a planar flame with the expansion coefficient $\Theta = 8$ and the activation energy $E/RT_b = 8.75$

Characteristic thickness of the reaction zone L_R where temperature is close to the final value, $(T_b - T)/T_b \propto T_b/E$, is much smaller than the total flame thickness $L_R \propto L_f T_b/E << L_f$. By this reason in the reaction zone the term with the first derivative of temperature in Eq. (9.3.5) may be neglected compared to the term proportional to the second derivative and the equation of thermal conduction takes the form

$$\frac{\kappa}{C_f}\frac{d^2T}{dz^2} + \frac{\rho_u T_u}{\tau_R}\frac{T_b - T}{T_b}\exp(-E/T) = 0, \tag{9.3.9}$$

where κ_b is the coefficient of thermal conduction at the temperature $T = T_b$.

The boundary conditions outside the reaction zone can written as

$$\frac{dT}{dz} = (\Theta - 1)\frac{T_u}{L_f} \text{ for } \frac{T_b - T}{T_b} >> \frac{T_b}{E} \tag{9.3.10}$$

ahead of the reaction zone, and

$$\frac{dT}{dz}(T = T_b) = 0 \tag{9.3.11}$$

behind the reaction zone. In the solution (9.3.8) the eigenvalue U_f is "hidden" in the definition for the flame thickness $L_f = \kappa_b/\rho_u C_p U_f$ in the boundary condition (9.3.10). Multiplying (9.3.9) by the temperature derivative and using the Frank-Kamenetsky approximation for the Arrhenius function in the reaction zone

$$\exp(-E/T) \approx \exp(-E/T_b)\exp\left(E\frac{T - T_b}{T_b^2}\right), \tag{9.3.12}$$

after integration across the reaction zone we obtain the equation for the flame thickness

$$\frac{\kappa_b}{2C_P}\frac{T_u^2(\Theta-1)^2}{L_f^2}=\frac{\rho_u T_u}{\tau_R T_b}\frac{T_b^4}{E^2}\exp(-E/T_b). \tag{9.3.13}$$

Taking into account the definition for the flame thickness, $L_f = \kappa_b/\rho_u C_p U_f$, we obtain from (9.3.13) for the flame velocity

$$U_f=\left(\frac{2\kappa_b}{C_P\rho_u\tau_R}\right)^{1/2}\frac{\sqrt{\Theta}}{(\Theta-1)}\frac{T_b}{E}\exp\left(-\frac{E}{2T_b}\right). \tag{9.3.14}$$

In agreement with the result of Chap. 5 obtained from dimensional analysis, the flame velocity is proportional to the square root of thermal conduction and inversely proportional to the square root of the characteristic reaction time $\tau_b^{1/2} \propto \tau_R^{1/2}\exp(E/2T_b)$:

$$U_f \propto \left(\frac{\kappa_b}{C_P\rho_u\tau_R}\right)^{1/2}\exp\left(-\frac{E}{2T_b}\right), \tag{9.3.15}$$

The Zel'dovich-Frank-Kamenetsky formula (9.3.14) is in a very good agreement with the numerical solution of the eigenvalue problem (9.3.5)–(9.3.6). Particularly, for the flame with the expansion coefficient $\Theta = 8$ and the activation energy $E/T_b = 8.75$ the analytical formula (9.3.14) gives the dimensionless value $\ln(L_f/U_f\tau_R) = 14.2$, while the numerically calculated result is $\ln(L_f/U_f\tau_R) = 14.3$.

In a general case of a reaction of an arbitrary order, n, and Lewis number different from unity, the expression for the velocity of a planar flame becomes

$$U_f=\left(\frac{2n!}{(Le)^n\Theta}\frac{\kappa_b}{C_P\rho_u\tau_R}\right)^{1/2}\left(\frac{\rho_b}{\rho_R}\right)^{\frac{n-1}{2}}\left(\frac{\Theta T_b}{[\Theta-1]E}\right)^{\frac{n+1}{2}}\exp\left(-\frac{E}{2T_b}\right). \tag{9.3.16}$$

The formula (9.3.16) may be derived on the basis of the set of isobaric hydrodynamic equations and the equation of state using the same method as was used for derivation of Eq. (9.3.14).

According to Eq. (9.3.16) velocity of a flame depends on the pressure in a fuel as $U_f \propto P^{(n/2)-1}$: the flame propagates slower in a fuel of higher pressure in the case of a reaction of the first order, while flame velocity increases with pressure for a reaction of the third order. For flames with second order reactions flame velocity is independent of the fuel pressure. Also the flames of higher order reactions are more sensitive to the increase of the initial temperature.

9.4 The Model of a Discontinuous Flame Front

A useful approach, used as a starting point for analytical study, is the model of a discontinuous flame front. The model of a discontinuous flame front is based on the fact that in many scases typical length scales of flame dynamics are much larger than the flame thickness. In this limit a flame front can be treated as a surface of zero thickness separating the fresh fuel and the burned products. The position of the flame front may be chosen as the position of one of the isotherms describing the inner structure of the flame since all isotherms of a flame are parallel to each other with a high accuracy. Dynamics of the flame front within the approach of a discontinuous flame front can be obtained by solving hydrodynamic equations outside the flame in the flows of the fresh fuel and in the flow of the burned gas. Solutions upstream and downstream the flame front should to be matched at the front using the conservation laws. In general the approach of a discontinuous flame front allows taking into account inexplicitly a finite thickness of the flame and transport processes since the flame thickness may enter the conservation laws at the flame front as an external parameter.

Because of the small flame velocities we can use the isobaric approximation and treat the upstream and downstream flows as being incompressible, assuming the constant densities $\rho = \rho_u$ and $\rho = \rho_b$ for the fuel and the burned matter, respectively. A flow of an ideal incompressible gas is described by the equations

$$\nabla \cdot u = 0, \tag{9.4.1}$$

$$\frac{\partial u}{\partial t} + (u \cdot \nabla)u = -\frac{1}{\rho}\nabla P. \tag{9.4.2}$$

Let a flame front be described, for example, by a function $F(\mathbf{x}, t) = z + U_f t = 0$, which means a planar stationary flame front moving in the negative direction. The boundary conditions at the flame front of zero thickness are the conservation laws of the mass flux and two components of the momentum flux

$$\rho\left(\mathbf{n} \cdot \mathbf{u} + \frac{1}{N}\frac{\partial F}{\partial t}\right)\Big|_{-}^{+} = 0, \tag{9.4.3}$$

$$\mathbf{n} \times \mathbf{u}|_{-}^{+} = 0, \tag{9.4.4}$$

$$\left[P + \rho\left(\mathbf{n} \cdot \mathbf{u} + \frac{1}{N}\frac{\partial F}{\partial t}\right)^2\right]\Big|_{-}^{+} = 0, \tag{9.4.5}$$

where $(\ldots)|_{-}^{+}$ represents the change of a quantity across the flame front from the fuel to the burned matter, $\mathbf{n} = \nabla F/|\nabla F|$ is the unit normal vector directed towards the products of burning and $N = |\nabla F|$. The conservation laws have to be complemented by the evolution equation of a flame front that determines the flame velocity with respect to the fuel. In general, the flame velocity depends on the flame shape and thickness, but in the simplest model of a discontinuous front of zero thickness we assume that the flame velocity is a constant considered as an external parameter. Then, the evolution equation can be written as

$$\mathbf{n} \cdot \mathbf{u}_{-} + \frac{1}{N}\frac{\partial F}{\partial t} = U_f, \tag{9.4.6}$$

where $\mathbf{u}_-$ is the fuel velocity just ahead of the flame. The conservation laws (9.4.4), (9.4.5) together with the evolution equation (9.4.6) may be also interpreted as the condition of continuity of streamlines at a flame front.

Unlike the conservation laws (9.4.3)–(9.4.5) the evolution equation (9.4.6) cannot be obtained self-consistently in the scope of the discontinuity model, but has to be added as a supplementary external condition. If we want to obtain the evolution equation rigorously, then we have to consider the continuous inner structure of the flame front and integrate the equation of thermal conduction across the front.

9.5 The Darrieus-Landau Instability of a Thin Flame Front

Flames propagate rather seldom as a planar stationary front. Usually the flame acquires a curved shape and sometimes transition to a turbulent regime of propagation occurs, which is accompanied by considerable amplification of the flame velocity. While the observed transition to the turbulent regime may be accounted for the interaction of the flow with rough tube walls, the curved shape of a flame in tubes with smooth, slip and adiabatic walls requires another explanation. A flame front may become spontaneously curved because of the hydrodynamic instability discovered by Darrieus and Landau (DL instability, Darrieus, 1939; Landau, 1944), which is the main reason of curved shape of the flame observed in many experimental studies.

The DL instability is inherent to all flames in gaseous mixtures since the instability is related to the gas expansion in exothermal reactions, and it appears to be very universal, explaining instability of the ionizing waves, ablation wave in the scheme of a laser fusion etc. In the approach used by Darrieus and Landau a flame front is treated as a surface of zero thickness separating the fresh fuel and the burnt matter and propagating with a constant velocity U_f with respect to the fuel. In this approach the velocity U_f is considered as an external parameter prescribed to the flame front. Since in the reference frame of the planar flame front the unperturbed solution is stationary, we can look for perturbations of a planar front in the form

$$\tilde{F}(x,t) = \tilde{F}(x)\exp(\sigma t + iky), \qquad (9.5.1)$$

where y is the coordinate along the flame front, $k = 2\pi/\lambda$ is the perturbation wave-number, λ is the wavelength of the perturbation, σ is the perturbation growth rate. Our purpose is to find the dispersion relation that is the growth rate as a function of the wave number $\sigma = \sigma(k)$. In the case of an infinitely thin flame front the expansion coefficient is the only dimensionless parameter of the stability problem. Therefore, on the basis of simple dimensional considerations one should expect the instability growth rate to be in the form $\sigma = \Gamma U_f k$ with a numerical factor depending on the expansion coefficient $\Gamma = \Gamma(\Theta)$. In the case of instability the real part of the growth rate is positive $\mathrm{Re}[\sigma] > 0$. In general, the perturbations (9.5.1) should be taken in a three-dimensional (3D) form $\tilde{F}\exp(\sigma t + i\mathbf{k}\cdot\mathbf{x})$, but there is no difference between 2D and 3D configurations for the linear problem.

Let us first consider the flame instability in a way similar to that we considered the Rayleigh-Taylor instability of incompressible liquid. First, in obtaining the dispersion relation for the growth rate of the flame instabilities we will be following the L. Landau method. We consider stability of a surface of the flame front, which is at rest in the co-moving co-ordinate system, separating the flow of the unburned gas upstream (u_1) and of the burned gas downstream (u_2). Since the flow is essentially subsonic, the flow can be treated as incompressible and potential on both sides from the flame front. The equations for small perturbations of the velocity, $u' << u$ and pressure, $P' << P$ are:

$$\nabla\cdot\mathbf{u}' = 0, \quad \frac{\partial \mathbf{u}'}{\partial t} + (\mathbf{u}\nabla)\mathbf{u}' = -\frac{1}{\rho}\nabla P'. \tag{9.5.2}$$

Let the x-axis is directed along the normal to the surface of the flame. Taking into account that the flow velocity is constant from the both sides of the flame front, and applying the operator ∇ to both sides of the second equation in (9.5.2), we obtain

$$\nabla^2 P' = \Delta P' = \frac{\partial^2 P'}{\partial x^2} + \frac{\partial^2 P'}{\partial y^2} = 0, \tag{9.5.3}$$

Substituting to (9.5.3) $P_1' = f(x)\exp(-i\omega t + iky)$ we obtain solution for f(x)

$$f(x) = Ae^{kx} + Be^{-kx}. \tag{9.5.4}$$

Upstream at $x<0,\ u = u_1,\ \rho = \rho_1$ we must take $B = 0$, so that at $x<0$

$$P_1' = Ae^{-i\omega t + iky + kx}. \tag{9.5.5}$$

Substituting (9.5.5) into (9.5.2–3) we obtain

$$-i\omega u_{1x}' + ku_1u_{1x}' = -A\frac{k}{\rho_1}e^{-i\omega t + ky + kx},\quad u_{1y}' = iu_{1x}'. \tag{9.5.6}$$

Thus

$$u_{x1}' = A\frac{k}{\rho_1(ku_1 - i\omega)}e^{-i\omega t + kx + iky}. \tag{9.5.7}$$

Since we are interested in unstable solutions when $\mathrm{Im}\{\omega\} > 0$, we introduce $\sigma \equiv -i\omega$ in follows. The unstable solutions then correspond to real positive σ.

There is another solution of the Euler equation (9.5.2) corresponding to the uniform equation with $P' = 0$

$$\frac{\partial u'_{1x}}{\partial t} + u_1 \frac{\partial u'_{1x}}{\partial x} = 0. \tag{9.5.8}$$

However, the solution of (9.5.8) $u'_{1x} = C\,e^{\sigma t + iky - \frac{\sigma}{u_1}x}$ is unlimited for real positive σ at $x \to -\infty$, so that $C = 0$.

Thus, upstream at $x<0$ we have

$$P'_1 = Ae^{-i\omega t + iky + kx}, \tag{9.5.9}$$

$$u'_{x1} = -A\frac{k}{\rho_1(\sigma + ku_1)}e^{\sigma t + kx + iky}, \tag{9.5.10}$$

$$u'_{x1} = -A\frac{ik}{\rho_1(\sigma + ku_1)}e^{\sigma t + kx + iky}. \tag{9.5.11}$$

In a similar way, downstream, at $x>0$, we obtain solutions

$$P'_2 = Be^{-i\omega t + iky - kx}, \tag{9.5.12}$$

$$u'_{x2} = \left(B\frac{k}{\rho_2(\sigma - ku_2)}e^{-kx} + Ce^{-\frac{\sigma}{u_2}x}\right)e^{\sigma t + iky}, \tag{9.5.13}$$

$$u'_{y2} = \left(B\frac{-ik}{\rho_2(\sigma - ku_2)}e^{-kx} + C\frac{i\sigma}{ku_2}e^{-\frac{\sigma}{u_2}x}\right)e^{\sigma t + iky}, \tag{9.5.14}$$

where solution of the uniform equation of type (9.5.8) was include in (9.5.13–14).

Let $\zeta = \zeta(y,t) = D\exp(\sigma t + iky)$ be a displacement of the flame surface caused by small perturbations. At the surface of the flame front considered as a discontinuity surface must be fulfilled the following conditions at $x = 0$. Pressure is continues, i.e.

$$P'_1(0) = P'_2(0). \tag{9.5.15}$$

The condition that the tangential velocity component is continuous reads

$$u'_{y1} + u_1\frac{\partial \zeta}{\partial y} = u'_{y2} + u_2\frac{\partial \zeta}{\partial y}. \tag{9.5.16}$$

Another condition for perturbations is that the flame velocity in the direction normal to the flame surface remains unchanged

$$u'_{x1} - \frac{\partial \zeta}{\partial t} = u'_{x2} - \frac{\partial \zeta}{\partial t} = 0. \tag{9.5.17}$$

Substituting solutions (9.5.9)–(9.5.11) and (9.5.12)–(9.5.14) into the boundary conditions (9.5.15)–(9.5.17) at $x = 0$, we obtain a system of four uniform equations for the coefficients A, B, C, D. With the use of equation $\rho_1 u_1 = \rho_2 u_2$, one can obtain

$$A = B = -\frac{\sigma}{k}\rho_1(\sigma + ku_1)D, \tag{9.5.18}$$

$$C = \sigma^2 \frac{u_1 + u_2}{u_1(\sigma - ku_2)} D. \tag{9.5.19}$$

The last two equations after simple algebra are reduced to

$$\sigma^2(u_1 + u_2) + 2\sigma u_1 u_2 + k^2 u_1 u_2 (u_1 - u_2) = 0. \tag{9.5.20}$$

In the flame $\rho_1 > \rho_2$, and since from the continuity equation, $\rho_1 u_1 = \rho_2 u_2$, and $u_2 > u_1$, both roots of the quadratic equation (9.5.20) are real. Introducing $\Theta = \rho_1/\rho_2 = u_2/u_1$, we obtain

$$\sigma = \pm k u_1 \frac{\Theta}{\Theta + 1}\left\{\sqrt{\Theta + 1 - 1/\Theta} - 1\right\}. \tag{9.5.21}$$

Existence of the positive root in (9.5.21) means instability of the flame front. Since ω is pure imaginary and has no real part, the perturbations in the flame growth locally, which means that they are not travel along the flame front. Formally the obtained solution means that the flame front is absolutely unstable against perturbations of any wavelength: as larger wave number (shorter the wave length) as larger σ is and as faster perturbations growth.

At the beginning of this section it was said that from dimensional analysis the instability growth rate should be $\sigma = \Gamma(\Theta)U_f k$. From the Eq. (9.5.21) we found the dependence on the expansion coefficient, $\Gamma(\Theta) = \frac{\Theta}{\Theta+1}\left\{\sqrt{\Theta + 1 - 1/\Theta} - 1\right\}$, which shows that there is no instability if there is no gas expansion across the flame front, $\Theta = 1$, and the growth rate only slightly changes for real flames: from $\Gamma = 1.437$ for $\Theta = 6$, till $\Gamma = 2.092$ for $\Theta = 10$. An important conclusion is that this kind of hydrodynamic instability of a wave front is universal and is due to a gas expansion across the front. It should be noticed that in the derivation of the instability we did not specify that the wave is combustion one.

It is obvious that the flow upstream, incoming to the flame front, is potential. It steams from the fact that the flow is potential far away from the flame front, and according to the Kelvin theorem of conservation of circulation it remains potential everywhere at $x < 0$. Using formulas (9.5.10), (9.5.11), (9.5.13), (9.5.14) and (9.5.18), (9.5.19) we obtain for the velocity components

$$u'_{x1} = \sigma D e^{-i\omega t + kx + iky}, \tag{9.5.22}$$

$$u'_{y1} = i\sigma D e^{-i\omega t + kx + iky}, \tag{9.5.23}$$

$$u'_{x2} = \left(\frac{\sigma^2(u_1 + u_2)}{u_1(\sigma - ku_2)} e^{-\frac{\sigma}{u_2}x} - \frac{\sigma u_2(\sigma + ku_1)}{u_1(\sigma - ku_2)} e^{-kx}\right) D e^{\sigma t + iky}, \tag{9.5.24}$$

$$\omega_2 = i\frac{\sigma^2(u_1 + u_2)u_1}{ku_2^2}(\sigma + ku_2)D\,e^{\sigma t - \frac{\sigma}{u_2}x + iky} \neq 0 \tag{9.5.25}$$

Then, it is easy to calculate vorticity $\boldsymbol{\omega} = \nabla \times \mathbf{u}$ upstream and downstream. Ahead of the flame front we obtain

$$\omega_1 = \frac{\partial u'_{y1}}{\partial x} - \frac{\partial u'_{x1}}{\partial y} = 0, \tag{9.5.26}$$

as it was expected. However behind the flame, at $x > 0$, at the combustion products

$$u'_{y2} = i\frac{\sigma^3(u_1 + u_2)}{u_1 u_2^2(\sigma + ku_2)}De^{\sigma t - \frac{\sigma}{u_2}x + iky} \neq 0. \tag{9.5.27}$$

This means that the instability of a flame front, its bending and corrugation, generate vorticities in the combustion products behind the flame front. Thus, the flow behind the flame will be turbulized even if turbulization caused by the wall friction in the case of smooth walls is weak.

9.6 The Linear Theory of Instability of the Thin Flame Front

In view of the fundamental importance of the hydrodynamic flame instability and its influence on flame propagation and combustion regimes, we will consider another approach to the problem. In the previous section solving the problem we used the condition (9.5.17), which means an assumption that the flame velocity in the direction normal to the flame surface remains unchanged. It must be noticed that condition (9.5.17) does not steam from the system of the hydrodynamic equations, and must be considered as an "assumption", which as it will be shown below is correct.

Since the problem is linear, the perturbations of all hydrodynamic values (which we will denote as $\tilde{\rho}$, $\tilde{u}_x$, $\tilde{u}_y$, $\tilde{P}$) are proportional to the perturbations of the flame front in the form (9.5.1). Small perturbations ahead of the flame and behind the flame are satisfied the linearized equations of an ideal non-viscous gas

$$\frac{d}{dx}\tilde{u}_x + ik\tilde{u}_y = 0, \tag{9.6.1}$$

$$\rho\sigma\tilde{u}_y + \rho u_x\frac{d}{dx}\tilde{u}_y + ik\tilde{P} = 0, \tag{9.6.2}$$

$$\rho\sigma\tilde{u}_x + \rho u_x\frac{d}{dx}\tilde{u}_x + \frac{d}{dx}\tilde{P} = 0, \tag{9.6.3}$$

where $\rho = \rho_f$, $u_x = U_f$ in the upstream flow, $\rho = \rho_f/\Theta$, $u_x = \Theta U_f$ in the downstream flow of the burned matter. The density is not perturbed in the upstream and downstream flows, so that we can put $\tilde{\rho} = 0$. If the solutions in the upstream and the downstream are obtained, then they have to be matched at the flame front using the conservation laws of the mass and momentum fluxes. The linearized conservation laws (9.4.3)–(9.4.5) at the flame front can be written as

$$(\rho\tilde{u}_x - \sigma\rho\tilde{F})\Big|_-^+ = 0, \tag{9.6.4}$$

$$(\tilde{u}_y + iku_x\tilde{F})\Big|_-^+ = 0, \tag{9.6.5}$$

$$(\tilde{P} + 2\rho u_x\tilde{u}_x)\Big|_-^+ = 0. \tag{9.6.6}$$

The linearized condition of a constant flame velocity with respect to the fuel, which is the linearized evolution equation (9.4.6), takes the form

$$\tilde{u}_{x-} = \sigma\tilde{F}. \tag{9.6.7}$$

The linearized equations of an ideal gas flow (9.6.1)–(9.6.3) must be also supplemented by the boundary conditions at infinity, far ahead of the flame front at $x = -\infty$, and far behind the flame, at $x = \infty$. The boundary conditions are the condition that solution of the problem is regular at infinity, so that all perturbations should vanish at infinity. This means that solution of Eqs. (9.6.1–3) should be in the form, $\tilde{u}_x, \tilde{u}_y, \tilde{P} \propto \exp(\mu x)$, with $\mathrm{Re}[\mu] > 0$ in the upstream flow and $\mathrm{Re}[\mu] < 0$ in the downstream flow. Substituting perturbations in this form into Eqs. (9.6.1–3) we find that solution upstream the flame ($x < 0$) consists only of a potential mode $\tilde{u}_x \propto \exp(kx)$, and the following relations hold ahead of the flame front

$$i\tilde{u}_{y-} = -\tilde{u}_{x-}, \tag{9.6.8}$$

$$\tilde{P}_- = -\rho_f U_f\left(1 + \frac{\sigma}{U_f k}\right)\tilde{u}_{x-}. \tag{9.6.9}$$

In the downstream flow ($x > 0$) two modes are possible: another potential mode $\tilde{u}_{xp} \propto \exp(-kx)$ satisfying the relations

$$i\tilde{u}_{yp} = \tilde{u}_{xp}, \tag{9.6.10}$$

$$\tilde{P}_p = -\rho_f U_f\left(1 - \frac{\sigma}{U_f k}\right)\tilde{u}_{xp}, \tag{9.6.11}$$

and a vortex mode $\tilde{u}_{xp} \propto \exp\left(-\frac{\sigma}{\Theta U_f}x\right)$, for which

$$i\tilde{u}_{yv} = \frac{\sigma}{\Theta U_f k}\tilde{u}_{xv}, \quad \tilde{P} = 0. \tag{9.6.12}$$

Obviously, the vortex mode is related to the drift of vorticity, produced by the curved flame. There is no vortex mode upstream since we assumed that there are no external sources of perturbations except the flame front itself. Therefore the fuel flow being irrotational at infinity pertains this property according to the Thomson circulation theorem until it passes through the perturbed flame. Unlike the vortex perturbations the potential modes are related to the degenerate sound perturbations for the case of an infinitely large sound speed, therefore the potential modes take place both upstream and downstream the flame front. Taking into account the structures of the potential mode in the burned matter and the vortex mode one obtains that perturbations in the downstream flow satisfy the equation

$$k\frac{\tilde{P}_+}{\rho_f U_f^2} + ik\tilde{u}_{y+} - \frac{\sigma}{\Theta U_f}\tilde{u}_{x+} = 0. \tag{9.6.13}$$

Relations in the upstream flow (9.6.8–9), in the downstream flow (9.6.13) and the boundary conditions at the perturbed flame front (9.6.4)–(9.6.6) constitute the complete algebraic set of equations, which allows one to determine the dispersion relation. Resolving the set of algebraic equations we obtain the dispersion relation

$$\sigma^2 + \frac{2\Theta}{\Theta + 1}U_f k\sigma - \Theta\frac{\Theta - 1}{\Theta + 1}U_f^2 k^2 = 0 \tag{9.6.14}$$

and the growth rate of the perturbations

$$\frac{\sigma}{U_f k} = \Gamma(\Theta) = \frac{\Theta}{\Theta + 1}\left(\sqrt{\Theta + 1 - 1/\Theta} - 1\right). \tag{9.6.15}$$

Equation (9.6.15) shows that the growth rate is real and positive for all expansion coefficients, $\Theta > 1$, i.e. for all flames in gaseous fuel mixtures. The factor $\Gamma(\Theta)$ increases with the increase of the expansion coefficient Θ, so that the larger the ratio of the fuel density and the density of the burned matter, the stronger the DL instability. If the density of the burning matter changes only slightly, $\Theta - 1 << 1$, then the instability growth rate becomes as small as $\sigma/U_f k = \frac{\Theta - 1}{2}$. Though the formulas (9.6.14–15) have been obtained in a 2D configuration, they also hold for a three dimensional geometry with the only replacement of k by the absolute value of the wave number vector $|\mathbf{k}|$. Thus, small perturbations imposed at an infinitely thin flame front grow exponentially in time, bending the flame front with the instability growth rate given by (9.6.15). This hydrodynamic flame instability is known as the DL instability.

The DL theory predicts growth of all perturbations: the shorter the perturbation wavelength, the faster it grows. Since there is no characteristic length scale in scope of the DL theory, the limit for the growth of instability must be imposed on the solution indirectly. Since the solution has been obtained assuming a zero flame thickness, then it holds only for perturbations of a long wavelength $\lambda >> L_f$ (or $kL_f >> 1$). Therefore on the basis of the DL theory one should expect that the DL instability results in 2D and 3D structures at a flame front with a typical length scale in the range of the flame thickness L_f. However experimental studies show that the characteristic length scale of the instability is about two orders of magnitude larger than the flame thickness, which cannot be explained by the DL solution.

9.7 Thermal Stabilization of the DL Instability

In the previous sections a flame front was treated as a surface of zero thickness separating the fresh fuel and the burnt matter and propagating with a constant velocity U_f with respect to the fuel. In this approach the velocity U_f is considered as an external parameter prescribed to the flame front and it was assumed that instability of a flame does not change velocity of the flame in the direction normal to the flame surface. Self-consistent approach to the problem must include the processes, which define velocity of the flame itself. As we discussed, these are thermal conduction and diffusion.

For the sake of simplicity, we consider a flame propagating in non-viscous fuel with a constant coefficient of thermal conduction, unit Lewis number and zero gravitational acceleration. We shall consider the flow, which consists of the hydrodynamic zones of a uniform flow upstream and downstream and the flame front itself composed of the heating zone and the reaction zone. In this approach we may treat the DL solution in the uniform flows of the fuel and the burned matter as the solution in the hydrodynamic zone, while the other zones have been replaced by the discontinuous flame front. In the hydrodynamic zone all variables change on some scale R of about the perturbation wavelength $R \propto \lambda = 2\pi/k$. If, for example, we consider flame in a tube, then the tube diameter gives the typical hydrodynamic length scale R. The typical length scale of the heating zone is of the order of the flame thickness L_f, and while of the reaction zone it is $L_R << L_f$.

Since the solutions in the hydrodynamic zone given by Eqs. (9.6.8–9) and (9.6.13) are known, then in order to solve the problem of flame stability for the case of a small but finite flame thickness we need to integrate the linearized hydrodynamic equations (9.1.1–2)–(9.2.6–7) across the flame front. The linearized hydrodynamic equations are

$$\sigma\tilde{\rho} + \frac{d}{dz}(\tilde{\rho}u_x + \rho\tilde{u}_x) + ik\rho\tilde{u}_y = 0, \tag{9.7.1}$$

$$\sigma\rho\tilde{u}_y + \rho u_x \frac{d}{dz}\tilde{u}_y = -ik\tilde{P}, \tag{9.7.2}$$

$$\sigma\rho\tilde{u}_x + \sigma v_x\tilde{\rho} + ik\rho u_x\tilde{u}_y + \frac{d}{dz}(\tilde{P} + 2\rho u_x\tilde{u}_x + u_x^2\tilde{\rho}) = 0, \tag{9.7.3}$$

$$\frac{\kappa}{C_p}\frac{d^2\tilde{T}}{dx^2} - \rho_f U_f \frac{d\tilde{T}}{dx} + \frac{dQ_R}{dT}\tilde{T} = (\rho\tilde{u}_x + \tilde{\rho}u_x)\frac{dT}{dx} + \rho\sigma\tilde{T} + \frac{\kappa}{C_p}k^2\tilde{T}, \qquad (9.7.4)$$

where perturbations of density are related to temperature perturbations according to the isobaric equation of state

$$\tilde{\rho} = -\rho_f \frac{T_f}{T^2}\tilde{T}. \qquad (9.7.5)$$

It is convenient to introduce the dimensionless variables $\zeta = x/L_f$, $\theta = T/T_f$, $\tilde{m} = \frac{\rho\tilde{u}_x + \tilde{\rho}u_x}{\rho_f U_f}$, $\tilde{w} = i\tilde{u}_y/U_f$, $\tilde{p} = \frac{\tilde{P} + 2\rho u_x \tilde{u}_x + \tilde{\rho}u_x^2}{\rho_f U_x^2}$, and the scaled parameters $K = kL_f$, $S = \frac{\sigma}{U_f k}$. The values, $\tilde{m}, \tilde{w}, \tilde{p}$ have the meaning of perturbations of the mass flux and two components of the momentum flux. In the dimensionless variables the equation of thermal conduction for the stationary flame becomes

$$\frac{d^2\theta}{d\zeta^2} - \frac{d\theta}{d\zeta} - \Theta_R(\theta) = 0, \qquad (9.7.6)$$

the Zel'dovich – Frank-Kamenetsky solution for the temperature distribution is reduced to

$$\theta = \begin{cases} 1 + (\Theta - 1)\exp\zeta, & \zeta < 0 \\ \Theta, & \zeta > 0 \end{cases} \qquad (9.7.7)$$

and the linearized equations take the form

$$\frac{d\tilde{m}}{d\zeta} = SK\frac{\tilde{\theta}}{\theta^2} - K\frac{\tilde{w}}{\theta}, \qquad (9.7.8)$$

$$\frac{d\tilde{w}}{d\zeta} = -SK\frac{\tilde{w}}{\theta} + K\tilde{p} - 2K\theta\tilde{m} - K\tilde{\theta}, \qquad (9.7.9)$$

$$\frac{d\tilde{p}}{d\zeta} = -SK\tilde{m} - K\tilde{w}, \qquad (9.7.10)$$

$$\frac{d^2\tilde{\theta}}{d\zeta^2} - \frac{d\tilde{\theta}}{d\zeta} - \frac{d\Theta_R}{d\theta}\tilde{\theta} = \tilde{m}\frac{d\theta}{d\zeta} + SK\frac{\tilde{\theta}}{\theta} + K^2\tilde{\theta}, \qquad (9.7.11)$$

where $\Theta_R = \Theta_R(\theta)$ stands for the scaled reaction terms similar to $Q_R = Q_R(T)$. The boundary conditions for the linearized equations (9.7.10)–(9.7.11) are that all perturbations vanish at infinity, which may be written in the form of perturbation modes in the uniform flows far ahead of the flame front and far behind the flame front. In addition to the three modes (9.6.9)–(9.6.13) the set of Eqs. (9.7.8)–(9.7.11) has two more independent modes related to the equation of thermal conduction (9.7.11). The thermal conduction modes go to zero on small length scales of about the flame thickness or less (less than unity in the dimensionless variables), therefore in the hydrodynamic zone these modes correspond to the condition $\tilde{\theta} = 0$. The last condition written in the dimensional variables gives $\tilde{p} = 0$. The other boundary conditions in the hydrodynamic zone for (9.6.9)–(9.6.13) in the dimensionless variables become

$$\tilde{w}_{-} = -\tilde{m}_{-}, \tilde{p}_{-} = -(S-1)\tilde{m}_{-} \tag{9.7.12}$$

in the upstream flow and

$$\tilde{w}_{+} + \tilde{p}_{+} - (S + 2\Theta)\tilde{m}_{+} = 0 \tag{9.7.13}$$

in the downstream flow.

We shall consider the solution to the Eqs. (9.7.8)–(9.7.11) in the long-wavelength approximation, $K << 1$. To do this we first write down (9.7.11) in the form

$$\hat{\Theta}_R\left[\tilde{\theta}\right] = \tilde{m}\frac{d\theta}{d\zeta} + KS\frac{\tilde{\theta}}{\theta} + K^2\tilde{\theta}, \tag{9.7.14}$$

where the operator $\hat{\Theta}_R$ is defined as

$$\hat{\Theta}_R\left[\tilde{\theta}\right] \equiv \frac{d^2\tilde{\theta}}{d\zeta^2} - \frac{d\tilde{\theta}}{d\zeta} - \frac{d\Theta_R}{d\theta}\tilde{\theta}, \tag{9.7.15}$$

so that $\hat{\Theta}_R\left[\frac{d\tilde{\theta}}{d\zeta}\right] = 0$.

It can be shown that the first and second terms in the right-hand side of (9.7.14) are of the same order, i.e. in the limit $K << 1$ the solution of (9.7.14) in zero order of K is

$$\tilde{\theta} = \zeta_T\frac{d\theta}{d\zeta}, \tag{9.7.16}$$

which means that the flame is shifted in the ζ direction on the scaled distance ζ_T. Using formula (9.7.16) for the perturbed temperature, we obtain the integrals of Eqs. (9.7.8–10) with the accuracy of the first order terms in $K << 1$:

$$\tilde{m} - \tilde{m}_{-} = KS\zeta_T\frac{\theta - 1}{\theta}, \tag{9.7.17}$$

$$\tilde{w} - \tilde{w}_{-} = -K\zeta_T(\theta - 1), \tag{9.7.18}$$

$$\tilde{p} = \tilde{p}_{-}. \tag{9.7.19}$$

Obviously, when related to the downstream the integrals (9.7.16)–(9.7.19) coincide with the linearized conservation laws (9.6.4–6) with $\tilde{F} = -\zeta_T L_f$.

To obtain the supplementary boundary condition on the perturbed flame velocity with the same accuracy we have to obtain the relation between the perturbed mass flux $\tilde{m}_{-}$ and the shift ζ_T. Substituting $\tilde{m}$ in the right-hand side of (9.7.14) and taking into account the terms of the first order in K, we find

$$\hat{\Theta}_R\left[\tilde{\theta}\right] = (KS + \tilde{m}_-/\zeta_T)\tilde{\theta}. \qquad (9.7.20)$$

The function $\tilde{\theta} = \zeta_T \frac{d\theta}{d\zeta}$ is the eigenfunction of Eq. (9.7.20) for the eigenvalue $KS + \tilde{m}_-/\zeta_T = 0$; it is the unique eigenvalue for which instability can develop ($Re[\sigma] > 0$). This gives the supplementary condition

$$KS\zeta_T + \tilde{m}_- = 0. \qquad (9.7.21)$$

When written for the dimensional variables the supplementary condition (9.7.21) gives the condition of constant flame velocity with respect to the fuel, which was assumed in the simplified model in previous sections.

The integrals (9.7.17)–(9.7.19) related to the downstream together with the boundary conditions (9.7.12–13) and the supplementary condition (9.7.21) yield the dispersion relation for the perturbation growth rate within the accuracy of the first order terms in the expansion in K

$$S = \Gamma(\Theta) = \frac{\Theta}{\Theta + 1}\left(\sqrt{\Theta + 1 - 1/\Theta} - 1\right). \qquad (9.7.22)$$

Thus, the first order term in the expansion in K corresponds to the DL solution of the flame stability problem. This is the first order term in the long wavelength approximation.

The second order term in the expansion in K describes the thermal stabilisation of the DL instability. To do this we substitute the first order solution Eqs. (9.7.17)–(9.7.19) into original set of Eqs. (9.7.8–10) and perform the straightforward integration

$$\tilde{m} = \tilde{m}_- + KS\zeta_T \frac{\theta - 1}{\theta} - K \int_{\zeta_-}^{\zeta} \frac{\tilde{w}_0}{\theta} dy, \qquad (9.7.23)$$

$$\tilde{w} = \tilde{w}_- - K\zeta_T(\theta - 1) - KS \int_{\zeta_-}^{\zeta} \frac{\tilde{w}_0}{\theta} dy + K \int_{\zeta_-}^{\zeta} \tilde{p}_0 dy - 2K \int_{\zeta_-}^{\zeta} \theta\tilde{m}_0 dy \qquad (9.7.24)$$

$$\tilde{p} = -(S - 1)\tilde{m}_- - KS \int_{\zeta_-}^{\zeta} \tilde{m}_0 dy - K \int_{\zeta_-}^{\zeta} \tilde{w}_0 dy, \qquad (9.7.25)$$

where ζ_- is some point in the hydrodynamic zone ahead of the flame front, $\tilde{m}_0$, $\tilde{w}_0$, $\tilde{p}_0$ are the first order terms in the expansion of $\tilde{m}$, $\tilde{w}$, $\tilde{p}$ in powers of K, defined by (9.7.17)–(9.7.19), and the boundary conditions (9.7.12) are taken into account. If the temperature distribution for the planar flame is known $\theta = \theta(\zeta)$, then the integrals (9.7.23)–(9.7.25) provide the linearized conservation laws at the perturbed flame front of finite thickness. The supplementary condition for a flame of finite thickness is calculated as an eigenvalue of the linearized equation of thermal conduction similar to the problem of planar flame propagation. Substituting $\tilde{m}$ from (9.7.23) into the right-hand side of (9.7.24) we find

$$\hat{\Theta}_R[\tilde{\theta}] = \left[(1 - K\zeta_-)\frac{m_-}{\zeta_T} + KS\right]\tilde{\theta} + K^2H(\zeta)\tilde{\theta}, \tag{9.7.26}$$

where we introduced $H(\zeta) = 1 - S\zeta - (S+1)J_1(\zeta)$ and the integral

$$J_1(\zeta) = \int_{\zeta-}^{\zeta} (1-\theta)/\theta \, dy, \tag{9.7.27}$$

Equation (9.7.26) can be reduced to an equation with a Hermitian operator by means of the substitution $\tilde{\theta} = \tilde{\theta}_H \exp(\zeta/2)$. If we treat the term $K^2H(\zeta)\tilde{\theta}$ as a small correction and use the standard techniques of perturbation theory, then the supplementary boundary condition to within terms of order K^2 can be written as

$$(1 - K\zeta_-)\tilde{m}_- + (KS + K^2\langle H\rangle)\zeta_T = 0, \tag{9.7.28}$$

where

$$\langle H\rangle = \frac{\int_{-\infty}^{\infty} H(\zeta)\left(\frac{d\theta}{d\zeta}\right)^2 \exp(-\zeta)d\zeta}{\int_{-\infty}^{\infty} \left(\frac{d\theta}{d\zeta}\right)^2 \exp(-\zeta)d\zeta}. \tag{9.7.29}$$

Substituting the integrals across the flame front, Eqs. (9.7.23)–(9.7.25), into the conditions in the hydrodynamic zone, (9.7.12–13) and taking into account the supplementary condition (9.7.28) one can find the dispersion relation of the DL instability for a flame of finite thickness, provided that the temperature distribution for the planar stationary flame $\theta = \theta(\zeta)$ is known. With the Zel'dovich-Frank-Kamenetsky temperature distribution, the dispersion relation of the DL instability for the case of a flame with the unit Lewis number and a constant coefficient of thermal conduction is

$$S^2 + \frac{2\Theta}{\Theta+1}\left(1 + \frac{\Theta\ln\Theta}{\Theta-1}K\right)S - \Theta\frac{\Theta-1}{\Theta+1}K^2 \left[1 - K\left(1 + \Theta\ln\Theta\frac{\Theta+1}{(\Theta-1)^2}\right)\right] = 0. \tag{9.7.30}$$

In the case of a flame front of zero thickness the obtained dispersion relation goes over to the DL dispersion relation (9.6.15), and in the dimensional variables the dispersion relation takes the form

$$\sigma^2 + \frac{2\Theta}{\Theta+1}\left(1 + \frac{\Theta\ln\Theta}{\Theta-1}kL_f\right)U_f k\sigma - \\ -\Theta\frac{\Theta-1}{\Theta+1}U_f^2k^2\left[1 - kL_f\left(1 + \Theta\ln\Theta\frac{\Theta+1}{(\Theta-1)^2}\right)\right] = 0. \tag{9.7.31}$$

From here, the rate of the growing mode of instability is

$$\sigma = \Gamma U_f k\left(1 - \frac{k\lambda_c}{2\pi}\right), \tag{9.7.32}$$

where Γ is the coefficient found in the theory of the thin flame instability (9.6.15), and λ_c is the cut-off wavelength for which the instability is suppressed by thermal conduction. For the case of unit Lewis number and the constant thermal conduction coefficient the cut-off wavelength is given by the formula

$$\lambda_c = \frac{2\pi L_f}{(\Theta-1)}\left(\Theta - 1 + \Theta\ln\Theta\frac{(\Theta+1)}{(\Theta-1)}\right). \tag{9.7.33}$$

For the expansion coefficients typical for laboratory flames, $\Theta = 8/10$, the cut-off wavelength is $\lambda_c \approx 20L_f$, and it depends slightly on thermal expansion. The wavelength of the fastest growing perturbations is approximately twice larger than the cut-off wavelength, $\lambda = 2\lambda_c \approx 40L_f$, and much greater than the flame thickness. In the case of small expansion coefficients, $\Theta - 1 << 1$, we come to the asymptotic value of the cut-off wavelength, $\lambda_c = 4\pi L_f/(\Theta - 1) \to \infty$ for $\Theta \to 1$.

The numerical solution for the scaled growth rates of the DL instability, $\sigma L_f/U_f$ versus the scaled wave number, kL_f for the expansion coefficients $\Theta = 8, 6, 3$ is depicted in Fig. 9.3.

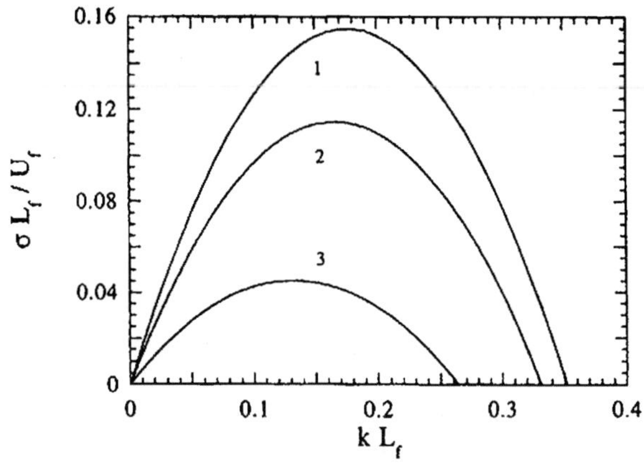

Fig. 9.3 The scaled growth rates of the LD instability versus the scaled wave number for the expansion coefficients $\Theta = 8, 6, 3$ shown by curves 1,2,3 respectively

9.8 Flame Instability in a Gravitational Field

If a flame propagates upwards in a gravitational field, then in addition to the DL instability there are conditions for development of the Rayleigh-Taylor (RT) instability. Because of the RT instability the interface between the two gases bends and the heavy matter becomes mixed with the light matter. In the case of the upward propagating flame the unburned fuel plays the role of the heavy gas, while the burned products act like a light gas, and DL instability is amplified by the RT instability. Contrary to this, if a flame propagates downward, then the DL instability is weakened and can be totally suppressed.

Let us consider the instability growth rate of an infinitely thin flame propagating upwards in a gravitational field. The problem is similar to that in Sect. 9.6 including the linearized equations of an ideal fluid. The only difference is related to the linearized equation of the conservation law for the normal component of the momentum flux, Eq. (9.6.6), which now becomes

$$\left(\tilde{P} + 2\rho u_x \tilde{u}_x - \rho g \tilde{F}\right)\Big|_-^+ = 0. \tag{9.8.1}$$

Since the linearized equations of an ideal incompressible gas do not change in the presence of gravity, then the relations in the uniform flows ahead of the flame (9.6.7–9) and behind the flame (9.6.13) remain the same. Resolving the relations in the uniform flows together with the new boundary conditions at the flame front we obtain the dispersion relation

$$\sigma^2 + \frac{2\Theta}{\Theta+1}U_f k\sigma - \Theta\frac{\Theta-1}{\Theta+1}U_f^2 k^2 - \frac{\Theta-1}{\Theta+1}gk = 0 \tag{9.8.2}$$

with the instability growth rate of an infinitely thin flame propagating upwards

$$\sigma_0 = \sqrt{\frac{\Theta-1}{\Theta+1}gk + \frac{\Theta^2+\Theta-1}{(\Theta+1)^2}\Theta U_f^2 k^2} - \frac{\Theta}{\Theta+1}U_f k. \tag{9.8.3}$$

The obtained dispersion relation coincides with the dispersion relation for the DL instability if the gravity acceleration is equal to zero. In the opposite case of a large acceleration, $U_f^2 k/g << 1$, or for perturbations of a very long wavelength ($k \to 0$) the instability growth rate goes over to the expression for the RT instability

$$\sigma_{RT} = \sqrt{\frac{\Theta-1}{\Theta+1}gk}. \tag{9.8.4}$$

It is seen from (9.8.3) that the resulting flame instability in a gravitational field is stronger than both the DL instability and the RT instability.

To find the dispersion relation of instability of a flame of finite thickness in a gravitational field, one has to solve the set of linearized hydrodynamic equations similar to Sect. 9.7 with the one extra term in Eq. (9.7.3) related to the gravitational field

$$\sigma\rho\tilde{u}_x + \sigma v_x\tilde{\rho} + ik\rho u_x\tilde{u}_y + \frac{d}{dz}\left(\tilde{P} + 2\rho u_x\tilde{u}_x + u_x^2\tilde{\rho}\right) = g\tilde{\rho}. \tag{9.8.5}$$

Respective changes should be also made in the corresponding dimensionless equation

$$\frac{d\tilde{p}}{d\zeta} = -SK\tilde{m} - K\tilde{w} - G\frac{\tilde{\theta}}{\theta^2}, \tag{9.8.6}$$

where the dimensionless number $G = gL_f/U_f^2$ (the inverse Froude number) shows the relative role of the gravitational field. The integration of (9.8.6) within the first order terms in K leads to the variations of the perturbed normal momentum flux inside the flame

$$\tilde{p} - \tilde{p}_- = G\zeta_T\frac{\theta - 1}{\theta}. \tag{9.8.7}$$

Integration similar to that performed in the previous section leads to the dispersion relation

$$\sigma^2 + \frac{2\Theta}{\Theta+1}\left(1 + \frac{\Theta\ln\Theta}{\Theta-1}kL_f\right)U_fk\sigma$$
$$- \Theta\frac{\Theta-1}{\Theta+1}U_f^2k^2\left[1 - kL_f\left(1 + \Theta\ln\Theta\frac{\Theta+1}{(\Theta-1)^2}\right)\right] - \frac{\Theta-1}{\Theta+1}gk = 0. \tag{9.8.8}$$

with the instability growth rate

$$\sigma = \sigma_0 - \frac{kL_f}{2}\frac{\Theta U_f k}{(\Theta+1)\sigma_0 + \Theta U_f k}$$
$$\left[\frac{2\Theta\ln\Theta}{\Theta-1}\sigma_0 + \left(\frac{\Theta+1}{\Theta-1}\Theta\ln\Theta + \Theta - 1\right)U_f k\right], \tag{9.8.9}$$

where σ_0 is the instability growth rate of an infinitely thin flame.

The dispersion relation (9.8.8) describes the RT instability of an upward propagating flame of finite thickness for the case of a unit Lewis number and constant coefficient of thermal conduction. The growth rates of the flame instability given by formula (9.8.9) and obtained from numerical solution of the stability problem are depicted in Fig. 9.4. As one can see, the larger is gravitational acceleration the larger is the maximal growth rate, while the cut-off wavenumber λ_{RT} shifts to the region of shorter wavelength. At the same time the maximal growth rate is shifted towards the long wavelength perturbations in comparison with $\lambda_{RT}/2$.

In the opposite case of a downward propagating flame the gravity plays stabilizing role. Unlike the thermal stabilization inherent to the short

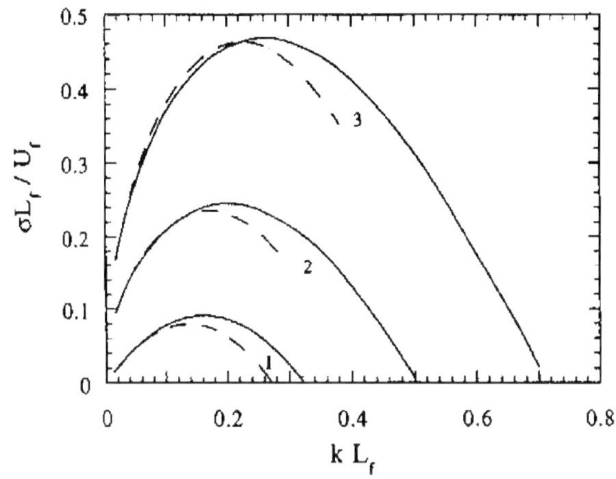

Fig. 9.4 The scaled growth rate of the DL instability for a flame of $\Theta = 5$ propagating upwards in a gravitational field for $gL_f/U_f^2 = 0,1,3$ shown by dashed curves 1,2,3 respectively, corresponding to Eq. (9.8.9): the solid curves correspond to the numerical solution

wavelength perturbations the gravitational stabilization concerns the perturbations of a long wavelength. The joint effect of the stabilizing gravity and thermal conduction may lead to complete suppression of the flame instability. To investigate the gravitational stabilization it is convenient to present the dispersion relation (9.8.8) in the form

$$\sigma^2 + \frac{2\Theta}{\Theta+1}\left(1 + \frac{\Theta \ln \Theta}{\Theta - 1} kL_f\right) U_f k\sigma + \Theta\frac{\Theta-1}{\Theta+1}\frac{U_f^2 k}{k_c}(k-k_1)(k-k_2) = 0, \tag{9.8.10}$$

where $k_c = 2\pi/\lambda_c$ is the cut-off wave number in the absence of gravity with the cut-off wavelength determined by (9.7.33), and the marginal wave numbers k_1 and k_2 show the stability limits in the case of stabilizing gravitational acceleration

$$k_1 = \frac{k_c}{2}\left(1 - \sqrt{1 - \frac{4|g|}{\Theta U_f^2 k_c}}\right), \quad k_2 = \frac{k_c}{2}\left(1 + \sqrt{1 - \frac{4|g|}{\Theta U_f^2 k_c}}\right). \tag{9.8.11}$$

Complete stabilization of the DL instability takes place when the marginal wave numbers coincide $k_1 = k_2$, which holds for the gravitational acceleration

$$|g_c| = \frac{\Theta}{4} U_f^2 k_c. \tag{9.8.12}$$

The dependence of the maximum instability growth rate on the gravity acceleration for upward propagating flame with a unit Lewis number and constant coefficient of thermal conduction is shown in Fig. 9.5.

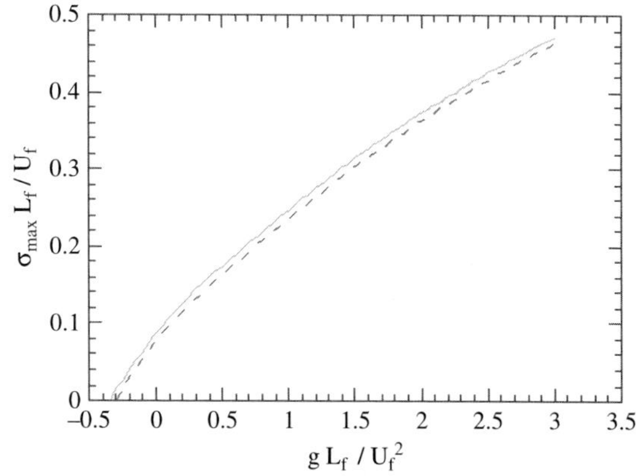

Fig. 9.5 The scaled maximum critical growth rate of the flame instability for the flame propagating upwards versus the scaled gravity acceleration; dashed line – Eq. (9.8.10); solid line is the numerical solution

The numerically calculated instability growth rates in the case of a stabilizing gravitational field and a flame with the expansion coefficient $\Theta = 5$ are shown in Fig. 9.6. The numerical solution for such a flame gives the value of the critical gravitational field $g_c L_f/U_f^2 = -0.35$, which is close to the analytical estimate $g_c L_f/U_f^2 = -0.31$. The marginal perturbation wave numbers k_1 and k_2 found numerically for different values of the gravitational acceleration are compared to the analytical results in Fig. 9.7.

For an arbitrary Lewis number and for the transport coefficients varying with temperature the dispersion relation of the flame instability can also be derived in a close form, though the calculations become more tedious and cumbersome. It is if practical interest that in the presence of acoustic or shock waves a flame can be either stabilized or destabilized, since the acoustic and shock waves produce an effective "gravitational" acceleration.

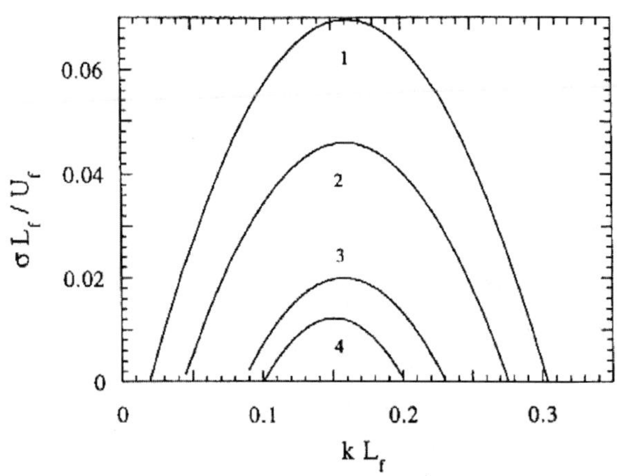

Fig. 9.6 The scaled growth rate of the LD instability for a flame of $\Theta = 5$ propagating downwards in a gravitational field for $gL_f/U_f^2 = 0.1;0.2;0.3;0.33$ shown by curves 1,2,3,4 respectively

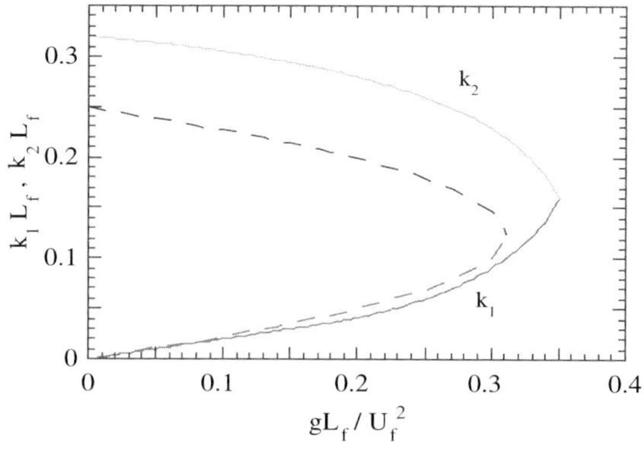

Fig. 9.7 The marginal wave numbers for a downward propagating flame versus the scaled gravitational acceleration; dashed line – analytical calculation, solid line – numerical solution

9.9 Thermal-Diffusive Instability

In the previous sections we considered the so-called hydrodynamic or Darrieus-Landau instability of a planar flame, which is an inherent feature of any flames in gaseous mixtures since the instability is related to the thermal gas expansion in exothermal reactions. There is another type of a flame instability, which is related to the disparity between molecular and thermal diffusivity. While the DL instability is inherent feature of any flames, the occurrence of the diffusive instability depends on the composition of the mixture. Unlike the DL instability that is characterized by relatively large scale and can be smoothed in relatively small-scale systems, the diffusive modes of instability manifest themselves in the emergence of small-scale irregularity recombining cellular structure of a flame front and can be relatively easily observed under normal laboratory conditions. The hydrodynamic and diffusive instability may well coexist, yet in theoretical studies they often analyzed separately.

Physics of flame stabilization by the thermal conduction and destabilization by diffusion can be understood qualitatively from a simple consideration. Let the Lewis number $Le = D/\kappa < 1$, i.e. thermal conductivity ($\kappa = \lambda / C_P \rho$) is the domination process, and diffusion (D) is not important. In this case when the flame front is slightly bended, the convex parts of the front loss more heat relative to the concave parts, so that the convex parts will be slackened while the concave parts accelerated thus smoothing the bending and suppressing the development of the instability. Contrary to this if $Le = D/\kappa > 1$ and diffusion is the dominating process, then the diffusion of a fresh fuel is larger at the convex parts and less at the concave parts of the flame front, increasing the reaction rate at the convex parts and, thus, enhancing the bending of the flame.

The dispersion relation, i.e. the dependence of the instability growth rate on wavelength of small perturbations, can be easily obtained in the case of a chemical reaction with large activation energy. In the case, when $Z = \frac{E}{2RT_b} \frac{(T_b - T_u)}{T_b} >> 1$, one can consider the reaction zone as a surface of zero thickness. Then, the equations for the adiabatic flame are reduced to the equation of thermal conduction and diffusion, which in a co-moving with the flame co-ordinate system are

$$\frac{\partial T}{\partial t} + u_1 \frac{\partial T}{\partial x} = \kappa \left(\frac{\partial^2 T}{\partial x^2} + \frac{\partial^2 T}{\partial y^2} \right), \tag{9.9.1}$$

$$\frac{\partial a}{\partial t} + u_1 \frac{\partial a}{\partial x} = D \left(\frac{\partial^2 a}{\partial x^2} + \frac{\partial^2 a}{\partial y^2} \right). \tag{9.9.2}$$

The unperturbed solution for the temperature profile and concentration are

$$T^0 = T_u + (T_b - T_u) e^{u_1 x/\kappa}, \ a^0 = a_0(1 - e^{u_1 x/D}), \text{ for } x<0, \tag{9.9.3}$$

$$T = T_b, \ a^0 = 0, \text{ for } x>0. \tag{9.9.4}$$

The boundary condition required that

$$C_P(T_b - T_u) = Qa_0$$

While for Le = 1, and correspondingly $\kappa = D$, chemical reaction completed at the adiabatic flame temperature T_b, in the case of perturbed flame with $\kappa \neq D$ temperature at the surface of chemical reaction zone can differ from T_b.

Since in the reference frame of the planar flame front the unperturbed solution is stationary, we can look for perturbations of a planar front against small perturbations bending the front in the form

$$\tilde{F}(x, t) = \tilde{F}(x) \exp(\sigma t + iky), \tag{9.9.5}$$

where y is the coordinate along the flame front, $k = 2\pi/\lambda$ is the perturbation wave-number, λ is the wavelength of the perturbation, σ is the perturbation growth rate.

For the case of long wavelength perturbations, $kL_f << 1$, or $\lambda >> L_f$, and large activation energy, high Zel'dovich number $Z = \frac{E}{2RT_b} \frac{(T_b - T_u)}{T_b} >> 1$, the result of solution of the thermal-diffusive instability problem can be presented in the form

$$\sigma = kU_f(kL_f)Z(Le - 1). \tag{9.9.6}$$

Thus, for Le > 1 the hydrodynamic instability of a flame is enhanced by the diffusive instability.

The growth of the flame instability given by Eq. (9.9.6) can be viewed as a linear dependence of the flame velocity on curvature of the flame front surface that is $\partial^2 x_f/dy^2$. Then, from the phenomenological consideration suggested by G.H. Markshtein the flame velocity depends on the curvature of the flame front as

$$U_f = U_{f0}(1 - L_{Mr}/R_f), \tag{9.9.7}$$

where U_{f0} is the normal velocity of a planar flame, R_f is the radius of curvature of the flame front and L_{Mr} is the Markstein length.

If one considers, for example, velocity of a cylindrically expanding flame for the case of $\mathrm{Le} = 1$, then compare equation for temperature (the energy equation) written in rectangular and cylindrical coordinate, one finds that velocity of the cylindrically expanding flame is

$$U_{cyl}(r = R) = U_{f0} - \frac{\kappa}{R} = U_{f0}\left(1 - \frac{L_f}{R}\right). \tag{9.9.8}$$

In this case the Markstein length is the width of the flame.

9.10 Nonlinear Stabilization of the Flame Instability

Let us consider a space filled by a combustible gas mixture. Let burning was ignited in some location. A flame can propagate form the location where it was ignited either being confined by the walls of a vessel, or as a freely propagating, for example, radially expanding from the point of ignition flame. Both of these configurations correspond to typical experimental situation of a laboratory flames propagating in either closed or open tubes, or ignited in the point and forming a radially expanding "fireball". The overall picture of flame propagation is described by the system of hydrodynamic equations, which take into account thermal conduction and the energy release. In any cases, the flame instability is the principle mechanism, which governs the shape and the velocity of propagating flame. As usually possibility to obtain an analytical solution to the problem is very attractive since it allows having a deeper insight for wide variety of all the problem parameters, which is difficult to obtain even with the most powerful computer simulation. Though the problem can be simplified if the flame can be treated as a surface of discontinuity, separating the fuel and the burning products, even with this simplification analytical description of a curved flame dynamics remains quite complex mathematical problem. Curved flame propagation is one of the most important and difficult issues in the combustion theory. Despite considerable efforts, its closed theoretical description is still lacking.

As it was pointed out one of the main reasons for a flame to lose its planar configuration is the DL instability of a planar flame front. Because of the DL instability small perturbations of a flame front are growing and bending the front. Development of perturbations at the nonlinear stage has been an important issue of many discussions. A general belief supported by experimental observations of flame dynamics in tubes is that the DL instability leads to the formation of a curved flame front and in a certain circumstances may cause the flame turbulization, though the flame turbulization is rather caused by another reasons such as rough boundary conditions at the tube walls, etc. The result of the development of the flame front instability is that a flame front acquires a curved stationary shape in the nonlinear stage of the DL instability. The

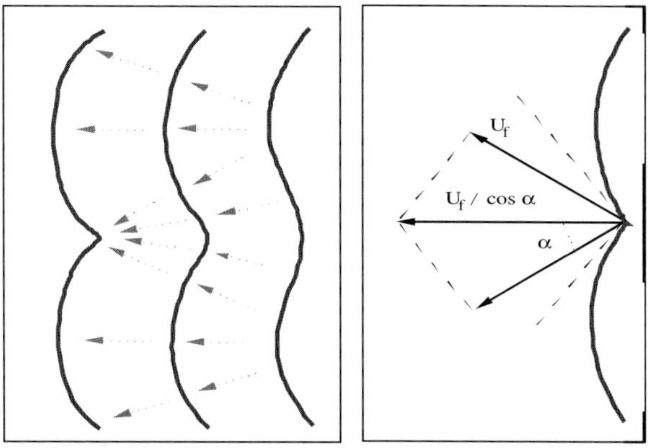

Fig. 9.8 The nonlinear mechanism of flame stabilization

physical mechanism of the nonlinear stabilization was proposed by Ya. Zel'dovich (1966) on the basis of a simple model of an infinitely thin flame and is illustrated schematically in Fig. 9.8. According to the Huygens principle the convex parts of a flame front diverge, while concave parts converge until angle points (cusps) are formed at the front. As it can be seen in Fig. 9.8 the angle points propagate faster than the smooth parts of the flame, the velocity increase at the cusps balances the growth of amplitude of the humps because of the DL instability and the flame acquires a curved stationary shape. In reality a flame front has finite thickness and the cusps are smoothed by thermal conduction.

An important effect related to the nonlinear stage of the DL instability is velocity amplification of a curved flame in comparison with planar one. Since the flame surface area increases with the development of the curved flame shape, the flame consumes more fuel per unit time and propagates faster. In the simplest model of an infinitely thin flame the amplification of the flame velocity can be estimated as the increase of the flame surface area. The amplification of the curved flame velocity may be also understood in the simple model shown in Fig. 9.8: in this model the propagation velocity of the curved flame is determined by the velocity of the cusps, which is larger than the normal flame velocity.

9.11 Curved Stationary Flames in Tube

The dynamics of premixed flames is one of the most fundamental problems in combustion theory. It is well established both theoretically and experimentally that the plane front of propagating flame in a gaseous combustible mixture is intrinsically unstable against small perturbations bending the flame front, for example, hydrodynamic instability such as the Darrieus-Landau (DL) instability arising at the flame front. The DL instability is inherent to all flames in gaseous mixtures since it is related to the gas expansion in exothermal reactions.

Nearly 100 years ago Mallard and Le Chatelier have found that the speed of flame in the mixture of methane and air increases progressively with the increase in diameter of the containing tube. Coward and Hartwell performed the first detailed experimental studies of the shape of the flame front and the flame speed depending on tubes diameter. They studied different mixtures of methane with air and measured velocity of the stationary flame propagating in tubes of diameters from 2.5 cm till 24 cm. They have written that "The flame is tilted but its shape is comparatively simple. In none of many photographs taken did the nodular flame appear, and when occasionally the flame had two heads the next snap-shot showed the simple form." In fact nodular flames can be observed in wide tubes for upward propagating flames, when the Rayleigh-Taylor instability affects the flame in addition to the Darrieus-Landau instability. A nodular flame structure is well documented experimentally but during nonstationary phase of flame propagation. More recent experiments on flame dynamics in tubes have shown that typical shape of a flame front is a curved flame.

On the linear stage of the DL instability small perturbations of a plane flame front grow exponentially with the instability growth rate depending on the perturbation wavenumber. Due to the DL instability a planar flame front becomes spontaneously curved, wrinkled by small perturbations and later on nonlinear stage, acquires a curved or corrugated shape. The saturation of the growth of perturbations is due to nonlinear effects, which leads to the formation of stationary curved flames and the increase of the velocity of flame propagation. Since the area of wrinkled or corrugated flame grows the flame consumes more fuel per unit time, and the flame speed undergoes a noticeable amplification. Outcome of the DL instability at the nonlinear stage may be wrinkling of the initially planar flame front, which may lead also (though more often it is thermal-diffusive instability) to the stationary cellular structure of the freely propagating flames shown in Fig. 9.9, or to a smooth curved shape of the flame. If the cell size is large enough compared to the cut-off wavelength, then the cellular flame in turn may become unstable against the DL instability on a smaller scale. As a result of this secondary DL instability a fine structure arises on the larger cells.

9.12 Nonlinear Equation for Stationary Flames

An analytical solution to the problem is very attractive allowing a deeper insight for wide variety of all the problem parameters. We will provide here a concise overview of the nonlinear theory of the flame evolution.

The difficulties encountered in trying to obtain a closed analytical description of flame propagation are first of all conditioned by the fact that the process is essentially nonlinear and nonlocal. One of the essential difficulties in analytical treatment of flames is the virtual impossibility to solve the flow equations

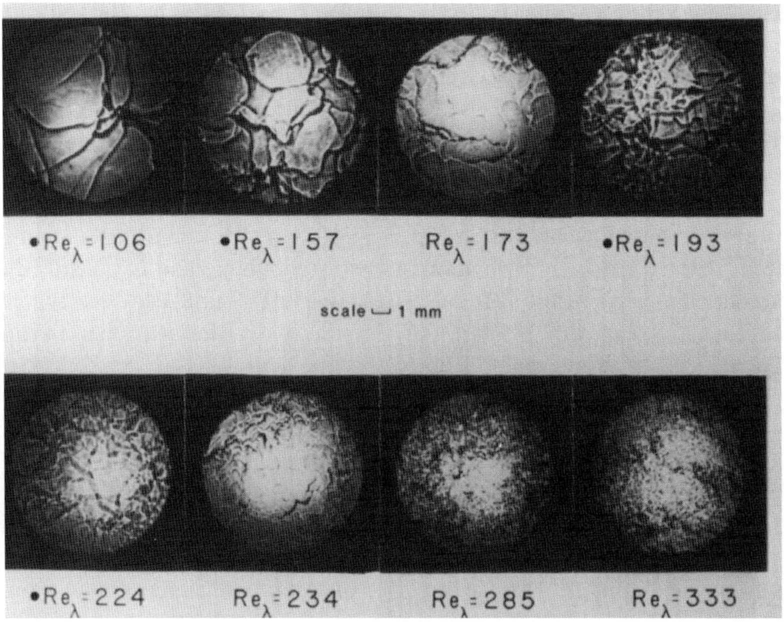

Fig. 9.9 Cellular structure of the flame front propagating in tube

governing dynamics of the burning gaseous mixture. The most important features of non-locality of the flame propagation is the vorticities produced by the curved flame downstream, which highly complicate the downstream flow structure. The flame induced gas flows from both side of the flame both upstream and downstream strongly affect the flame front structure itself. The latter means that the non-locality of equations governing flame propagation is more complex than that encountered in the linear theory and described by the Darrieus-Landau theory. In particular, the local relation between pressure and velocity field upstream, expressed by the Bernoulli equation, no longer holds for the flow variables downstream. Only in the case of small gas expansion $(\Theta - 1) < 1$ can the problem be treated both perturbatively and locally, since then the amplitudes of perturbations remain small compared to their wavelengths at all stages of development of the DL-instability and the flow is potential both upstream and downstream in the lowest order in $(\Theta - 1) < 1$. It should be stressed that the above-mentioned inconsistency for finite values of Θ resides in the equations of motion themselves. From the mathematical point of view perturbation analysis is inapplicable because the assumptions of weak nonlinearity and stationarity of the flame contradict each other. In particular, it is completely irrelevant to the problem of formation of the stationary flame configurations, and it requires an adequate nonperturbative description of flames with arbitrary values of the flame front slope.

The main reason underlying the complexity of the problem is the hydrodynamic instability of zero-thickness flames. In view of this, evolution of the flame front cannot be prescribed in advance and should be determined in the course of joint analysis of the flow dynamics outside the flame front and the heat diffusion inside. In general, nonlinear interaction of different perturbation modes under the smoothing influence of thermal conduction leads to the formation of a steady curved flame front configuration with the curvature radius of the order $20L_f$. This estimate for the curvature radius steams from the linear theory of the DL-instability, where it corresponds to the cut-off wavelength for the front perturbations $\lambda_c \approx 20L_f$. The exponential growth of unstable modes is ultimately suppressed by the nonlinear effects. It is clear that since the governing equations in the case of arbitrary gas expansion do not contain small parameters, so the DL instability can only be suppressed by the nonlinear effects. The stabilizing role of the nonlinear effects is clearly seen in numerical experiments on the flame dynamics, where even in narrow tubes the observed flame slope is about 1.5.

The equation describing flame propagation can in essence be derived for an arbitrary coefficient of gas expansion in a closed form, though the equation is quite complex for the analysis. The corresponding calculations are rather tiresome and tedious, therefore we shall restrict our consideration below giving the main ideas without detailed derivation. For the sake of simplicity, let us consider two-dimensional problem. A flame front can be described by the function $F \equiv z - f(x,t) = 0$, which satisfies the evolution equation (9.4.6)

$$\mathbf{n} \cdot \mathbf{u}_- + \frac{1}{N}\frac{\partial F}{\partial t} = U_f, \tag{9.12.1}$$

where $\mathbf{u}_-$ is the fuel velocity just ahead of the flame front. However, the unknown flow itself results from the flame front evolution, so that $\mathbf{u}_-$ in Eq. (9.12.1) is some function of F. The relation between $\mathbf{u}_-$ and F can be found solving the equations of ideal hydrodynamics ahead and behind the flame front and matching these solutions at the flame front. It can be done considering flame as a front of zero thickness, as well taking into account small but finite thickness of the flame, L_f.

For equal coefficients of fuel diffusion and thermal diffusivity, Le = 1, and for a small but finite thickness L_f of the flame front such procedure allows to obtain a nonlinear equation describing evolution of a curved flame. In general case of arbitrary value of Θ analysis is difficult, but it is considerably simplified for weak gas expansion. In this case at the third order of expansion in $(\Theta - 1)$ the equation reads

$$\Theta(f')^2 + (\Theta - 1)\hat{H}(f') - 2\Theta W = (\Theta - 1)\frac{\varepsilon\lambda_c}{2\pi}f''. \tag{9.12.2}$$

Here we used the non-dimensional variables for co-ordinates: $(\eta, \xi) = (x/D, z/D)$, and for the velocity: $(w, u) = (u_x/U_f, u_z/U_f)$; $f' = df/d\eta$. $W \equiv V - 1$ is the actual increase of the nondimensional velocity of a curved flame front, with the average flame velocity across the tube section $V = \frac{1}{D}\int_0^D N d\eta$, where $N = \sqrt{1 + (f')^2}$, and $\varepsilon = L_f/D$.

The expression for the cut-off wavelength is

$$\lambda_c = \frac{4\pi L_f}{\Theta - 1}\left(1 + 3\frac{\Theta - 1}{2}\right), \tag{9.12.3}$$

which is the first order approximation of the exact value of the cut-off wavelength in the linear theory of the DL instability.

For the flame propagating in a tube with ideally slip adiabatic walls the solution to the Eq. (9.12.2) with the boundary conditions at the walls $f'(\eta = 0) = f'(\eta = 1) = 0$ can be obtained using so-called method of pole decomposition.

$$f(\eta) = -\varepsilon\frac{3\Theta - 1}{\Theta}\sum_{k=1}^{3M} \ln \sin\left[\frac{\pi}{2D}(\eta - \eta_k\right].$$

In dimensional variables it gives for the velocity of curved stationary flames propagating in a rectangular channel of width D the following expression

$$U_w = U_f + 4U_m\frac{M\lambda_c}{2D}\left(1 - \frac{M\lambda_c}{2D}\right), \tag{9.12.4}$$

where $M = \mathrm{Int}\{D/\lambda_c + \frac{1}{2}\}$ with $\mathrm{Int}\{x\}$ denoting the integer part of x, and the maximal flame velocity increase depends on the gas expansion coefficient as (here U_f stands for the normal velocity of a planar flame), and the maximum velocity increase is

$$U_m = \frac{(\Theta - 1)^2}{8\Theta^2}U_f. \tag{9.12.5}$$

It is noteworthy that for the adiabatic boundary conditions at the ideal walls $f' = df/d\eta = 0$ for $\eta = 0, 1$ mean that the flame front touches the walls at the right angle, which is a direct consequence of the thermal conduction taken into account in right hand part of Eq. (9.12.2) by the term proportional to $\varepsilon = L_f/D$. If one considers a very wide tube and neglects the thermal conduction, then the flame front would touch one of the walls at a sharp angle leading to formation of a stagnation zone in the flow close to the wall behind the flame. In terms of Eq. (9.12.2) the statement of the problem with neglected thermal conduction implies $\varepsilon = 0$, so that the order of Eq. (9.12.2) decreases; one of the boundary conditions at the wall is excessive and the flame front does touch the wall at a sharp angle. However in the complete form of Eq. (9.12.2) with small but non-zero $\varepsilon = L_f/D$ thermal conduction always smoothes the sharp angle inside a thin "boundary layer" of the thickness of about ε. Similarly, the discontinuity surface between the main flow behind the flame and the stagnation zone close to the wall reduces to a thin (but continuous) region of large velocity gradients.

It follows from (9.12.4) that for sufficiently wide tubes and accordingly for sufficiently small ε, curved stationary flames of different shapes and velocities corresponding to solutions with different numbers of poles are possible. The number of poles does not have clear physical interpretation, however it can be shown that for a given value of ε all but one of these solutions are unstable and do not realize. Indeed, one can expect the flame velocity in a tube to be determined by a solution with a number of poles that provides the largest flame velocity for a given tube width. For a given value of ε the largest flame velocity corresponds to the solution with the number of poles $M = \mathrm{Int}\left[\frac{1+\varepsilon}{2\varepsilon}\right]$.

The dependence of the scaled flame velocity increase on the inverse tube width given by the Eq. (9.12.4) is depicted in Fig. 9.10 by dashed line, where marks show the results of numerical studies of the flame propagating in the two-dimensional tube.

It represents a combination of the parabola pieces with the maximum at the points corresponding to the tube widths $D = M\lambda_c/2$, with $M = 1, 2, 3, \ldots$. An important feature of the solution is the existence of a maximum velocity of a curved stationary flame that cannot be exceeded with increase of the tube width. Thus, outcome of the DL instability at the nonlinear stage results in a smooth curved stationary shape of a flame front for a flame propagating in a tube of width larger than the first critical tube width $D_{c1} = \lambda_c/2$. Note, that the width of an ideal tube determines half of the maximal possible perturbation wavelength and the perturbations of the wavelength shorter than λ_c are suppressed by thermal conduction.

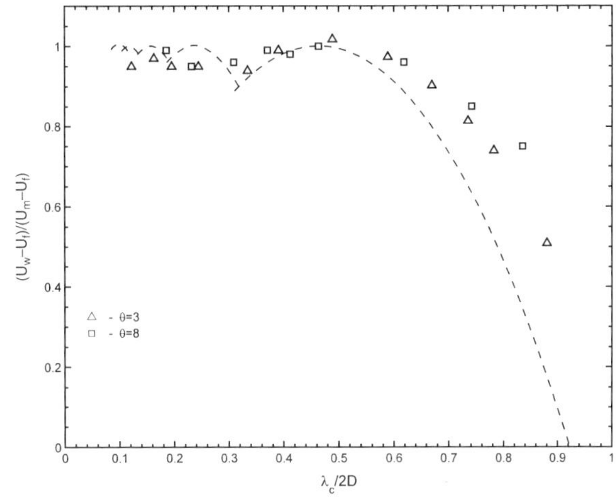

Fig. 9.10 The flame velocity amplification versus the inverse tube width

9.13 Stability Limits of a Curved Stationary Flame in Tubes

The velocity of a flame propagating in tubes is one of the most important issues in combustion science. While normal velocity of a planar flame U_f is determined by thermal and chemical fuel parameters, the resulting velocity of flame propagation U_w depends also on the flame shape: the more curved and wrinkled the flame front, the faster it propagates. In the absence of external turbulence a curved flame shape results from intrinsic flame instabilities such as the Darrieus-Landau instability inherent to any premixed flame in a gaseous fuel. It follows from the linear theory of the DL instability that a planar flame front is bent by two-dimensional and three-dimensional perturbations, if the perturbation wavelength exceeds the cutoff wavelength λ_c determined by thermal conduction and finite thickness of the flame front, L_f. For most of the laboratory flames the cutoff wavelength is considerably larger than the flame thickness, $\lambda_c \approx 20L_f$.

Consider a flame propagating in tubes of different widths and for different values of the expansion coefficient assuming that the walls are ideally slip and adiabatic, i.e. there is no heat losses to the walls. The flame is assumed to be initiated as a planar one and starts propagating with the normal flame velocity, U_f. Because of the ideal boundary conditions at the walls the tube width D determines a half of the largest possible wavelength of permitted perturbations, i.e. the perturbation wavelength may take the values $\lambda = 2D/n$, $n = 1, 2, 3$, etc. For perturbations of a wavelength shorter than the cut-off wavelength development of the DL instability is suppressed by the thermal conduction. Therefore, if a tube is narrow, so that $D < D_{C1} = \lambda_c/2$, then all permitted perturbations belong to stable part of the dispersion relation (9.7.32). In agreement with the theory, the numerical simulations have shown that flame evolution in narrow tubes $D < D_{C1}$ always leads to a planar flame front, even if rather large amplitude initial perturbations are imposed the flame front returns to the planar configuration. We shall refer to D_{C1} as the first critical tube width.

In wider tubes $D > D_C1$ the hydrodynamic instability develops and results in a stationary curved shape of a flame front. As the amplitude of the first perturbation harmonic with the wavelength $\lambda = 2D$ grows, it induces due to the nonlinear interaction the growth of harmonics with shorter wavelengths $\lambda = D$, $2D/3$ etc. Since harmonics with high wave-numbers are stable, the growth of perturbation slows down and stops when the final amplitudes corresponding to the stationary curved flame front are achieve. Numerical simulations of flame dynamics in tubes of moderate width with ideally slip and adiabatic walls have shown that at the nonlinear stage the DL instability results in a smooth curved stationary flame shape, which may be described as a hump directed towards the fuel as is shown in Fig. 9.11. Maximum flame velocity increase in nondimensional variables, $W_m = (U_m - U_f)/U_f$ versus the gas expansion coefficient is shown in Fig. 9.12 by solid line, where filled circles are from numerical simulation of the problem. The velocity U_w of a curved

Fig. 9.11 The isotherms showing shape of the flame propagating in a tube of width $D = 2D_c$, $\Theta = 5$

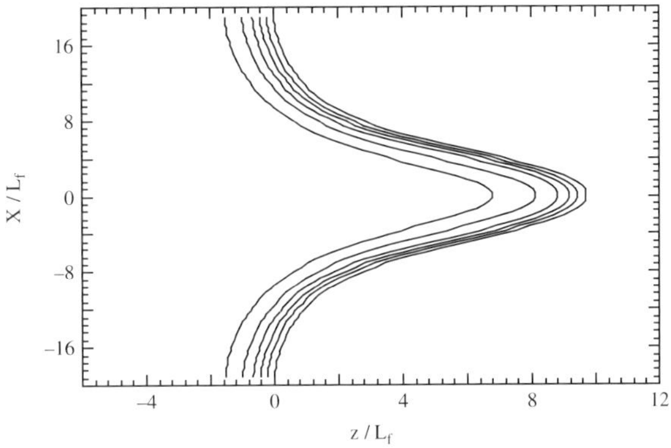

stationary 2D flame in a tube with ideal walls found in numerical experiments as a function of the scaled inverse tube width λ_c/D is in a very good agreement with the analytical solution (9.12.4) for flames in tubes of moderate widths $0.3 < D_{C1}/D < 1$. As it is seen from Fig. 9.10 the calculated in numerical simulation values of the flame velocity fit the same parabola shaped dependence of the flame velocity on the inverse tube width as the analytical formula (9.12.4). The maximum velocity is achieved for the tube width $D \approx 2D_{C1}$ corresponding to the perturbations with the wavelength $\lambda = 2\lambda_c$ that have also the maximal growth rate at the linear stage of the DL instability.

The amplification of the flame velocity due to the curved flame shape corresponds to the increase of the flame surface area (increase of the flame length in the 2D configuration) for moderate expansion coefficients ($\Theta < 5$), but for larger expansion coefficients $\Theta > 5$, these two values deviate one from another by about 30%, which is presumably coupled to the nonlinear terms related to vorticity production behind a curved flame.

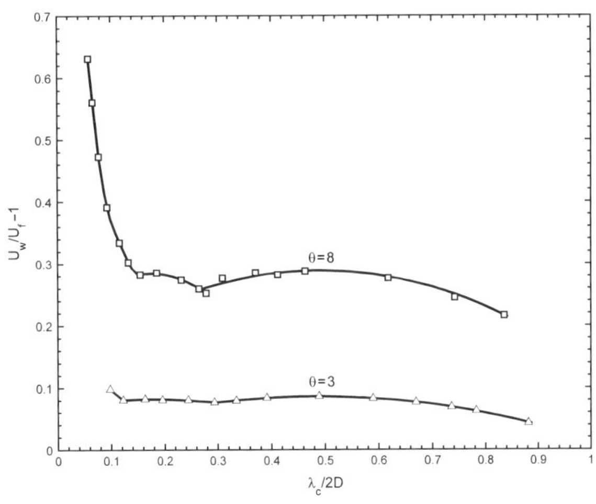

Fig. 9.12 Dependence of the scaled flame velocity increase on the inverse tube width. Squares and triangles corresponds to calculations for the gas expansion coefficient coefficients $\Theta = 8$ and $\Theta = 3$

According to the analytical solution (9.12.4) the flame velocity tends to its maximal value U_m with the increase of the tube width, and for sufficiently wide tubes the flame front velocity should be almost equal to U_m. However this is not what is observed in reality. The dependence of the flame front velocity on the inverse tube width presented on Fig. 9.12 clearly indicates the existence of two bifurcation points, which corresponds to the primary $D > \lambda_c/2$ and secondary $D > 3.4\lambda_c$ DL instabilities. The squares and triangles in Fig. 9.12 show the calculated velocity increase for flames with expansion coefficients $\Theta = 8$ and $\Theta = 3$ versus the inverse scaled tube width obtained in the numerical simulations. As one can see the velocity of the curved stationary flame increases with increase of the tube width until it reaches a local maximum in agreement with the solution (9.12.4). The nonlinear theory predicts the local maximum for the tube width equal to the doubled first critical value $D = \lambda_c$, which also corresponds to the largest instability growth rate in the linear theory. In tubes of a moderate width $\lambda_c/2 < D < 3.4\lambda_c$ the DL instability leads to a smooth curved stationary flame front. The shape of the flame front represents a hump directed towards the fresh fuel mixture.

If the tube width increases, the radius of curvature of the curved stationary flames increases too and the stabilizing influence of the curved flame shape weakens: the flame front resembles locally a planar flame and the DL instability should occur on a new scale. This secondary DL instability can lead to additional wrinkling of the front and to additional increase of the flame velocity. The result of the secondary DL instability is the abrupt change in the dependence of the flame velocity for the tube width near the point $D_{C2} \approx 3.4\lambda_c$. Contrary to what one could expect, in tubes of width $D > 3.4\lambda_c$ the flame velocity increases progressively with increase of the tube width, while the flame shape remains smooth being slanted in the tube and without formation of a new cusps or nodular flame structure.

The solutions to the stationary nonlinear equation (9.12.2) describe dynamics of curved stationary flames in tubes of an arbitrary width including very wide tubes of a width much larger than the cut-off wavelength of the DL instability. However, in very wide tubes the stationary curved flames like those shown in Fig. 9.11 do not happen in reality. For a tube width much larger than the cutoff wavelength the radius of curvature close to the hump of these flames goes to infinity with increase of the tube width, so that such flames coincide locally with a thin planar flame front. The stabilizing influence of the curved shape weakens and, consequently, a curved stationary flame presumably becomes unstable against perturbations of a small scale, much smaller than the curvature radius of the flame but larger than the cutoff wavelength. The numerical simulation of a flame in wide tubes show that long-living "many-humps" or nodular flames are possible in wide tubes, but only during nonstationary stage as it is shown in Fig. 9.13.

In the stationary mode only relatively simple "single-hump" or slanted flame front are stable. Such a simple flame shape can be understood from analysis of the flame front evolution. In tubes of moderate width the initial perturbation grows and turns into a single-hump structure that propagates without

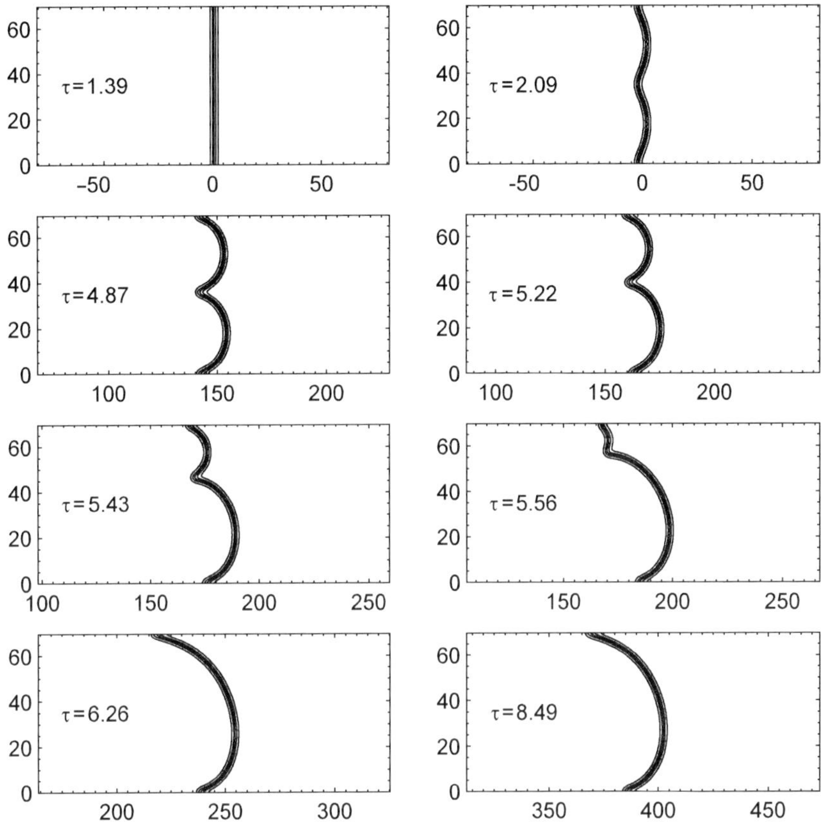

Fig. 9.13 Numerical simulation of the temporal evolution of the flame propagating in a rectangular channel of width $D = 4\lambda_c$, $\Theta = 8$. The isotherms are shown at different time instants with the last plot corresponding to the stationary flame

changing. Contrary to this in wide tubes the nonstationary "intermediate" "many-hump" configuration of the flame front can occur, and transition from such a wavy solution to a convex one looks like pushing out of one hump by another. This tilted shape of the flames propagating in wide tubes is consistent to those observed in experiments.

A simple qualitative explanation of such shape of the flames is the following. When the flame front is slanted to the direction of the tube axis, it becomes stretched and the distance between different points of the front increases with time. As a result, perturbations at the flame front become stretched too and their wavelength increases, which leads to decrease of the instability growth rate. Besides, the perturbations have finite time to develop since any localized perturbation gets suppressed as soon as it is transferred to the walls by the tangential component of the flow field inherent to slanted flame. Instead of forming a new cusp on the flame front the perturbation vanishes when it reaches the wall due to the imposed boundary conditions. For the flame front strongly inclined to the wall, the tangential transfer of the perturbations is faster then the time required for instability to grow. Formally this corresponds to the stability limit – loss of stability of the solution (9.12.4). A critical value of the tube width $D_{c2} \approx 3.4\lambda_c$ obtained from 2D numerical simulation can be considered as the limit for an applicability of the nonlinear theory. The described scenario of the secondary DL instability consists in appearance of the extra cusps arising near the center of the tube, which are transferred along the flame front to the tube walls. This causes additional stretching of the cusp while the front retains its original shape becoming more slanted in the tube with comparatively simple shape. It is worth to note that the new shape of the flame front leads to considerable increase of the flame velocity in comparison with the velocity of curved stationary flames in tubes of smaller width. The primary DL instability causes increase in the flame velocity by the factor about $U_w/U_f = 1.2/1.3$. For wider tubes the second critical tube width D_{c2} is approximately four times larger than the first critical tube width D_{c1}, for which the DL instability overcomes the stabilizing influence of thermal conduction.

9.14 Spherically Expanding Flames

A spherical flame spreading out from an ignition source is one of the most fundamental manifestations of combustion. While such flames are quite feasible in the laboratory, in the regime of well-developed hydrodynamic instability a spherical flame spreading out from the ignition source appears not as the outward propagating smooth spherical front but as a multiple-scale surface resembling a cauliflower with the structure quite similar to the Koch fractal structure. Since the surface area of a wrinkled flame increases with the wrinkling development compared to a smooth unperturbed spherical flame front, the flame consumes more fuel per unit time and propagates faster.

As this was discussed, in the regime of hydrodynamic instability the growth of perturbation is limited due to nonlinear effects, and the outcome of the DL instability at the nonlinear stage may be wrinkling of the initially smooth flame front. For the case of the freely propagating flames this wrinkling may lead to cellular structure of a flame, and for the case of the flame propagating in tubes a smooth stationary curved shape is usually observed. It should be emphasized that the morphology of flames propagating in tubes appears to be essentially different from that occurring in spherically expanding wrinkled flames. A planar flame front propagating in a tube becomes spontaneously curved due to the DL instability and the instability of the flame front results in formation of a single hump, whose scale is controlled uniquely by the tube radius. Contrary to this, in the case of the spherically expanding flames, the humps merging at the flame front is balanced by the overall expansion, and while propagating, flame is wrinkled by the additional hierarchies of humps of all sizes. The available experimental data indicate that expanding wrinkled flames spreading out from the ignition source in the regime of well-developed hydrodynamic instability undergo considerable acceleration with the temporal dependences of the average flame radius being similar for all flames:

$$R(t) \propto At^{\alpha}, \tag{9.14.1}$$

where R(t) is the average radius of the wrinkled flame ball and A is some empirical constant. This regime of the flame propagation can be associated with the development of a fractal structure on the flame front with total surface of the spherical flame

$$S = 4\pi R^2 \frac{\alpha A^{1/\alpha}}{\Theta U_f} R^{(\alpha-1)/\alpha}. \tag{9.14.2}$$

Interpreting this regime as the development of a fractal structure of the flame front, it was found that the experimentally measured velocity of flames, $U_w = dR/dt = \alpha A R^{(\alpha-1)/\alpha}$, is consistent with the expression (9.14.1) for $\alpha = 3/2$.

The power law behavior (9.14.1) of freely propagating flames can be substantiated as result of the Darrieus- Landau hydrodynamic instability of the expanding flame. When the curvature of the expanding flame becomes large compared to the cut-off wavelength, then the flame front resembles a locally planar one and in turn becomes unstable against the DL instability on a smaller scale. As a result of this secondary DL instability, a fine structure arises on the larger scales. If the largest instability length scale exceeds the cut-off wavelength by many orders of magnitude, then cascades of the secondary instabilities consisting of small humps imposed on larger humps may form hierarchies of humps at the flame front with structure similar to the fractal structure. This picture is consistent with what is seen in experimental studies, which indicate that the initially smooth front of the freely expanding spherical flames undergoing wrinkling due to well-developed hydrodynamic instability on the large scales and expand with a noticeable acceleration. The process is similar to the Koch curve formation, which is constructed as a cascade of triangles. Numerical simulations of spherically expanding flames is a difficult task since it requires a huge computational recourses, however, two-dimensional simulation of cylindrically expanding flame is possible and its extrapolation to 3D case confirmes the picture of formation a fractal structure and observed dependence $R(t) \propto At^{\alpha}$ of the flame acceleration with $\alpha = 3/2$.

Let us consider a curved stationary flame as one step of the cascade similar to the Koch triangles. If a flame is treated as an infinitely thin discontinuity surface, then the increase of the flame velocity is equal to the increase of the total flame surface area on one step of the cascade. Let us assume that a step of the cascade k decreases the size of triangles by a factor $b = L_k/L_{k+1}$, which results in the increase of the total length of the curve by a factor $\beta = S_k/S_{k+1}$. The overall cascading process is limited from above by a largest triangle size λ_{max} and by the smallest triangle size λ_{min} from below, so that the total number of the steps in the cascade is $N = \ln(\lambda_{max}/\lambda_{min})/\ln b$, and the total length of the curve resulting in the development of cascade is

$$S_N = S_0\beta^N = S_0(\lambda_{max}/\lambda_{min})^d, \tag{9.14.3}$$

where $d = \ln(\beta/b)$ is the fractal excess, and $1 + d$ is the fractal dimension in the two-dimensional case, while it is $2 + d$ in the case of three-dimensional geometry.

To evaluate the increase of the length scale for a fractal generator (the factor b) we can assume that the increase of the flame surface is directly associated with the corresponding increase of the flame velocity. Then, $b = D_{C2}/D_{C1}$, where D_{C1} and D_{C2} are the tube widths for the first and secondary stability limits of a curved stationary flame propagating in tubes. It follows from the results of previous section that for the first cascade $b = D_{C2}/D_{C1} \cong 3.4$. The corresponding increase of the flame surface on one step cascade is the velocity amplification $\beta = U_w/U_f \cong 1.28$, for 2D problem. In this case, for 2D circular flame (cylindrically expanding flame) we obtain for the fractal excess

$$d_{2D} = \ln(\beta_{2D})/\ln(b_{2D}) \approx 0.20 \tag{9.14.4}$$

The corresponding fractal dimension is $1 + d = 1.20$, and the exponent in the expression (9.14.1) is $\alpha = 1/(1 - d) = 1.25$. This means that we can expect that in 2D geometry the radially expanding cylindrical flames are accelerating according to

$$R_{2D} \propto At^{1.25}, \tag{9.14.5}$$

The fractal dimension of the radially expanding cylindrical flames obtained from direct numerical simulation of the complete system of Navier–Stokes equations including one-step chemical kinetics, shown in Figs. 9.14—9.16 corresponds to power-law acceleration with exponent 1.25 in agreement with theoretical evaluation of fractal dimension for 2D geometry. The sequences of expanding flame front obtained in the numerical simulations are shown in Fig. 9.14.

A representative plot of the stream lines, flow field and vortexes generated by wrinkled flame front is shown in Fig. 9.15.

It is seen from Fig. 9.14 that formation of the wrinkling cells is limited from below. In all simulations the smallest hump size is of the order of $\lambda_C \approx 20L_f$, which is the cut-off wavelength of DL instability according to the theoretical estimate. The perturbation size is also limited from above since perturbations with a length scale larger than $2\pi R/n_c$ are suppressed due to the finite curvature of the flame front, where n_c is the critical number of a spherical harmonic. Only the perturbations with sufficiently large harmonic number are unstable. It

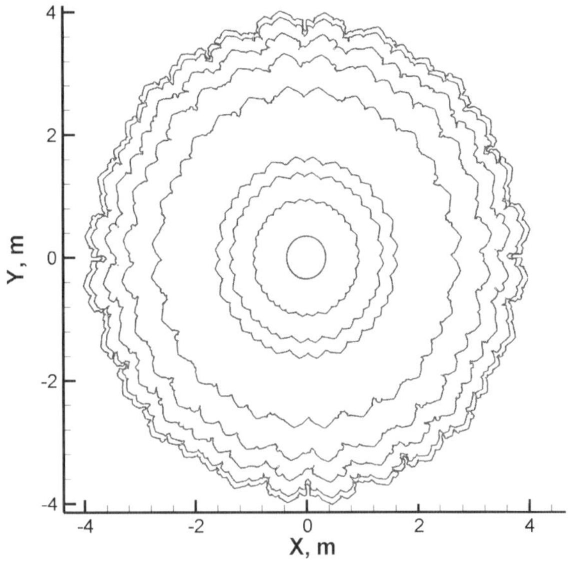

Fig. 9.14 Formation of a fractal structure of the radially expanding flame front (numerical modeling)

should be noted that if perturbations grow slower compared to the rate of the radial flame expansion, which corresponds to linear temporal dependence R(t), then the flame front does not acquire fractal structure and remains smooth. As a result of development the DL instability the initially smooth flame front becomes wrinkled and the front surface area increases during expansion exhibiting cusps that multiply and grow in amplitude. Evolution of the flame radius

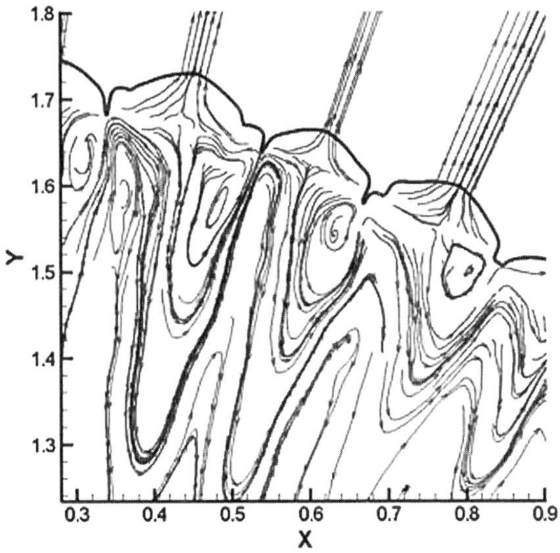

Fig. 9.15 The flow field and vortexes generated by wrinkled flame front (numerical modeling)

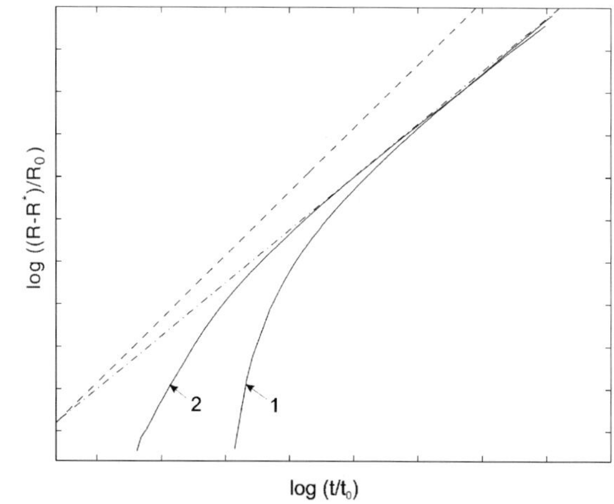

Fig. 9.16 Radius of the radially expanding flames vs time (numerical simulation): curves 1,2. *Dashed-doted line* corresponds to the power-law 5/4; *dashed line* – 3/2

obtained in the numerical simulation is shown in Fig. 9.16 for two different initial conditions. It is seen that the average flame radius approaches asymptotically the power law (9.14.5).

To extrapolate the obtained results from the 2D to 3D geometry of spherically expanding flames we should take into account that sum of two solutions of the nonlinear equation for the flame front is also solution to the equation. Then we conclude that velocity amplification for spherical flame should be twice of velocity amplification for cylindrical flame, e.g. for spherically expanding flame $\beta_{2D} = U_w/U_f \cong 1.56$. Then, for the spherically expanding flame $d = \ln \beta_{2D}/\ln b \cong 0.36$. This gives fractal dimension 2.36, which is close to fractal dimension 2.33 found in the experimental studies for $R(t) \propto t^{3/2}$.

It should be emphasized that the fractal regime of flame propagation represents an intermediate asymptotic of the flame dynamics, which is achieved after some transition period and based on assumption that the flame instability caused solely by the hydrodynamic instability. The regime is achieved at a large radius compared to the flame thickness, so that we can expect that it would depend on Markstein number, though transition to self-similar regime is rather smooth and it would be difficult to determine the critical point experimentally. The critical flame radius evaluated from numerical calculations is $R_c \simeq 1.8$ m, which corresponds to the critical Reynolds number $Re_c \approx U_f R_c/\nu = 2.8 \cdot 10^4$, that is within the experimentally measured critical Reynolds numbers in the range 10^3–10^6.

9.15 Flame in a Gravitational Field: Theory of Rising Bubbles

Velocity of curved stationary flames resulting from the development of the DL instability may increase by the factor 1.2–1.5 in comparison with the planar flame velocity in the case of 2D flames and about two times more in the case of

3D flames. However experimental studies of the flame propagating in tubes show that velocity of curved stationary flames may exceed these values considerably. It has been found, for example, that a flame in the mixture 10% CH_4 and air propagates in a horizontal tube of diameter 90 cm with the velocity 245 cm/s exceeding considerably the normal velocity of the flame, which is about 43 cm/s. Such a strong increase of the flame velocity both in vertical and horizontal tubes cannot be explained solely by bending of the flame due to the DL instability, but it can be related to the flame stretching in a nonuniform flow induced ahead of the flame due to noslip boundary conditions at the walls, or due to the wall roughness, or due to influence of a gravity.

The wall friction can change drastically dynamics of the flame propagating in a tube. Both, the DL and thermal-diffusive instability results in bending the flame front, its wrinkling and consequently in the increase of the flame surface and the flame acceleration. The accelerating flame acts as a semi-transparent accelerating piston. This means that a weak shock is generated ahead of the flame. When shock is formed, it generates a flow behind it, i.e. ahead of the propagating flame. The unburned gas ahead of the flame, which has been motionless before the flame started accelerating, is now evolved in motion. The self-similar solution says that such flow consists of advancing shock followed by rarefaction wave where velocity dropped to the velocity of the gas adjusted to the flame front. Such a flow in a tube with adhesive walls generates a boundary layer. If width of the tube is D and characteristic velocity is U, the Reynolds number is $Re = UD/\nu$, then the width of the boundary layer is $\delta \simeq D/(Re)^{1/2}$. In general, the width of the boundary layer is small and does not affect considerably the flame. The Poiseuille flow with the parabolic distribution of the velocity (zero at the walls and maximum at the axis) will be established when the width of the boundary layer became about the width of the tube. This will occur at the distance $L \propto D^2U/\nu \simeq D \cdot Re$. The flame does not feel nonuniform flow ahead immediately, but only when nonuniform flow is shifted to the flame by convection. When the flame enters the domain of the nonuniform flow, it mimics the shape of the velocity field, became curved and additionally accelerates. The whole picture depends on the tube width: for a small tube diameter only nonuniformity of the flow will be responsible for the flame acceleration, but for wide enough tubes DL and thermal diffusive instability may enhance acceleration. This picture is even more pronounced for the case of a flame propagating from the closed end of the tube, where the flow with the velocity $(\Theta - 1)\, U_f$ is required by the boundary conditions at the closed end of the tube.

An additional flame wrinkling can also be related to the RT instability of a flame and to influence of the gravitational field. The linear stage of the RT instability development at the flame front propagating upwards has been considered previously. The nonlinear stage of the instability is more complex problem, though it can be qualitatively described as a bubble rising. A flame front propagating in a cold combustible mixture converts it to hot products of burning of lower density, which tend to move upwards in a gravitational field. We can reasonably assume that the rising burned light products form a bubble with the flame front being the surface of the bubble. Let us evaluate how the velocity of the rising bubble depends on the gravitational acceleration, g. From two values: the gravitational acceleration g and the bubble size of radius R can be formed a unique value of a velocity dimension, $U_b \propto \sqrt{gR}$, which should be the velocity of the rising bubble with the accuracy of a numerical factor of the order of unity. The last one can be found from detailed calculations. The bubble velocity depends also on the geometry of a bubble, which may be characterized as open bubbles or as closed bubbles shown in Fig. 9.17. The result can be different also for two-dimensional or three-dimensional cases.

As an example, let us evaluate velocity U_b of an open 2D bubble formed by a light gas rising upwards in an ideal tube of width R filled by a liquid. Let density of the light gas is negligible compared to the density of the liquid, and we assume it zero. In the reference frame of the bubble ($x = 0$, $z = 0$ being the top of the bubble) the fluid at infinity flows towards the bubble with the uniform velocity $u_z = U_b$, rounds the bubble and forms a thin jet. At the surface of a massless bubble the Bernoulli equation is

$$\frac{1}{2}u^2 + gz = 0. \qquad (9.15.1)$$

At the rigid walls at $x = (0, R)$ the x-component of the velocity is zero, $u_x = 0$. Since there are no external sources of vorticity the flow must remain potential everywhere. In order to obtain the bubble velocity we need to solve the Laplace equation for the velocity potential $\nabla^2\varphi = 0$ in the whole flow with the boundary conditions at the walls and at the bubble surface. This is rather difficult mathematical problem, which has not been solved yet. Let us consider an approximation for the velocity potential and the stream function for the incoming flow

$$\varphi = U_b z + U_b \frac{R}{\pi} \exp\left(-\frac{\pi z}{R}\right) \cos\left(\frac{\pi x}{R}\right), \qquad (9.15.2)$$

$$\psi = -U_b x + U_b \frac{R}{\pi} \exp\left(-\frac{\pi z}{R}\right) \sin\left(\frac{\pi x}{R}\right). \qquad (9.15.3)$$

The corresponding velocities are

$$u_x = -U_b \exp\left(-\frac{\pi z}{R}\right) \sin\left(\frac{\pi x}{R}\right), \qquad (9.15.4)$$

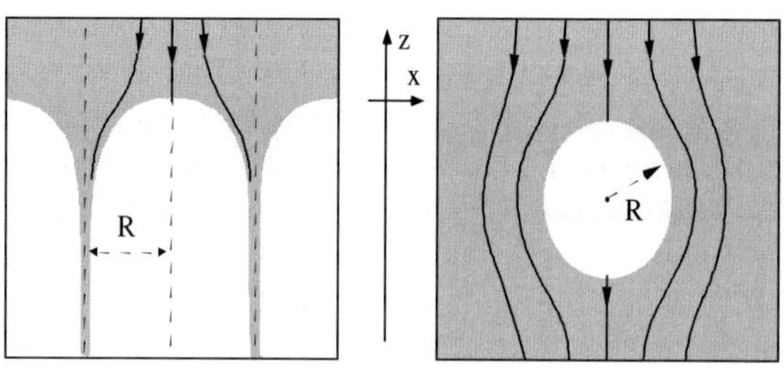

Fig. 9.17 Configurations of open and closed bubbles

$$u_z = U_b\left[1 - \exp\left(-\frac{\pi z}{R}\right)\cos\left(\frac{\pi x}{R}\right)\right]. \quad (9.15.5)$$

From the equation $\psi = 0$ for the streamline that passes through the stagnation point at the bubble top steams that the bubble surface near the top is given by the formula $z = -\pi x^2/6R$. Substituting (9.15.4–5) into the Bernoulli equation and taking into account the expression for the bubble shape close to the top we find for the velocity of the bubble

$$U_2 b = \sqrt{gR/3\pi} = 0.33\sqrt{gR}. \quad (9.15.6)$$

The velocity of three-dimensional bubbles is larger: for the open bubbles, $U_{3b} = 0.51\sqrt{gR}$, and for the closed bubbles, $U_{3b} = 0.67\sqrt{gR}$.

In general the ratio of the burned gas density in a bubble, ρ_b and the fresh fuel density, ρ_f is non-zero, $\rho_b/\rho_f = 1/\Theta$. Correction for the finite density ration can be taking into account. Let us assume that the gas in the bubble is at rest, then the pressure inside the bubble is $P = \rho_b gz$. From the Bernoulli equation (9.15.1), taking into account the pressure balance at the surface of the bubble we have

$$\frac{1}{2}\rho_f u^2 + \rho_f gz = \rho_b gz. \quad (9.15.7)$$

As a result, in the formula for the bubble velocity term g should be replaced by $g(\Theta - 1)/\Theta$, so that the Eq. (9.15.6) becomes

$$U_b = 0.33\sqrt{(\Theta - 1)gR/\Theta}. \quad (9.15.8)$$

It should be noticed that, in the case of non-zero bubble density the stationary flow is unstable also against the Kelvin-Helmholtz instability of the heavy matter sliding along the surface of the light matter. As a consequence, the nonlinear stage of the RT instability is often accompanied by development of mushroom shaped structures in the jets of the falling heavy matter and by mixing of the heavy and light gases.

9.16 Flame in Horizontal and Vertical Tubes

When a flame propagates upwards in vertical tubes in a gravitational field we deal with the configuration of open bubbles. Since according to (9.15.6) the bubble velocity increases with the tube radius we can expect that for a sufficiently wide tube the velocity of the light bubble of burned matter, which may be much larger than the normal velocity of a planar flame front, determines the flame velocity. Thus, if $gR/U_f^2 >> 1$, the problem of flame propagation does not include the process of burning and fuel consumption but just velocity of rising bubble formed by the flame. If one considers solution of the complete set of the equations of flame dynamics in a strong gravitational field, then the bubble velocity represents the principal term of the expansion of the flame velocity into power series of the small parameter $U_f^2/gR << 1$.

Typical shape of the axisymmetric curved flames with the expansion coefficient $\Theta = 5$ in a cylindrical tube with the radius $R = 2R_c$ obtained in numerical experiment, which is similar to the shapes of the curved flames observed experimentally in vertical tubes, is presented in Fig. 9.18 for different values of the dimensionless acceleration $gL_f/U_f^2 = 0, 1, 3$. It is seen that the flame front curvature increases with the increase of the dimensionless parameter gL_f/U_f^2: for larger acceleration, $gL_f/U_f^2 = 3$, which corresponds to $gR/U_f^2 = 75$, the shape of the flame looks like a rising semi-infinite bubble. The only important difference of the flow caused by the rising flame bubble from the bubble flow is the absence of the infinitely long jets of the falling heavy matter. The heavy fuel of the jets is consumed by the flame front at some distance from the bubble top due to the "fire-polishing" effect, and the larger the dimensionless gravity acceleration, the longer the fuel jets.

The velocity of a convex flame rising in a cylindrical tube is considerably larger than the velocity of a planar flame. The velocity of convex flames increases with the increase of the gravity acceleration, but not as fast as the velocity of an inert bubble $\propto \sqrt{g}$, which may be explained by the influence of specific flame properties. With account of both the flame properties and the effects of bubble motion the velocity of an upward propagating flame can be approximated as

$$U_w = \sqrt{U_{DL}^2 + U_b^2} = \sqrt{U_{DL}^2 + \beta^2 \frac{\Theta - 1}{\Theta} gR}, \tag{9.16.1}$$

where U_{DL} is the velocity of the curved flame for the case of zero gravity and the numerical factor β is of the order of unity, $\beta_{3D} = 0.51$ and $\beta_{2D} = 0.35$. The expression (9.16.1) provides rather likely realistic values for the flame velocity compared to what is seen in experiments.

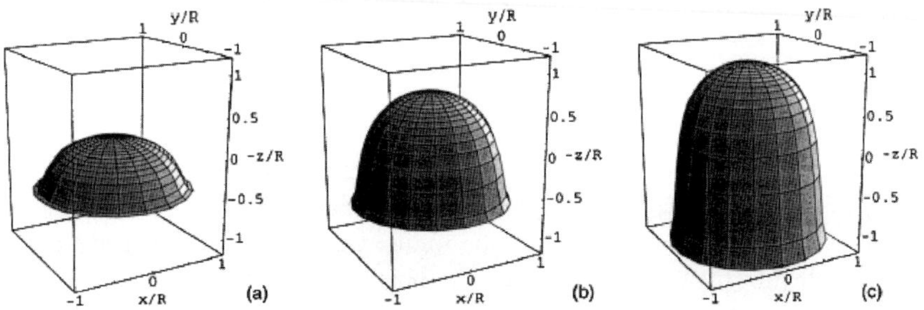

Fig. 9.18 The shape of the convex flame propagating upward in a cylindrical tube for different values of the dimensionless acceleration: $gL_f/U_f^2 = 0,1,3$; correspondingly: (**a**), (**b**), (**c**)

It is obvious from the Eq. (9.16.1) that influence of gravity and the bubble formation on the flame velocity is more important for slow flames. For flames propagating in the tubes with radius of about $R \approx 3$ cm, the gravity is important for slow flames with velocities $U_f < 20$ cm/s such as propane flames. On the contrary, the effect of gravity is negligible for fast hydrogen-oxygen flames with the velocities about 10 m/s.

Flames propagating in horizontal tubes of large diameters is also influenced by the gravity effects similar to flames moving upwards: the hot burned gas of smaller density tends to occupy the upper part of the tube, while the heavy fuel extends along the lower part. As a result the flame front acquires a curved shape and the flame propagates with larger velocities than the velocity of a planar flame. Since for most of the laboratory flames the fuel density exceeds the density of the burning matter about 10 times, so that the model of a massless bubble of the products of burning is quite satisfactory, and if the gravity effects dominate, a good approximation of the flame velocity can be obtained assuming the model of an inert massless open bubble moving in a horizontal channel.

Consider a stationary flow caused by a 2D open bubble propagating in a horizontal channel of width R as shown schematically in Fig. 9.19. A characteristic feature of a flame in horizontal tubes is the sharp angle $\alpha_b = \pi/3$ at the top of the flame front, which is clearly seen in the snapshot photographs of a flame in horizontal tubes. The sharp angle can be clearly distinguished on the bright part of the flame front, where effects of thermal conduction and losses to the walls do not influence the flame structure. From dimension considerations it follows that the velocity of the bubble is $U_b = \beta\sqrt{gR}$. The numerical coefficient β can be found form the solution of the problem in a way similar to (9.15.2)–(9.15.5). Similar to the case of upward propagating flames we can construct an approximate solution for the flow around the bubble and match this solution with the solution at the bubble top. The final result gives $\beta = 0.43$. Thus, a massless bubble in a horizontal 2D channel of width R propagates with the velocity $U_b = 0.43\sqrt{gR}$. In the case of non-zero bubble density the velocity of the bubble is $U_b = 0.43\sqrt{(\Theta - 1)gR/\Theta}$. The obtained velocity of a horizontally moving bubble is quite close to the velocity of open bubbles rising upwards in ideal 2D vertical tubes.

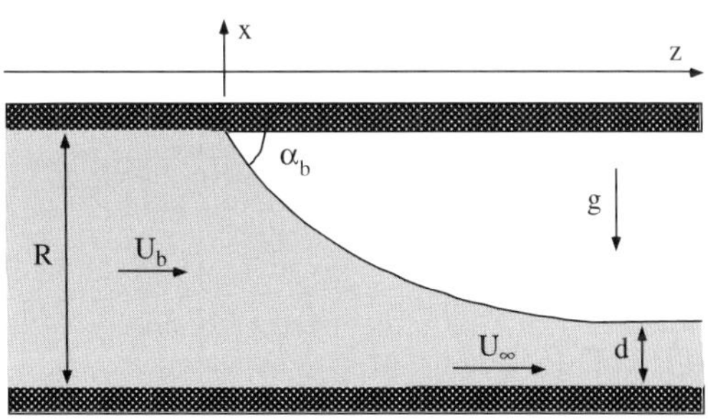

Fig. 9.19 Schematic configuration of an open bubble in 2D horizontal channel

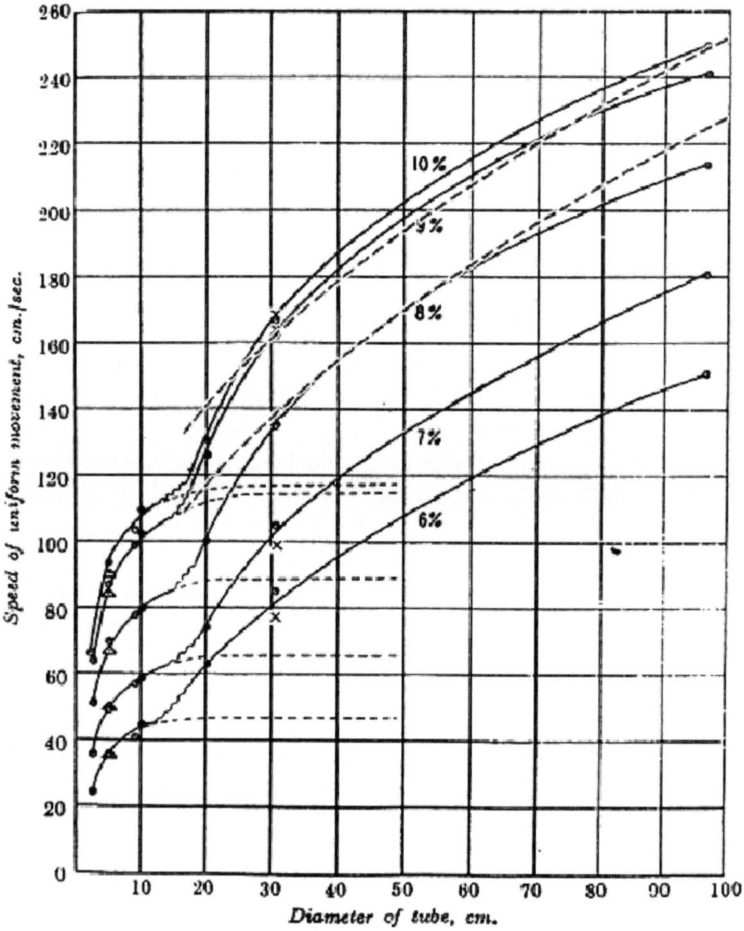

Fig. 9.20 Flame velocity in the methane-air mixtures versus the tube diameter

Effect of the buoyancy is more pronounced for slower flames and becomes noticeable for the tube widths exceeding $D \approx 5.4U_f^2/g$. This means that for slow flames, like in 6% CH_4 with the normal velocity $U_f = 6\,cm/s$ the gravity will already influence the flame velocity for the tube widths larger than $5.4U_f^2/g = 0.2\,cm$, and for faster flames, like in 10% CH_4, with $U_f = 43\,cm/s$, for the tubes $D > 10\,cm$.

For example, Coward and Hartwell (1932) found that velocity of the curved flame propagating in a wide horizontal tube (diameter 90 cm) in the mixture 10%CH_4 was $U_w = 245\,cm/s$, which considerably exceeds the normal velocity $U_f = 43\,cm/s$. The velocity of the light 2D bubble of burned products in the tube of such diameter can be estimated as $U_b = 0.43\sqrt{gD(\Theta - 1)/\Theta} \approx 120\,cm/s$. Taking into account that 3D bubbles propagate with larger velocity,

$U_{3b} \approx 0.64\sqrt{gD(\Theta - 1)/\Theta} \approx 180\,\text{cm/s}$, we obtain for the flame speed expression $U_W = 1.7U_f + U_{3b}$, which gives values close to the experimental measurements. This dependence of the flame velocity on the tube widths is shown in Fig. 9.20 by the dashed lines, where solid lines are from the experiments of Coward and Hartwell. An additional increase of the flame front velocity may be due to wall roughness (noslip boundary conditions) which is not taken into consideration in the present analysis.

9.17 Flame in a Closed Burning Chamber

A flame in a closed burning chamber is the most typical manifestation of combustion, and therefore flame dynamics in closed tubes is an important subject of combustion study. Particularly, flame propagation in a closed chamber is a key point for understanding most of industrial applications, such as e.g. combustion in engines. One of the most interesting problems is interaction of a flame with pressure waves, which can be viewed as a flame in an effective time-altering gravity field. When a flame front acquires a curved shape, the interaction of acoustic waves with the flame front becomes more effective, which makes easier the spontaneous reaction in the fresh fuel ahead of the flame front. Even in the case of ideal walls there are many specific factors that may influence development of the DL instability in closed tubes. Flames propagating in an unbounded fuel or in open tubes consume fuel in a nearly isobaric regime, since expansion of matter in the burning process is balanced by the flow of the burnt matter away from the flame front. The situation is quite different for flames propagating in closed tubes. In the latter case expansion of burning matter causes adiabatic compression of the fuel ahead of the flame. Compression of the fuel is accompanied by the pressure and temperature build-up, so that close to the end of burning the pressure and the temperature of the unburned fuel may considerably increase

$$P_1 = P_0\left(1 + \frac{\gamma Q}{C_P T_0}\right) = (\gamma\Theta_0 + 1 - \gamma)P_0, \tag{9.17.1}$$

$$T_1 = (\gamma\Theta_0 + 1 - \gamma)^{1-1/\gamma}T_0, \tag{9.17.2}$$

where the initial and final fuel parameters in a closed tube are denoted by the label 0 and 1, respectively.

For $\Theta_0 = 8$, $\gamma = 1.4$, for example, the final pressure is $P_1 \approx 11\,P_0$, exceeding by order of magnitude the initial pressure, and the final fuel temperature and density are $T_1 = 1.95\,T_0$, $\rho_1 = 5.47\,\rho_0$.

The increase of fuel pressure and temperature alters all characteristics of the flame such as the normal flame velocity, the flame thickness, the expansion coefficient, etc. Though the fuel parameters change in time, these changes are rather slow compared to the typical time scale of the flame L_f/U_f. Therefore there is enough time for the internal flame structure to adjust to the change in external conditions, so that the flame can be considered as a freely propagating quasi-isobaric flame with some new values of the fuel pressure and temperature. According to Eqs. (5.4.19–20), dependence of the normal flame velocity on the initial pressure and temperature implies different behavior for flames with different order of the reaction. Practically, the normal flame velocity changes negligibly for a flame of the first order reaction, however, for the third order reaction the normal flame velocity at the end of burning increases by an order of magnitude.

The flame thickness decreases considerably to the end of burning for flames of both the third order reaction and the first order reaction, though in the former case the decrease is much stronger. For a flame with $n = 1$ the decrease of the flame thickness happens mostly due to increase of the fuel density. On the contrary, in the case of a third order reaction $n = 3$ both amplifications of the fuel density and the normal flame velocity contribute to the decrease of the flame thickness. The scaled flame velocity and the flame front thickness are shown in Fig. 9.21 versus the flame front position in a tube of length $L = 500\ L_{f0}$ for $\Theta = 5$, $\gamma = 1.4$ for the first, $n = 1$ and for the third, $n = 3$ order reactions. Eventual decrease of the flame thickness in a closed tube makes competition of perturbation modes on the nonlinear stage of the DL instability more complicated: decrease of the flame thickness leads to the decrease of the cut-off wavelength λ_c, so that perturbation modes, that have been stable at the beginning of burning, may become unstable later. Flame dynamics in a closed chamber may be also affected by flame acceleration or deceleration that reaches significant values in short tubes and leads to strong suppression or enhancement of flame instabilities.

Note, that velocity of the flame propagating in a closed tube differs considerably in laboratory coordinate system from the normal flame velocity. In the reference frame of a planar flame front the fuel moves to the front with velocity U_f and the burned gases are drifted away with the velocity ΘU_f. At the beginning the burned products are at rest, the unburned fuel is pushed forward

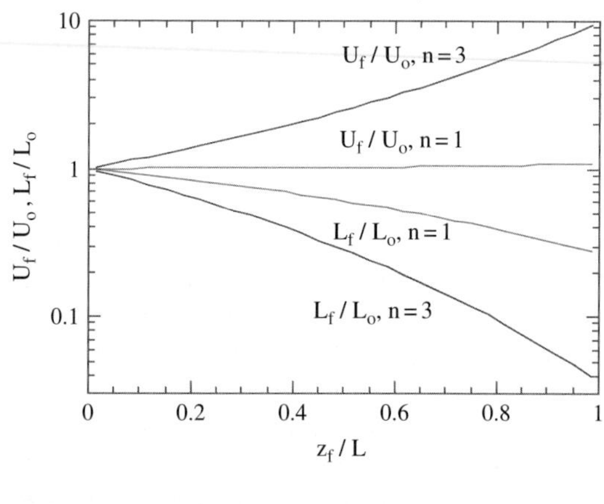

Fig. 9.21 The scaled velocity U_f/U_f0 and the flame thickness L_f/L_0 depending on the flame position in the closed tube for the flames of the first, $n = 1$ and third $n = 3$ order reactions

and the velocity of flame propagation in the laboratory coordinate system is $U_l = \Theta_0 U_{f0}$. When the flame comes to the opposite end of the tube, the fuel is at rest, the burned matter is pushed backward and the flame velocity relative to the walls equals the normal flame velocity. This means that the flame undergoes acceleration or deceleration, while travelling through the closed tube. The effective acceleration of the flame in a tube of length L may be estimated as

$$g = \frac{U_1^2 - \Theta_0^2 U_0^2}{2L}, \qquad (9.17.3)$$

and it turns on in addition the Rayleigh-Taylor instability of the flame front, so that the hydrodynamic instability of the flame can be either enhanced or weakens by the RT instability.

9.18 Flame Quenching and Flammability Limits

Not all the combustible mixture is capable of sustaining combustion waves. An addition of diluent gases in sufficient quantity may limit the flammability of the mixture. These diluents may consist of excess of fuel or excess of oxidants or it can be inert gases. For the stoichiometric mixture the combustible mixture being ignited reaches a sufficiently large temperature, moves into the explosive regime and supports the flame propagation. However, being diluted to the sufficient lean or rich concentration mixtures being ignited are not able to reach a sufficiently large temperature and not capable of sustaining combustion waves. The ability of the given fuel-oxidizer mixture ratio to support reaction propagation is related to the chemical rate of the given mixture ratio and the rate of heat generated in the reaction, and the flammability limit is defined by a competition between the rate of heat generated in the reaction and the external rate of heat loss by the flame. For some limiting mixture ratio when the mixture is too rich or too lean, its heat release is not sufficient to sustain the self-supported reaction propagation. With the increased dilution of the combustible mixture, when the mixture ratio falls into the lean or rich ends of the concentration spectrum, the burning temperature decreases and as a consequence also decreases the combustion wave speed. As regarding the fuel-oxidizer mixture ratio the flammability limit is defined by the minimum concentration of a combustible substance that is capable to support self-sustained propagation of the reaction wave. In real conditions the flammability limit is determined by the competition between the rate of heat generated by reaction and the external rate of heat loss by the flame. As a result of the external heat losses (heat losses to the walls, radiation, etc.) the flame temperature becomes lower than adiabatic flame temperature, and correspondingly the flame speed is smaller than normal flame velocity. Heat losses in turn depend on the velocity of the flame propagation. The leanest and richest concentrations, which just able to support flame, are called the lean and rich flammability limits.

A flame propagating in tubes is quenched if mechanisms of species diffusion and heat transfer, which are responsible for flame propagation, are affected by losses to the tube wall. Heat and momentum losses to the walls can reduce the reaction rate to the point when it can not sustain the propagating reaction. If flame propagates in a wide tube one can neglect by the heat losses to the cold wall. In narrow tubes the heat losses increase due to the increase of the transverse temperature gradient, and for some critical tube diameter – *quenching distance*, the self-supported flame propagation becomes impossible. Assuming that the reaction propagates in the deflagration regime as a plane flame in a tube of given diameter, and the heat losses to the tube walls, which are at the room temperature, will be the only mechanism responsible for the flame quenching, we can find critical tube diameter. For a simple case of two-dimensional problem this is the minimal distance between two parallel plates when reaction can propagates. The quenching distance corresponds to the distance between the plates, such that the rate of heat generated in the reaction is equal to the rate of heat losses to the walls. If the heat of the reaction per mole for the given stoichiometric mixture is Q_R, and τ_R is the reaction rate (mole/t $\cdot$ L^3), then the rate of heat generating in the flame is

$$\dot{q} \propto \tau_R (dS \cdot d) Q_R, \tag{9.18.1}$$

where dS is a surface element at the wall, d is the distance between the walls.

The rate of heat losses to the wall is

$$q_T = \lambda \frac{dT}{dx} dS. \tag{9.18.2}$$

Let T_q is the lowest temperature at which the flame still can propagate, and T_0 is the temperature of the cold wall. Approximating the rate of the heat losses to the wall as

$$q_T = \lambda \frac{dT}{dx} dS \cong 2\lambda \frac{T_q - T_0}{(d_q/2)} dS, \tag{9.18.3}$$

we obtain equation for the minimal distance between the wall when flame still can propagate (quenching distance):

$$4\lambda \frac{T_q - T_0}{d_q} dS \propto \tau_R (d_q dS) Q_R. \tag{9.18.4}$$

Taking into account that $Q_R = C_P(T_f - T_0)$, we obtain for the quenching distance

$$d_q^2 \propto \frac{4\lambda}{C_P \tau_R} \frac{T_q - T_0}{T_f - T_0}. \tag{9.18.5}$$

Taking into account that $U_f \propto \sqrt{\lambda/C_p\tau_R}$, we conclude that $d_q \propto U_f$. As larger is normal flame speed as larger the quenching distance for this particular mixture; the shorter is time for the heat losses to the wall. As initial temperature of the unburned mixture increases, the flame is able to pass through narrower channel. The quenching distance also depends of pressure as, $d_q \propto 1/P$.

Problems

9.1. Estimate the growth rate of the plane flame front instability form dimensional consideration. Write the equations for a linear problem of an infinitely thin flame front stability.

9.2. It is desired to stabilize flame propagating in a tube. How flame stabilization depends on the tube radius? Will it be stabilized for larger or smaller tube diameter? Is flame more stable propagating up or down in vertical tube? How it depends on gravity acceleration?

9.3. Flame propagating in a tube acquires a nearly parabolic shape due to heat losses to the tube wall. Will the change of the flame shape influence the speed of the flame?

9.4. In an experiment a plane flame is initiated near the closed end of a horizontal tube and propagates to an opposite open end in the first experimental investigation, and it is initiated near the open end and propagates to the closed end in another case. Determine velocity distribution in the flow ahead and behind the propagating flame in both cases. What is the flame velocity in laboratory co-ordinate system, if the normal flame velocity is U_f and the expansion coefficient $\Theta = \rho_b/\rho_u$ is known?

9.5. An experimentalist wants to measure the normal flame velocity for a given combustible mixture. In series of the studies he took mixture with different initial temperatures. What will be result of the experiment? How velocity of the flame and temperature of the burned products will be changed depending on the initial temperature?

9.6. Normal velocities of some planar flames at normal conditions are: $U_f = 5.0$, 50, 220, 520, 900, 1330 cm/s for the mixtures: 6%CH_4+Air, 5%C_3H_8+Air, H_2+Air, H_2+0.5O_2+ 0.5 N_2, C_2H_2+2.5O_2, correspondingly. What must be possible oncoming flow velocities for these mixtures to obtain a steady flame in the Bunsen burner? What is the angle of the Bunsen flame cone? Assume a two-dimensional version of the axisymmetric Bunsen burner. How result will change for cylindrically symmetric burner? How would you design the experiment for flame speed measurement using the Bunsen burner?

9.7. According to experimental studies the average radius of the outward propagating spherical flame moving through combustible mixture grows as $t^{3/2}$. This implies that the flame interface area grows faster than the square of its average radius, i.e. that the wrinkled interface assumes a fractal-like structure. Show that assumption about fractalization of the infinitely thin flame front is compatible with the 3/2 – power law of the experimental data.

9.8. A hydrocarbon-air mixture is ignited at the center of a soap bubble of diameter 3 cm. Assuming that flame starts propagating as a spherical ball with unburned gas temperature $T_u = 300$ K, and temperature of the products $T_b = 2000$ K. The velocity of the expanding spherical flame relative to the burned products is $U_b = 200$cm/s. What is the normal velocity of a laminar flame?

9.9. A stoichiometric fuel-air mixture forms a well-defined conical flame when burning in a Bunsen burner. The mixture is then made leaner. For the same flow velocity in the tube, how does the cone angle is changed?

9.10. Compare the velocity of a flame in stoichiometric hydrogen-air mixture with hydrogen-air lean and rich mixtures. In which case the flame speed is larger? Explain.

9.11. A laminar flame propagates in a horizontal tube 2 cm in diameter. The normal velocity of a planar flame is 43 cm/s. Due to buoyancy effect the flame front tilts at 30 degrees to the wall forming a shape of half parabola. Estimate the velocity of a flame, and the burning rate of the mixture in gram/s if initial density of the mixture is 0.015 g/cm^3, and the flow is assumed laminar.

9.12. The flame speed in hydrocarbon-air mixture is 43 cm/s, and the activation energy is 40 kal/mol. Assuming that the adiabatic temperature is 1800 K, and adding of the inert gas lower temperature up to 1500 K, how will change the flame speed in the diluted mixture?

9.13. It is desired to measure the laminar flame speed in the mixture of 11.6% CH_4 + air using the burner tube method. For such mixture the normal flame velocity, temperature and expansion ratio are $U_f = 25$ cm/ s, $T_b = 2100$ C, $\rho_u/\rho_b = 7.5$. What a tube diameter should be taken for the measurement?

Chapter 10
Regimes of Premixed Flames

Majority of our discussion is focused on premixed gas combustion, the combustion of gaseous reactants that are perfectly mixed prior to ignition. Premixed combustion requires that fuel and oxidizer to be completely premixed before combustion. Although premixed combustion may occur homogeneously throughout the volume, e.g. as an explosion, this mode is not typical. The most distinctive feature of premixed combustion is its ability to assume the form of a self-sustained reaction wave initiated in some location and then propagating either subsonically (deflagration) or supersonically (detonation) at a well-defined speed. Once fuel and oxidizer have been homogeneously mixed and heat source is supplied, it becomes possible for a reaction wave to propagate through the combustible mixture. There are several important applications of premixed combustion, the principal being spark-ignition engines, lean-burn gas turbines, household burners, recently renewed interest to pulsed detonation propulsion, and many others. Apart from their technological relevance, premixed combustion constitutes a truly fascinating dynamical system, displaying an amazingly rich variety of phenomena such as non-uniqueness of possible propagation regimes, their birth (ignition) and destruction (extinction), chaotic self-motion and fractal-like growth, and various hysteretic transitions.

10.1 Combustion Waves in Laminar Flow

Propagating combustion zone, which is called combustion wave, can be initiated in a combustible mixture by a hot wire, or an electric spark, or by some other ignition sources capable to initiate a self-sustained reaction wave. In this section the discussion will be restricted to laminar regimes of flame propagation, whereas the effects of turbulence on flame propagation will be considered in the next section.

An ignition source, which is in general a source of heat, raises temperature and initiates the reaction. However, a heat source such, for example, as an electric spark also can produce atoms and free radicals, which may act as chain carries in the chemical reaction. The flux of heat as well as the flow of

M.A. Liberman, *Introduction to Physics and Chemistry of Combustion*,
DOI: 10.1007/978-3-540-78759-4_10,

chain carries from the ignition source initiates chemical reaction in an adjacent layer of the combustible mixture, so that this layer itself becomes a source of heat and/or chain carries capable of initiating chemical reaction in the next layer, and so on. In this way a zone of reaction propagates through the mixture, and as we discussed this in the previous chapters, the speeds of the wave propagating due to mechanisms of either thermal conduction or diffusion are much less than sound speed.

Within the layer of the propagating burning wave, the rate of chemical reaction accelerates considerably with rising temperature according to the Arrhenius law, $W(T) \propto \exp(-E/RT)$, in the manner similar to a thermal explosion. Depending on the reaction mechanism, the intermediate chemical products, atoms, free radicals, may play a role as chain carries. They also may cause self-acceleration of the reaction if the chains are branched (see Chap. 3). Bearing in mind that the flow ahead and behind of the flame front is essentially subsonic, the flow can be treated as incompressible. The pressure difference across a flame front is very small by the same reason. Taking into account continuity equation, $\rho_u U_u = \rho_b U_b$, we find the pressure difference across the flame front

$$P_u - P_b = \rho_u U_f^2 \left(\frac{\rho_u}{\rho_b} - 1 \right) \tag{10.1.1}$$

There exist also so called cool flames, which are formed in burning hydrocarbon (benzene) and play important role in understanding autoignition, which is knocking regime of engine combustion. For the cool flames the increase of the free-radical concentrations is a specific mechanism of the global reaction acceleration. However, in the case of usual thermal or "hot" combustion, the temperature rise is the main and the most important factor in the reaction self-acceleration and the wave propagation compared to the increase of free radical concentration or the chain branching.

A combustion wave being initiated from a small point-like ignition source, such as electric spark, will propagate as a spherical expanding wave. In this case, the flux of the heat from the burned to unburned mixture is divergent, and this should have effect of diluting the heat that is transferred to the unburned gas, which may result in the wave to be slowed down or even quenched. However, it was established both theoretically and experimentally (Chap. 9) that spherical flames spreading out from the ignition source undergo considerable acceleration with the temporal dependence of the average radius of the wrinkled flame ball.

We consider the simplest configuration of a flame in the Bunsen burner shown schematically in Fig. 10.1a, b, which is a common experimental device used in laboratories. Gaseous fuel from the fuel supply enters through an orifice into the mixture chamber, where air is entrained from the outside through adjustable openings. The opening area of the fuel orifice may be adjusted by moving the needle through an adjustment screw into the orifice, thereby allowing the velocity of the jet entering into mixture chamber to be varied and the

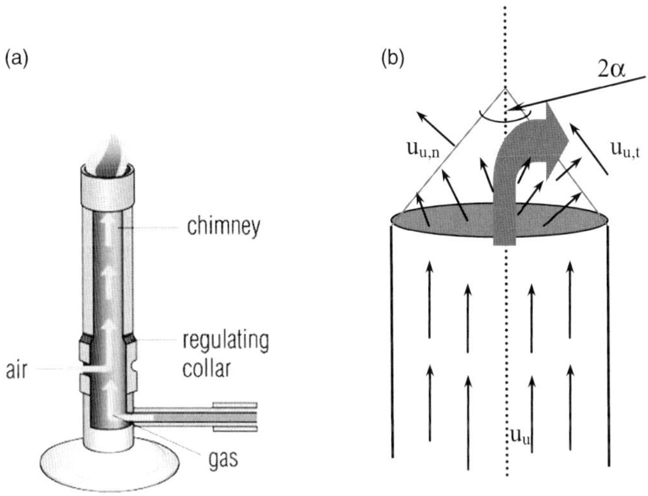

Fig. 10.1 (**a, b**) Bunsen burner

enter of the air and the air-fuel mixture to be optimized. Depending on the velocity of the jet, both laminar and turbulent steady premixed flames can be obtained on a Bunsen burner. If the velocity of the mixture is sufficiently large, the flow inside the Bunsen tube becomes turbulent and we obtain turbulent flame, if the velocity is low, the flame is laminar. Obviously, the laminar or turbulent flames are established if the speed of the flow running out the Bunsen tube is larger than the corresponding speed of laminar or turbulent flame. In this case a flame cone is formed on the Bunsen tube.

Let us find the shape of the flame in a Bunsen burner. For the sake of simplicity, we consider two-dimensional problem and calculate the angle of the flamer cone. Let angle of the flamer cone above the burner tube is 2α (Fig. 10.1a, b). The oncoming flow velocity of the unburned mixture $\mathbf{u}_u$ has normal component, $u_{u,n}$ and tangential component, $u_{u,t}$ to the flame front, which is the planar side of the cone (Fig. 10.2).

From continuity condition of the mass flow in normal direction to the flame, $\rho_u u_{n,u} = \rho_b u_{n,b}$, we find $u_{n,b} = u_{n,u}\rho_u/\rho_b$. Obviously, the tangential velocity component must be continuous, $u_{t,b} = u_{t,u}$. Since the whole flow is supposed to be steady, and the flame front is stationary, the normal velocity of the flow from the side of the unburned mixture is nothing, but the normal burning velocity of a planar flame, $u_{n,u} = U_f$. Taking into account that $u_{n,u} = |\mathbf{u}_u| \cdot \sin\alpha$, we obtain

$$\sin\alpha = U_f/|u_n|. \qquad (10.1.2)$$

This expression for the cone angle is valid also in case of turbulent flame if instead of U_f we will use the mean velocity of turbulent flame. Evidently, velocity of the flow in the Bunsen tube must be larger than the normal flame

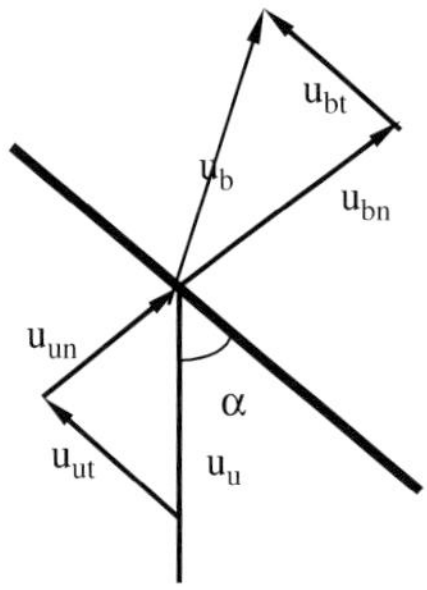

Fig. 10.2 Schematic configuration of flame and flow velocities in Bunsen burner

velocity. If we do know the flow velocity in the Bunsen tube, we can measure velocity of the flame measuring the cone angle.

Accurate solution to the problem of the flame shape in a Bunsen burner will require more calculations. Let us assume that at the round exit of the burner orifice with radius R_0 velocity distribution corresponds to the Poiseuille flow

$$u = U_0\left(1 - \frac{r^2}{R_0^2}\right), \tag{10.1.3}$$

where $u = u(r)$ is the velocity along the tube, U_0 is the maximal velocity on the axis.

The normal to the flame surface component of the velocity must be equal to the normal velocity of the flame:

$$u_n = U_f = |u| \cdot \sin\alpha = U_0\left(1 - \frac{r^2}{R_0^2}\right) \cdot \sin\alpha \tag{10.1.4}$$

For the flame surface given by the equation: $r = r_f(z)$, we have $tg\alpha = dr_f/dz$, and

$$\sin\alpha = \frac{dr_f/dz}{\sqrt{1 + (dr_f/dz)^2}}. \tag{10.1.5}$$

Substituting (10.1.5) in (10.1.4) we obtain ordinary differential equation for the flame shape, which can be written as

$$\frac{dy}{dx} = \pm\frac{v}{\sqrt{(1 - y^2)^2 - v^2}}, \tag{10.1.6}$$

where $y = r_f/R_0$, $x = z/R_0$, $v = U_f/U_0 < 1$.

The integral in (10.1.6) can be expressed in elliptical integral of the first and second kind, though further analysis requires more careful consideration of the physical process, such as thermal losses at the orifice edge and burning conditions at the nearest vicinity of the orifice. The reason is that since the flow velocity near the tube wall is low the heat losses to the burner rim are an important factor in stabilizing the flame at the top.

One may distinguish different colors within the Bunsen cone. Inside the cone is the dark zone, which is nothing but the unburned premixed gas before it enters the reaction zone. The outer zone, which is the flame front itself, is the luminous zone, typically less than 1 mm thickness. Here the temperature is the highest and here is where the reaction and heat release take place. Visible light and color of the luminous zone in the Bunsen cone are different depending on the used fuel and whether the mixture is fuel lean or fuel rich. For example, hydrogen-air flames are almost invisible. Color of the lean hydrocarbon-air flames is a deep violet due to formation of exited CH radicals, while due to excited C_2 molecules the tint of green color appears at the rich hydrocarbon-air flames. For high enough temperature of the burned gases the reddish tint may appear due to water vapor and CO_2 radiation. If mixture is very rich, then due to carbon particles appears an intense yellow radiation.

10.2 Regimes of Premixed Turbulent Combustion

Turbulence may considerably influence the propagation of combustion wave. If in case of a laminar flame the flow conditions do not alter the rate of energy release in the reactions or chemical mechanisms, situation can be different in presence of turbulence. In the turbulent flow there are fluctuating components of velocity, temperature, density, pressure and concentration, and it is possible that such fluctuating components affect the chemical reactions and the heat release rate. The flame structure depends upon the relation of characteristic time associated with the chemical reaction ($\tau_{Ch} = L_f/U_f$) and the characteristic time associated with turbulent velocity fluctuating components ($\tau_t = \lambda/u'$). If $\tau_{Ch} \ll \tau_t$, then the velocity fluctuating components do not influence essentially the chemistry, if the opposite condition is true $\tau_{Ch} \gg \tau_t$, then turbulence can affect the rate of chemical reaction, the energy release and the flame structure. The concept of turbulent reacting flow may have many different meanings and depends on the interaction range, which is governed by the overall character of turbulent flow. For example, in the case of very strong turbulent pulsations, the fluctuating temperature and concentrations may affect the chemical reactions and heat release, and then the combustion products may be rapidly mixed with reactants within a time interval shorter than the chemical reaction time. The case of strong influence of turbulence is called *stirred reactor* regime, and there is no meaning of a flame structure exists in such combustion. If external turbulent flow is relatively weak contrary to the previous example, which means that the turbulent flow is dominated by large-scale low intensity turbulence, then turbulence will affect the premixed flame so that it becomes a wrinkle one, either enhancing its intrinsic hydrodynamic instability, or producing wrinkle flame independently. In the latter case the flame remains locally a laminar one. As the intensity of turbulence increases, local laminar structure of the flame remains until scale of the smallest turbulent eddies is large compared to the laminar flame width. This regime of a quasi-laminar flame in presence of turbulence is called a laminar *flamelet* regime. With further increase of the turbulence intensity the laminar flame structure will be destroyed.

Based on this consideration we can introduce dimensionless parameter, which is Damköhler number

$$\mathrm{Da} = \frac{\text{turbulent time}}{\text{chemical time}} = \frac{\tau_t}{\tau_{Ch}} = \frac{U_f}{L_f}\frac{\lambda}{u'}. \tag{10.2.1}$$

If $\mathrm{Da} < 1$, turbulent pulsations of fluid induce change in the reactants composition within a time interval shorter than the chemical reaction on the scale of the flame thickness. In this regime, the turbulent eddies are nearly all embedded in the reaction zone, so that meaning of the "flame front" has no sense anymore.

Introducing the turbulent Reynolds number (see Chap. 7)

$$\mathrm{Re}_\lambda = \frac{u'\lambda}{\nu}, \tag{10.2.2}$$

and taking into account that by the order of magnitude the thickness of the flame front is $L_f \approx \nu/U_f$ (since thermal conductivity, $\kappa = k/\rho C_p$ equal to the kinematic viscosity ν), we can rewrite expression (10.2.2) as

$$\mathrm{Re}_\lambda = \frac{u'\lambda}{L_f U_f}. \tag{10.2.3}$$

Such defined the turbulent Reynolds number gives the measure of turbulence influence on the reacting flow. In the region, $\mathrm{Re}_\lambda = u'\lambda/L_f U_f < 1$, where $\lambda < L_f$ and $u' < U_f$ the flame is laminar. It is convenient to present different regimes of flame propagation using so-called Borghi diagram (Fig. 10.3), where vertical axis is non-dimensional ratio u'/U_f, and horizontal axis is non-dimensional ratio λ/L_f. Region of the laminar and wrinkled flames are located below $\mathrm{Re}_\lambda = 1$ and $u'/U_f = 1$. The line corresponding to the equation $\mathrm{Da} = 1$ $(\tau_t = \tau_R)$ separates the region of well-stirred reactor regime. If we introduce also dimensionless parameter – Karlovitz number, which characterize the ratio of the laminar flame scale in terms of the Kolmogorov's turbulent scale

$$\mathrm{Ka} = \left(\frac{\text{flame width}}{\text{Kolmogorov scale}}\right)^2 = \frac{L_f^2}{\lambda^2} = \frac{u'^2}{U_f^2}, \tag{10.2.4}$$

then the corrugated flames are situated below this line in the Borghi diagram.

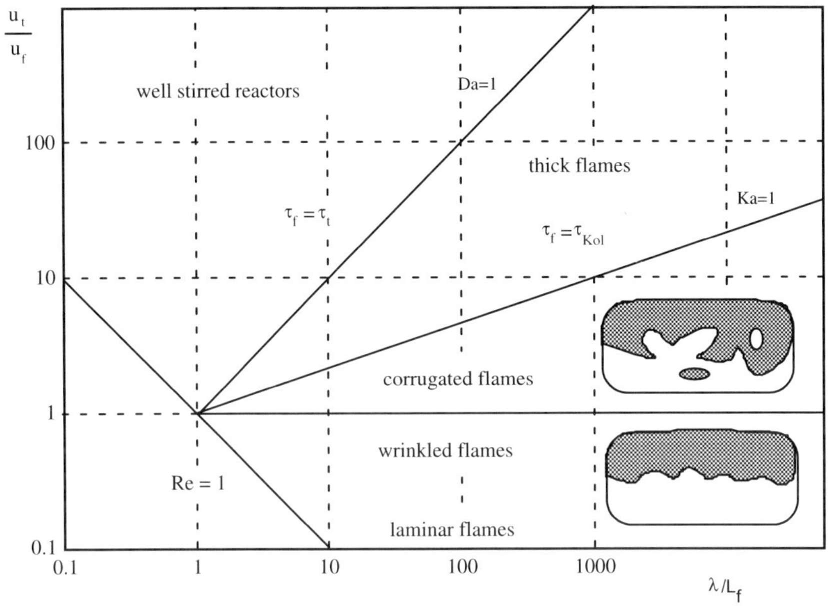

Fig. 10.3 Borghi diagram

The intermediate regime, between Ka = 1 and Da = 1, corresponds to thick flames, when intensity of turbulence is high so that turbulence changes the transport properties of the gas and weakens the flame, though flame still represents a configuration, when burned gas is separated from unburned mixture.

The most common regime is the flamelet regime, when thickness of the flame is much smaller compared to the smallest turbulent scale – Kolmogorov scale. These combustion regimes are typical for combustion in engines, industrial burners and gas turbines.

The flamelet regime of premixed turbulent combustion can be described using phenomenological scalar G-equation. Let us consider surface given by the scalar function $G(\mathbf{r}, t) = G_0$, such that the region of the burned gas corresponds to $G(\mathbf{r}, t) > G_0$, and the region of the unburned gas corresponds to $G(\mathbf{r}, t) < G_0$, where $\mathbf{r} = (x, y, z)$. Let $\mathbf{n} = -\nabla G/|\nabla G|$ is the unit normal vector to the front surface in the direction of the unburned gas. The time derivative of the function $G_0 = G(\mathbf{r}, t)$ is

$$\frac{dG_0}{dt} = \frac{\partial G}{\partial t} + \nabla G \cdot \frac{d\mathbf{r}_f}{dt} = 0, \tag{10.2.5}$$

where $\mathbf{r}_f$ is the vector coordinate at the surface.

It is obvious that the propagation velocity $d\mathbf{r}_f/dt$ of the front equals to sum of the flow velocity $\mathbf{u}$ and the component of the local burning velocity normal to the front

$$\frac{d\mathbf{r}_f}{dt} = \mathbf{u} + \mathbf{n} \cdot U_f = \mathbf{u} - \frac{\nabla G}{|\nabla G|} U_f, \tag{10.2.6}$$

where for normal vector we used expression $\mathbf{n} = -\nabla G/|\nabla G|$.

Substituting (10.2.6) into (10.2.5) we obtain equation for G

$$\frac{\partial G}{\partial t} + \mathbf{u} \cdot \nabla G = U_f |\nabla G|. \tag{10.2.7}$$

As an example of application of G-equation, let's calculate the angle of the flame cone in a Bunsen burner. We consider two-dimensional problem with angle of the flame cone above the burner tube being 2α, axis x being the symmetry axis directed along the central line of burner channel, and the burner rim is located at $x = 0, y = \pm d$, similar to the problem considered earlier in Sect. 10.1. For the steady flow with the oncoming vertical flow velocity of the unburned mixture u, the G-equation (10.2.7) takes the form

$$u \frac{\partial G}{\partial x} = U_f \sqrt{\left(\frac{\partial G}{\partial x}\right)^2 + \left(\frac{\partial G}{\partial y}\right)^2}. \tag{10.2.8}$$

For our purpose it is enough to consider only inner part of the flame cone – the region of the unburned mixture. Solution to the Eq. (10.2.8) we shall seek in the form

$$G = G_0 + x + F(y), \tag{10.2.9}$$

where G_0 is some constant.

Substituting (10.2.9) into (10.2.8) we obtain

$$u = U_f \sqrt{1 + \left(\frac{\partial F}{\partial y}\right)^2}. \tag{10.2.10}$$

Integrating (10.2.10) gives

$$F(y) = \frac{\sqrt{u^2 - U_f^2}}{U_f} (|y| - \text{const}). \tag{10.2.11}$$

Taking into account the boundary conditions at the burner rim: $F(x = 0, y = \pm d) = 0$, and substituting this into (10.2.9) we will find solution of the G-equation

$$G = G_0 + x + \frac{\sqrt{u^2 - U_f^2}}{U_f}\left(|y| - \frac{d}{2}\right). \qquad (10.2.12)$$

According to derivation of G-equation the flame is the surface $r = r(x_f, y_f)$, where $G(x, y) = G_0$, therefore for the flame surface we have

$$x_f = \frac{\sqrt{u^2 - U_f^2}}{U_f}\left(\frac{d}{2} - |y_f|\right). \qquad (10.2.13)$$

From this formula it is clear that flame may exist only if the flow velocity in the tube is larger than the normal flame velocity, and for the flame tip, which is located at $y = y_f = 0$ we find

$$x_{f0} = \frac{d}{2}\frac{\sqrt{u^2 - U_f^2}}{U_f}. \qquad (10.2.14)$$

From the geometrical consideration, it is clear that the flame angle is

$$\tan\alpha = \frac{d}{2x_{f0}} = \frac{U_f}{\sqrt{u^2 - U_f^2}}. \qquad (10.2.15)$$

It is easily seen that this expression is the same that we already obtained at the end of Sect. 10.1, $\sin\alpha = U_f/|u_u|$. It should be noted that this expression for the cone angle is valid independently of either a flame is laminar or it is turbulent.

10.3 Velocity of Turbulent Flame

One of the most important tasks of a turbulent premixed combustion model is calculation of the mean fuel consumption rate. In the flamelet regime, which is the most common for practical use in industrial burners, engines, etc., flamelet propagates normal to itself, and therefore its location depends on the flow field and is determined by interaction of the locally laminar flame with the oncoming external turbulent flow. The flamelet regime itself can be divided into regimes of wrinkled flamelets and corrugated flamelets (see Fig. 10.3). In the regime of wrinkled flames, when $u' < U_f$, u' can be interpreted as circumferential velocity of the large eddies, which bend the flame front but which are "too large" to convolute the front so that to form islands or multiply reaction sheets. The interaction between week turbulence and the flame front in the regime of wrinkled flames can be described using asymptotic methods for large activation energy. In particular, effective method is to include phenomenologically term describing the flame respond to the external turbulence in the nonlinear equation of Sect. 9.12. In this way the nonlinear equation takes into account both influence of the weak external turbulence and the DL instability. The values of the turbulent flame velocity obtained in this way are in a good agreement with experimental measurements and reproduce the limiting cases for the flame speed obtained for the case of zero thermal expansion of the fuel mixture.

Rough estimate for the turbulent flame velocity of a wrinkled flame can be obtained according Damköhler as

$$U_t = U_f(S_f/S_t), \tag{10.3.1}$$

where S_f denotes the overall area of the wrinkled flame, S_t is the area of the mean turbulent flame front, and U_f is the normal laminar flame velocity.

Assuming that $S_f/S_t = 1 + u'/U_f$, one obtains

$$U_t = (1 + u'/U_f) \cdot U_f. \tag{10.3.2}$$

For the case of large activation energy the flame speed is very sensitive to temperature change in a thin reaction layer caused by the flame stretch induced by interaction with turbulent eddies since it changes relative diffusion of heat and the reactant. Then we conclude that positive stretch increases the enthalpy and therefore the temperature in the reaction layer and this results in the flame acceleration if $Le < 1$. Correspondingly, the flame decelerates if flame stretch is negative and $Le < 1$ or if flame stretch is positive and $Le > 1$.

The regime of corrugated flames is much more difficult to analyze both analytically and using numerical modeling since in this regime $Ka < 1$, and according to (10.2.4) $u' > U_f$ so that the velocities of largest eddies are greater than the normal flame speed and this means that the largest eddies may cause considerable convolution of the flame front, in particular, splitting the burned islands.

As the turbulence intensity u' increases, a maximum in the turbulent flame velocity U_t is observed. An interaction between the flame and large-scale eddies of the turbulent flow field results in the extension of the flame interface and thereby in the burning rate enhancement. Yet, it has long been observed that for each gaseous combustible mixture there is a certain level of turbulence intensity at which the flame speed reaches it maximum, and further increase of the turbulence intensity leads to a drop of the flame speed followed by the flame disintegration and extinction. The flamelet model suggests a likely explanation of the phenomena. The positive stretch caused by a high intensity turbulent flow reduces the reaction rate. The sign of the stretch alternates along the flame front in a turbulent flow. Turbulent eddies bring reactants toward the flame, and this leads to steeper gradient at the flame front, which in turn causes larger diffusion losses. Finally, with increasing of the turbulence intensity, the finite rate of the chemical kinetics is unable to generate products as fast as reactants are delivered. As a consequence, the flame temperature drops, which further lowers the reaction rate. Finally this leads to extinguish of the flame.

10.4 Spontaneous Reaction in Nonuniform Combustible Mixture

The regimes of propagation of a self-sustained reaction wave propagating either subsonically (deflagration) or supersonically (detonation) at a well-defined speed, which were considered in previous chapters, are caused by the diffusion of heat or reactants or by the change in the reaction rate due to compression and temperature increase in a shock wave. Therefore, these regimes represent a thin layer of the reaction wave, which is a wave front propagating in a medium that can be considered as uniform compared to large gradients within the wave front. In many cases the combustible mixture can be considered as being uniform everywhere, except of the thin layer of the reaction wave. Such approach is based on the essential difference in chemical reaction and characteristic hydrodynamic time scales. Though in general local self-ignition of the reaction represents a thermal explosion, which is accompanied by generation of the pressure waves heating and compressing the surrounding fresh mixture, in the case when chemical reaction and hydrodynamic time scales are essentially different the motion of the matter resulting from thermal expansion of a reactive mixture is of no importance and it can be not taken into account. Self-ignition of the reactive mixture takes place in the first time in a layer with highest temperature corresponding shortest induction time of the reaction, highest temperature of the mixture etc. In this case the reaction propagates over all the volume in the form of reaction wave, which is either deflagration or detonation wave.

Let reaction occurs in a vessel of characteristic size L. If the reaction wave propagates in deflagration regime, then the characteristic reaction time is $\tau_R = L/U_f$, and the characteristic hydrodynamic time scale is $\tau_a = L/a_s$. Since the flame velocity is much less than the sound speed, then $\tau_a \ll \tau_R$. In this case sound waves equalize the pressure over the whole volume, and the process is isobaric with accuracy $\Delta P/P \propto M^2 \ll 1$. Correspondingly, the motion of the matter and related hydrodynamic effects are negligible. The opposite limiting case, $\tau_a \gg \tau_R$, corresponds to self-ignition of uniform reaction pocket, which is a thermal explosion. Maximum pressure jump corresponding to thermal explosion is of the order of $P_2/P_0 \propto T_b/T_0 \propto 10$, so that shock waves generated by thermal explosion are relatively weak and not able to initiate a detonation in the surrounding cold mixture; the corresponding Mach number is $M_{sh} \propto 2$. Typical strength of the shock wave initiating detonation corresponds to a shock wave with the Mach number $M_{sh} \propto 6/8$. Detonation is an example of coupled motion of a shock wave and a chemical reaction front, which corresponds to condition $\tau_a \cong \tau_R$.

It is obvious that initiation a flame requires that the size of the reactive pocket needed to ignite a flame must be of the order of the laminar flame thickness with its temperature being close to adiabatic flame temperature. If the initial temperature of the fuel mixture is less than the adiabatic flame temperature, then the initial reaction time is much longer than the reaction time in a flame. If in addition the initial temperature distribution is nonuniform, and there is a gradient of the initial temperature, and corresponding gradient of the induction time, then the layers of the combustible mixture will react independently according to adiabatic thermal explosion. The reaction wave and energy release in this case will be propagating according to the nonuniform character of the initial temperature distribution, which Zel'dovich, who considered the problem for the first time, called spontaneous propagation of the reaction, or spontaneous reaction wave.

For the reaction of the Arrhenius type the reaction time depends on the initial fuel temperature as

$$\tau_R = \tau_0 \cdot \exp(E/T), \tag{10.4.1}$$

where we use energy units and assume that activation energy is large, $E/T \gg 1$, τ_0 is constant of time dimension.

When reaction starts, the fuel concentration changes only slightly and energy release is small. The reaction is completed as an abrupt explosion within a short time interval about $\tau_{exp} \simeq (T/E)\,\tau_R$. For a nonuniform initial temperature distribution $T = T(r)$, the reaction region can be considered as a wave of reaction propagating with the velocity

$$u_{sp} = \left(\frac{d\tau_R}{dr}\right)^{-1} = \left(\frac{d\tau_R}{dT}\right)^{-1} \frac{\nabla T}{|\nabla T|^2} = -\frac{T^2}{E\tau_R}\frac{\nabla T}{|\nabla T|^2} \tag{10.4.2}$$

The reaction region can be considered as a wave with well-defined front if the thickness of the reaction zone $\Delta_{sp} = u_{sp}\tau_R$ is small compared to the length scale of the temperature variation, $L_T \cong T/|\nabla T|$. This condition always holds for the large activation energy, except the region of the temperature maximum, or if $\nabla T = 0$,

$$\Delta_{sp} = u_{sp}\tau_R = \frac{T^2}{E}\frac{\nabla T}{|\nabla T|^2} \ll \frac{T}{|\nabla T|} \tag{10.4.3}$$

For a nearly uniform temperature distribution corresponding to a very small temperature gradient ($\nabla T \to 0$), when the velocity of spontaneous reaction is much larger than the local acoustic velocity, a nearly constant volume thermal explosion takes place. In the opposite limiting case, when the velocity of spontaneous reaction is much smaller than the local acoustic velocity, which corresponds to a large temperature gradient, the explosion is almost isobaric, which corresponds to a deflagration regime of reaction wave propagation.

For an intermediate temperature gradient a detonation wave can be initiated due to the initial temperature gradient if the generated shock wave is synchronized with the energy released by the self-ignition, which occurs at different time at each place.

If a hot spot and as a consequence the nonuniform temperature profile was formed at the fuel mixture, for example, with the temperature distribution in the form

$$T = T_0\left(1 - \alpha\frac{r^2}{R_0^2}\right), \tag{10.4.4}$$

where T_0 is the maximum temperature at $r = 0$, then the velocity of the spontaneous reaction wave is

$$u_{sp} = \frac{T}{\tau_R E}\frac{R_0^2}{2\alpha r} \tag{10.4.5}$$

The velocity is infinitely large at the $r = 0$, but at this location the wave is not formed yet. The condition (10.4.3), when the reaction wave front is formed, for the temperature profile (10.4.4) can be written as,

$$\Delta_{sp} = u_{sp}\tau_R = \frac{T^2}{E}\frac{\nabla T}{|\nabla T|^2} \ll R_0 \;\; \Rightarrow \;\; r \gg \frac{T}{\alpha E}R_0. \tag{10.4.6}$$

At the beginning, when the spontaneous wave is formed, its velocity is

$$u_{sp} = \frac{T}{\tau_R E}\frac{R_0^2}{\alpha r} \le \frac{R_0}{\tau_R} \tag{10.4.7}$$

The synchronization between the traveling shock wave produced by the local volume explosion at high temperature region and the traveling energy source associated with the self-ignition of the spontaneous wave at lower temperature, which amplifies the shock wave and may induce detonation, can be written as

$$u_{sp}(r) \propto a_s(r). \tag{10.4.8}$$

The condition (10.4.8) determines the temperature profile, or in the case of linear temperature distribution the temperature gradient, for the spontaneous formation of a detonation wave.

10.5 The Deflagration to Detonation Transition

As was said a self-sustained wave of exothermic chemical reaction in a homogeneous combustible mixture propagates either as a subsonic deflagration or as a supersonic detonation wave. Normally, deflagration is initiated by a mild energy release, such as e.g. a spark, while detonation is provoked by shock waves via localized explosions. Detonations can be generated directly if a large amount of energy is rapidly released in the explosive mixture. Thus, both deflagration and detonation appear to be stable attractors each being linked to its own base of initial conditions. Yet, it has long been known that the initially formed deflagration undergoes acceleration accompanied by a rapid change in the velocity of the reaction wave following abrupt transition from deflagration to detonation mode.

Since the first description of the deflagration-to-detonation transition by two prominent French teams of Berthelot and Vieille (1882), and Millard and Le Chatelier (1883), more than century ago, a large number of experimental, theoretical and numerical studies have been performed to understand the basic physical mechanisms responsible for transition from deflagration to detonation and to determine the critical conditions that lead to or prevent the DDT in reactive mixtures. Understanding of deflagration to detonation transition (DDT) is of great importance since detonation represents the most sever form of explosion hazard produced in industrial accidents involving fuel-air explosions. Predicting the occurrence of DDT has practical importance e.g. for pulse-detonation engines and because of its destructive potential. It still remains a major concern and source of industrial hazards produced by DDT in H_2-air, C_2H_2-air and CH_4-air mixtures.

Experimental studies indicate that normally DDT is reluctant to occur in unconfined freely propagating flames, like spherical clouds with central ignition. Yet the transition to detonation occurs considerably easier in presence of walls and flow obstacles. In the case of unconfined freely propagating flames this can be attributed to the geometrical effects of expansion, which make smoother and weaker development of flame instabilities and correspondingly the flame acceleration. Also, shocks, which precede the detonation, might be weakened or turbulence might be damped by the expansion. All these effects lead to the preconditioned mixture becomes reluctant to explode. On the contrary, the presence of rough walls or obstacles produce dramatic acceleration of a flame within a short distance of 6–8 tube diameter. It was found that the following parameters are important: the intrinsic laminar flame speed, the thickness of the reaction zone, transverse dimension of the obstacles, obstructed area ratio.

How can detonation be formed in the absence of strong energy sources and associated blast wave? A detonation is a supersonic combustion wave across which the pressure and density increase. The wave speed is typically slightly larger 2000 m/s for hydrocarbon fuel-air mixtures and about 3000 m/s for hydrogen-oxygen mixture. An idealized model of detonation structure, the classical ZND model discussed in Chap. 8, is a planar shock wave followed by a laminar chemical reaction zone, where reaction is initiated thermally in the high temperature of the order of 1500 K produced by the leading shock. The spatial location of exothermal reaction moves at exactly the same speed as the leading shock, thus being coupled to the shock wave. However, in reality such a simple laminar scenario of detonation is unstable. Almost all self-sustaining gaseous detonations exhibit formation of the cellular structure with the characteristic cell size in between 10 and 100 times of the idealized reaction zone thickness. A key role of the transverse waves for self-sustaining propagation of the detonation and for formation of the cellular structure in a tube with rough walls has been first noted by Ya. B. Zel'dovich (1949). He noted that if the inner surface of the tube was coarse then the detonation velocity was considerably lower than that in smooth tubes. In case of a tube with rough walls the reflection of the shock from irregularities of the wall produces transverse waves, which in turn create Mach reflections, so that the temperature behind the Mach steam is high to form hot spots required for local ignition and rapid energy release. Calculations using the idealized ZND reaction zone thickness have shown that the cell size is approximately proportional to the reaction zone thickness. In particular, transition to detonation is possible if the detonation cell size is less than the tube diameter. Transitions to detonation were never observed in the experiments, where special measures were taken to damp the transverse pressure waves, which means that the transverse waves are important both to accelerate the deflagration wave and to sustain detonation.

The main features of qualitative understanding of transition to detonation have been acquired largely due to the comprehensive experimental studies of K. I. Schelkin, Ya. B. Zel'dovich and their co-workers, and later of A. K. Oppenheim and co-workers with their stroboscopic laser schlieren photography technique. A high-speed framing camera has been used by researches at McGill University to photograph sequences of schlieren images to form a cinematic impression of the behavior of flames and detonations in tubes. Their studies of DDT have shown that the main feature leading to the detonation onset, which is common for all cases even though the process of transition is in essence non-reproducible in its detailed sequence of events, is a moderate acceleration in the velocity of the combustion wave following by abrupt increase in the velocity and an explosive character of transition.

The deflagration-to-detonation transition exhibits two stages of evolution of the propagating flame corroborating majority experimental observations. During the first stage the flame propagating from the closed end of the tube acts as a permeable moving piston, which sets into motion the gaseous mixture within part of the flow between the flame front and the advancing shock wave emitted by the flame after its initiation. The velocity of the blast flow satisfies the boundary condition ($u_b = 0$) at the closed end in the burned gas, so that from the continuous equation velocity of the flow ahead is

$$u_u = (\Theta - 1)U_w. \qquad (10.5.1)$$

Propagating flames rarely have a smooth or laminar appearance. Many instabilities as well as interaction with upstream flow disturbances cause the flame surface to become highly distorted. The flame burning velocity, U_w increases due to the flame convolution caused by different factors, which involve, the flame front stretching due to interaction with nonuniform flow ahead of the flame caused by noslip or rough walls, interaction with turbulent eddies of the blast flow, hydrodynamic flame instability, etc. The increase of the flame burning velocity can roughly be estimated as the ratio of the actual flame area brush to that of the tube cross section. It can be several times larger than the incipient velocity of a planar flame but still much less than the sound speed. Accelerating flame together with the flow of the burned products acts as an accelerating piston generating pressure and shock waves in the flow ahead. At the initial stage the shocks emitted by the accelerating flame are too weak to provide noticeable heating of the unburened gas for further flame acceleration, but may have positive feedback effect enhancing turbulence-induced flame folding and further moderate flame acceleration. Duration of the first stage depends on the combustible mixture and initial and boundary conditions. The former may promote the spontaneous origin of a flame folding producing preheating zone suitable for further local and much faster self-accelerating of the reaction wave on the second stage. It was argued that hydraulic resistance may also be involved in the mechanism of deflargation-to-detonation transition.

Just before the detonation onset the flame acquires either a tulip-shaped front or spikes and folds developed at the flame front. Either the flame acquires a convex wrinkled shape, or concave tulip-shaped front depends on the initial pressure, flame thickness, and on the roughness of the channel walls. The second stage is an abrupt transition to detonation, which is accompanied by a very fast acceleration of the reaction wave. The onset of detonation is prompted by an explosion, which is localized either within the flame brush or in the boundary layer, or sometime by an explosion inside the region between the flame and advancing shocks, and as a rule resulted from merging of repeated shocks and the reaction wave. In the latter case the explosion spreads out from a point it originated and its reflection from the walls results in the transverse and refraction waves, while propagating front acquires the character of a self-sustained detonation. The distance at which combustion transforms into detonation is typically reaches tens of tube diameters.

Turbulence may play an important role in transition to detonation enhancing the flame acceleration. The result is an extended turbulent "flame brush" propagating with velocity several times larger compared to the normal velocity of a laminar flame. Many experimental studies of DDT in coarse tubes have been performed being very much influenced by the concept of the effect of turbulence on the deflagration to detonation transition. Yet, compared to what have been thought on earlier stage of the DDT studies, the principle mechanisms of the DDT triggering are accomplishable irrespectively of whether the bulk flow and the flame are laminar or turbulent. In particular, DDT was observed in very thin capillary tubes, where the flow is always laminar.

The experimental study of the detonation onset has shown that the detonation onset depends on the particular pattern of accelerating flame and on the action of the shock generated by the flame. A distinction should be made between DDT in rough tubes and smooth tubes since the wall roughness plays a significant role on both the propagation of deflagration and detonation. In a smooth tube where only the shear at the wall provides the means for turbulence generation, the deflagration speeds are typically an order of magnitude less than that at tubes with rough walls. In particular, it was observed that the roughness of the channel's walls and obstacles are considerably shorten time of the transition. Transition to detonation is also not so clearly marked by an abrupt increase in the reaction wave velocity.

More recent experimental studies of the Karlsruhe Research Center group (M. Kuznetsov, I. Matsukov and co-workers) using high speed shadow photography, have confirmed the main features of transition, helped to considerably clarify role of the boundary layer formation for the transition in smooth and coarse tubes, and shed light on the physical mechanism of transition to detonation. Experiments indicate that detonation often starts near the wall, especially in case of rough walls (N. Smirnov, 1995). It is likely due to viscous heating of the unburned mixture in the boundary layer (A. Merzhanov, 1996). If T_0 is temperature near the tube axis, and M_u is the flow Mach number ahead of the flame front, then the temperature excess in a stationary boundary layer $\Delta T_{wall} = T_0\left(1 + \frac{1}{2}(\gamma - 1)M_u^2\right)$ is negligibly small when flame starts; $M_u \simeq$ 0.2–0.3 at the beginning. However, velocity of the flow ahead of the flame exceeds considerably before the transition, and M_u became close to unity, so that temperature excess in the boundary layer becomes noticeable. This may considerably influence both extension and temperature of the preheating zone in the boundary layer, therefore favouring location of the transition somewhere near the wall (see below Sect. 10.6 Mechanism of deflagration to detonation transition). Besides, the effect is also accumulating. Thus, the distance of transition will be not some universal value but rather depends on the tube diameter.

The earlier stage of the flame propagation in $H_2 + O_2$ mixture before the transition is shown in the sequence of frames in Fig. 10.4. The flame-generated advancing shock and transverse pressure waves are clearly seen in the first frame. On the second frame the advancing shock is seen together with transverse shocks generated by accelerating corrugated flame. The flame in this photo is beyond the vision. The boundary layer formed behind the advancing shock is seen in these photographs as a white strip near the lower wall with its thickness gradually increasing in the direction to the flame.

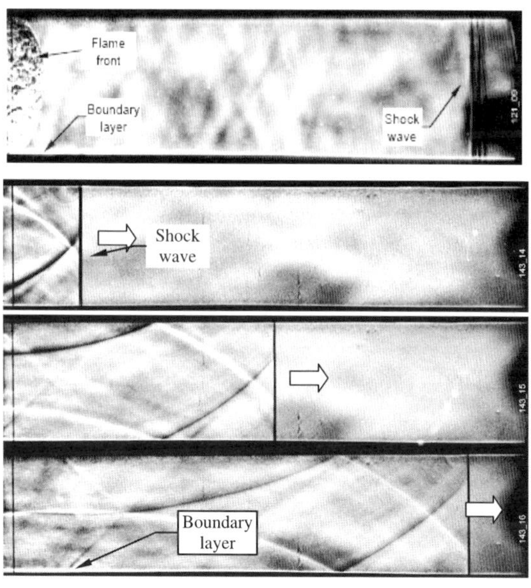

Fig. 10.4 Shadow photographs of early stage of flame propagation in a smooth tube filled with a hydrogen-oxygen: P = 0.75 bar, window at 210–440 mm from ignition point (Courtesy of Kuznetsov, M. and Matsukov, I. Private communication, 2007)

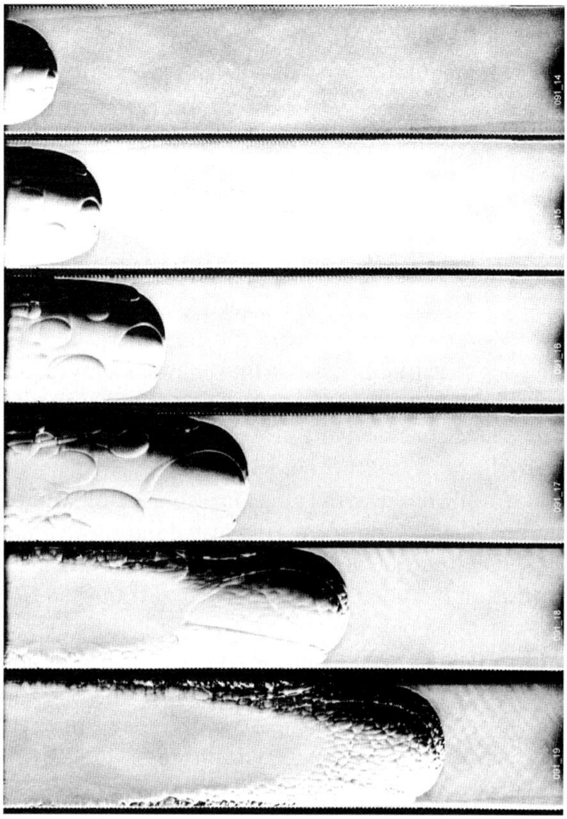

Fig. 10.5 Flame surface has cellular structure due to hydrodynamic instability. Cell size decreases with distance. Flame generates week shock waves ahead. $H_2 + O_2$, P = 0.6 bar (from: Kuzntsov, M., Matsukov, I., Alekseev, V., Breitung, W., Dorofeev, S., Effect of boundary layer on flame acceleration and DDT, Proc. 20th Int. Coll. On Dynamics of Expl. and Reactive Systems, Montrel, Canada, 2005)

Sequence of shadow photographs in Fig. 10.5 demonstrates typical scenario of the flame surface distortion due to interaction with nonuniform flow upstream, flame wrinkling and formation of cellular structure of the flame surface due to hydrodynamic instability.

For a larger roughness of the channel walls and for higher pressure the flame brush acquires a tulip shaped (V-shape) with the leading edges of petals at the wall, where intense turbulence and pressure waves are generated first from the boundary layer, as it is seen in sequence of shadow photographs in Fig. 10.6. The flame is becoming increasingly disturbed by vortices of the flow, being stretched in the upstream nonuniform flow, acquiring more and more corrugated shape and may pass finally to a turbulent flame brush. The process of positive feedback between the flame and its induced flow is one of the key themes in deflagration to detonation transition. The paradigm of this amplification process is the interaction of a flame with a vortex generated by the shear formed behind the advancing shock wave generated by the accelerating flame, or a flame wrapping by wakes behind the obstacles, as well as gas-dynamic feedback mechanism of flame acceleration. The flame acceleration may be caused also by the DL instability that also triggers a positive feedback process. Though the positive feedback may

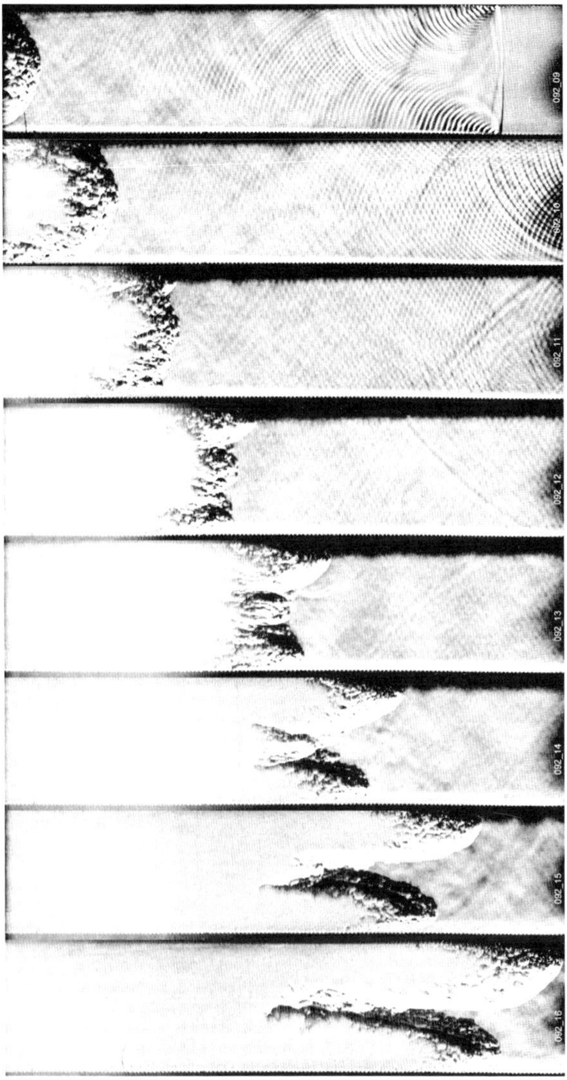

Fig. 10.6 Sequence of shadow photographs showing boundary layer ahead of accelerating flame propagating from the left to right. Boundary layers are seen as dark regions on the top wall and as lighter regions on the bottom wall of the channel. Wall roughness is 1 mm, $H_2 + O_2$, $P = 0.75$ bar (from: Kuzntsov, M., Alekseev, V., Matsukov, I., Dorofeev, S., DDT in smooth tube filled with hydrogen-oxygen mixtures, Shock Waves, vol. 14, pp. 205–215, 2005)

be not very pronounced in case of a tube with smooth walls, it strongly influences transition to detonation in the tubes with rough walls. For slow flames and in the case of rough walls this positive feedback may be the dominant process involved in the mechanism triggering transition to detonation.

The increase of the flame velocity due to stretching of the flame front by the nonuniform flow ahead can be estimated as being proportional to the increase of the flame surface.

Two different stages of the evolution of combustion wave velocity are seen in Fig. 10.7. During the first stage the flame accelerates, though its velocity

Fig. 10.7 Velocity evolution of the combustion wave for different initial pressure and roughness of the tube walls (from: Kuzntsov, M., Alekseev, V., Matsukov, I., Dorofeev, S., DDT in smooth tube filled with hydrogen-oxygen mixtures, Shock Waves, vol. 14, pp 205–215, 2005)

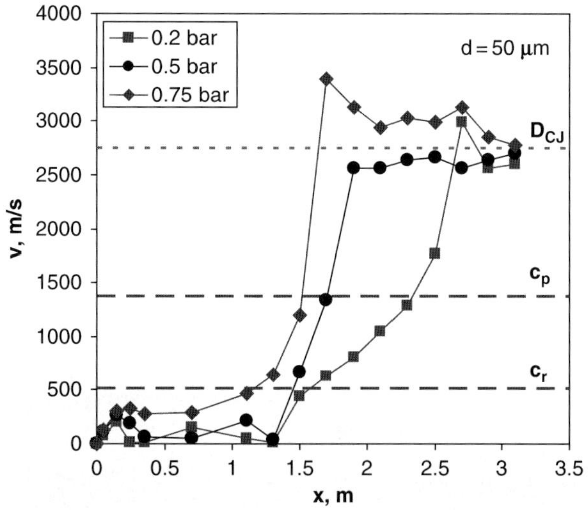

remains considerably subsonic. The nonuniform flow formed ahead of propagating flame, stretches the flame, so that the flame mimics its shape, which becomes similar to the velocity profile, and this increases the flame surface area accelerating the reaction wave and consequently the flow velocity ahead. As the flame accelerates and generates shock waves running ahead of the flame front, the mean flow and turbulent intensity between the flame and the leading shock increase. Estimating the flame velocity increase being proportional to the increase of the flame surface area and taking into account the above consideration, we obtain that the flame must accelerates exponentially in time

$$U_w \propto U_{f0} \exp(\alpha t). \tag{10.5.2}$$

This conclusion agrees with linear dependence V(x), which is seen in the experimental diagrams Fig. 10.7 and 10.9.

Of great importance for DDT is the transverse pressure and shock waves irradiated by the accelerating curved flame, which are essential ingredients for the onset of detonation and for the amplification of the blast waves by the preconditioned mixture. Obstacles, rugged walls and turbulence deform the flame front enhancing the rate of the instability development and correspondingly promoting development of folds in the flame front and increasing its acceleration.

Though the flame acceleration is an essential stage which promotes transition to detonation, it does not explain a mechanism of transition since it does not explain formation of a strong shock typical for a detonation or amplification of a shock. In fact the amplification of the shock wave and transition to detonation originate due to formation of the preheating zone in the unburned gas adjacent to the advancing deflagration front, which leads after extended

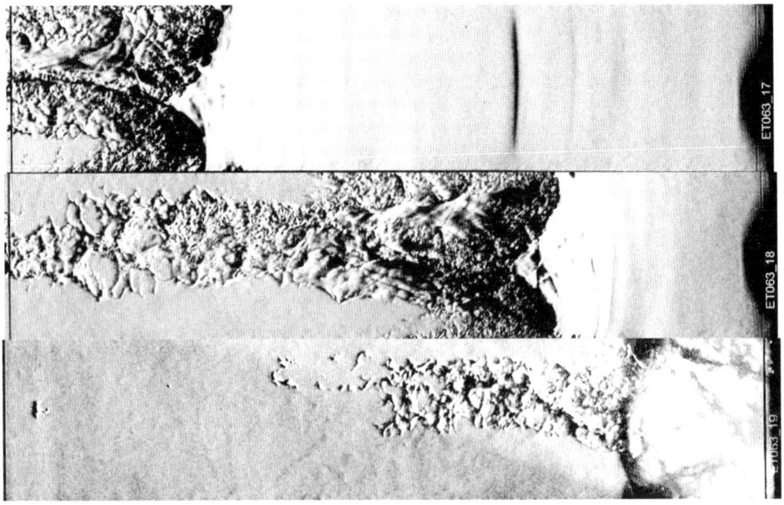

Fig. 10.8 Formation of a preheating zone adjusted to the flame and transition to detonation. Stoichiometric ethylene-oxygen flame, $P = 0.121$ bar, $U_{f0} = 5.2\,m/s$, $L_f = 0.61$ mm (Courtesy of Kuznetsov, M. and Matsukov, I. Private communication, 2007)

period formation of a proper preconditioned preheating zone to triggering an abrupt transition to detonation.

Note that after its initiation the flame acts as an accelerating piston and generates the first weak shock wave, which sets into motion the gaseous mixture within part of the flow between the flame and the advancing shock wave. Typical Mach number of the leading shock is about 1.5. Therefore the flame

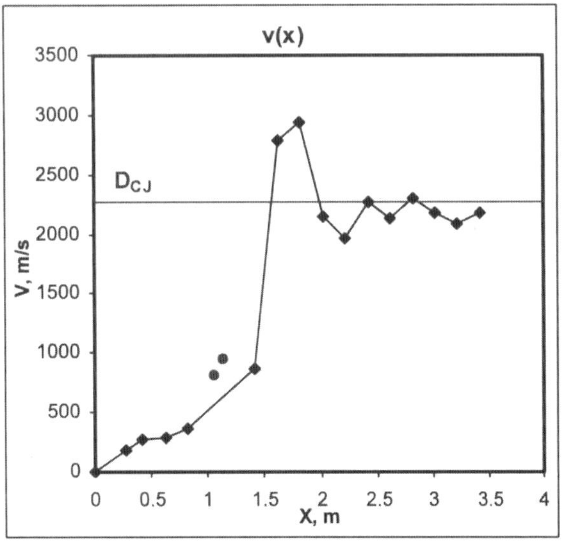

Fig. 10.9 V-x diagram of the combustion process for the conditions of Fig. 10.8 (Courtesy of Kuznetsov, M. and Matsukov, I. Private communication, 2007)

propagates form the very beginning through the mixture of higher than initial pressure and temperature, $P_2/P_0 \simeq 2.4$, $T_2/T_0 \simeq 1.24$. This means that even planar flame will propagate with the velocity noticeably higher than the intrinsic velocity of a planar flame in the initial mixture. Further distortion of the flame surface due to stretching by nonuniform flow upstream, wrinkling due to instability, interaction with turbulent eddies, etc., even more increase both the flame velocity and the velocity of the blast flow upstream.

There are different possible mechanisms of formation of the preheating zone required for triggering transition to detonation. The wall friction or resistance caused by the wall roughness lead to the reduction of the gas flow momentum, and this hydraulic resistance causes a gradual precompression and hence preheating of the unburned gas. A noticeable preheating certainly would require considerable increase in the flame and the blast flow velocities. Also hydraulic resistance causes preheating of the unburned gas of the whole flow between the leading shock and the flame, but not of the unburned gas adjusted to the propagating flame. Therefore, it is unlikely that the hydraulic resistance is a mechanism of the preheating zone formation in DDT events. More plausible is that the preheating zone is formed by compression and heating of the unburned gas by series of weak shocks generated by the accelerating flame when the flame, which already propagates with considerably increased velocity achieved during the first stage of acceleration, is additionally accelerated, for example, due to additional wrinkling. In some cases the preheating zone suitable for triggering detonation may be formed inside a deep and narrow fold within the flame brush originated from the flame instability (Sect. 10.7).

Formation of a preheating zone in the unburned gas adjacent to the advancing deflagration followed by triggering a detonation, first observed experimentally by M. Kuznetsov and I. Matsukov, is shown in sequence of shadow photo of Fig. 10.8. Here the higher temperature corresponds to the white curve strips ahead of the flame front. The measured length of the preconditioning zone is $13/15\,\mathrm{mm} \simeq (20/24)L_f$. In accordance to the above discussion the shape and visible structure of the preheating zone looks more as if it is formed by the sequence of shocks emitted by the accelerated curved flame rather than by the hydraulic resistance. In the latter case one should expect larger extension and higher temperature near the walls, ahead of the flame front petals.

Dependence of the reaction wave velocity on the distance along the tube for the experiment shown in Fig. 10.8 is presented in the V-x diagram in Fig. 10.9.

We can estimate a distance from the accelerating flame front to the location where shock wave is formed due to the flame acceleration using data of Fig. 10.9. With help of the formulas obtained in Sect. 6.15, the first shock corresponding to the first flame acceleration, which is the first linear part of V(x), less then approximately 0.5 m in Fig. 10.9, is formed at about (1.5–2)m ahead of the flame. The second flame acceleration corresponding to the distances from 1 m to 1.5 m is larger and takes place at much larger flame velocities. In this case the shock wave generated by the flame is formed at the distance (2–5)cm from the flame, which is close to the width of preheating zone 1.5 cm.

As it is seen from the V-x diagram in Fig. 10.9, transition to detonation takes place when the velocity of the reaction wave becomes equal to the local sound speed. For the experimental conditions of Figs. 10.8 and 10.9, the sound speed in the unburned gas of reactants is 324 m/s, and taking into account that temperature in the preheating zone is approximately 700 K, the local sound speed is 500 m/s. Overshooting of the detonation seen in Fig. 10.9 is because of the formed detonation wave initially propagates through the mixture precompressed by the leading shock.

When temperature and extent of the preheating zone become large enough, the local explosion followed by transition to detonation takes place. In the last frame of Fig. 10.8, which corresponds to the transition to detonation, one can see only the retonation wave propagating to the left.

10.6 Mechanism of Deflagration to Detonation Transition

Detonation represents a propagation of reaction, which is preceded by compression and heating of the gas by a strong shock wave. The pressure in this shock increases in about two orders of magnitude. How is such a wave formed during flame propagation when the flame velocity in the most rapidly burning mixture does not exceed (10–15)m/s and the pressure change was negligibly small? What is the mechanism of formation of such a strong shock wave which is capable of rising temperature high enough to ignite the reaction? How can detonation be formed in the absence of strong energy sources and associated blast wave?

For cogent insight to the mechanism of triggering detonation, and to elucidate connection between local formation of extended preheating zone ahead of the advancing flame it should be taken into account that triggering detonation requires generation of rather strong shock wave, which is strong enough to ignite burning. Typical strength of the shock wave must satisfy condition of strong coupling between the leading shock and chemical reaction wave, which means that the shock wave must be strong enough to raise temperature up to about 1500 K to initiate the reaction. Typical Mach number of the leading shock wave in Chapman-Jouguet detonation is $M_{sh} \simeq 6/7$, corresponding to temperature elevation and the pressure jump across the shock in the compressed combustible mixture

$$T_2/T_0 \cong \frac{2\gamma(\gamma-1)}{(\gamma+1)^2} M^2 \approx 1500\,K, \tag{10.6.1}$$

$$P_2/P_0 \cong \frac{2\gamma}{(\gamma+1)} M^2 = 40/50, \tag{10.6.2}$$

which is enough to induce exothermal chemical reaction.

The shock waves, generating by the accelerating flame in the unburned fuel during the first stage, are too weak with the Mach numbers about 1.5. Such shocks may only slightly raise temperature of the unburned mixture ahead of the flame front. It should be noticed that a maximum pressure jump, which can be produced by thermal constant volume explosion of a uniform reactive pocket, is about the ratio of the adiabatic flame temperature and the initial mixture temperature. The corresponding magnitude is $P_2/P_0 \sim T_{ad}/T_0 \propto 8/10$, which is about one order less than the value of pressure jump in a detonation stock wave (10.6.2).

In traditional attempts to explain the phenomenon the role of confinement was reduced exclusively to generation of turbulence. Only recently the idea was suggested that the flame acceleration and transition to detonation may occur for laminar flames, when it was realized that the preheating zone of the unburned gas adjusted to the advancing deflagration is always a precursor of a detonation triggering (M. Liberman and G. Sivashinsky, 2006). It was found that a proper extended preheating zone of the unburned gas adjusted to the advancing deflagration can be produced if a laminar flame propagates in a tube with adhesive or rough walls, as well as an appropriate local preheating zone can be produced by the influx of heat from the folded reaction zone formed due to the hydrodynamic flame instability if the fold induced by the instability is sufficiently narrow and deep.

The fruitful idea is that the reaction wave trajectory can be set-up by reactant and temperature gradients alone. The accelerating reaction wave generates pressure waves in the partially reacted gas in the region between the flame and precursor shock. If these pressure waves and the reaction waves move synchronously, then the pressure waves can be amplified while the reaction wave remains closely coupled, spontaneously forming a detonation. The idea of spontaneous initiating a detonation has been proposed by Ya. B. Zel'dovich (1970) and independently by J. H. S. Lee with collaborators (1978), who called it SWACER (Shock Wave Amplification by Coherent Energy Release) and demonstrated the detonation initiation by using photochemical excitation of a hydrogen-chlorine mixture. The SWACER mechanism requires a synchrony between the progress of the shock and the sequential release of chemical energy.

Since velocity of a deflagration is much less than the sound speed, the flow upstream and downstream is nearly isobaric. The reaction wave – flame accelerates if part of unburned mixture adjusted to the flame front is preheated and acts as a permeable piston emitting pressure waves. If the pressure waves produced by the accelerating reaction wave are synchronized with the further accelerating reaction wave and therefore considerably amplified and steepen into a strong shock wave then detonation can be initiated. This occurs if the preheating zone is formed ahead of the flame and the temperature profile within the flame front is modified so that the reaction wave accelerates and simultaneously a shallower temperature gradient is formed from the initial steep temperature profile of the deflagration wave. Then the mechanism of the reaction propagation changes from the diffusive deflagration to "spontaneous" regime of the reaction wave propagation (see Sect. 10.4). The detonation development in such scenario is similar to the situation with a prescribed nonuniform initial temperature profile, when the generated shock wave is synchronized with the energy released by the self-ignition occurring at different time and locations.

In numerous numerical calculations it was found that scenario of the DDT depends only slightly on the particular temperature profile in the preheating zone. Qualitatively, the picture is very similar independently on either temperature exponentially decreases from its maximum value T_m just ahead of the flame front, or it decreases linearly, or the stepwise temperature profile is taken within the preheating zone. For the one-dimensional numerical simulation of a flame propagating from the closed end of a tube, the initial spatial temperature distribution is taken as

$$T(0<x<x_f) = T_b, \qquad T(x_f<x<L) = T_m, \qquad T(L<x<\infty) = T_u, \quad (10.6.3)$$

where x_f is the initial position of the flame front set at $x_f \cong 25L_f$ from the closed end, L is the characteristic length of the preheating zone with temperature T_m.

The initial conditions for the concentration, pressure and flow velocity are specified as,

$$Y = (T_b - T)/(T_b - T_u), \qquad P(x>0) = P_u, \qquad u(x>0) = 0. \quad (10.6.4)$$

Initial distributions of the temperature, pressure and the rate of reaction in the flame are shown in Fig. 10.10. Due to the preconditioning the incipient flame speed exceeds the normal flame speed associated with the uniform initial state at x > L, and in the course of the subsequent evolution, depending on details of the initial data, the process settles either into the deflagrative or

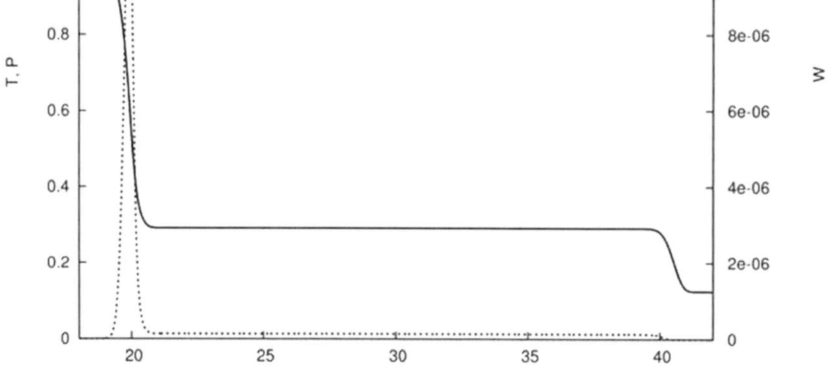

Fig. 10.10 Initial distribution of temperature, pressure and reaction rate for the deflagration wave with the preheating zone

detonative combustion. From the beginning the whole process is isobaric, flame starts to propagate, and pressure is nearly constant everywhere.

Temporal evolution of the temperature, reaction rate, pressure, velocity, and density profiles shown in Figs. 10.11 and 10.12 illustrate the described scenario

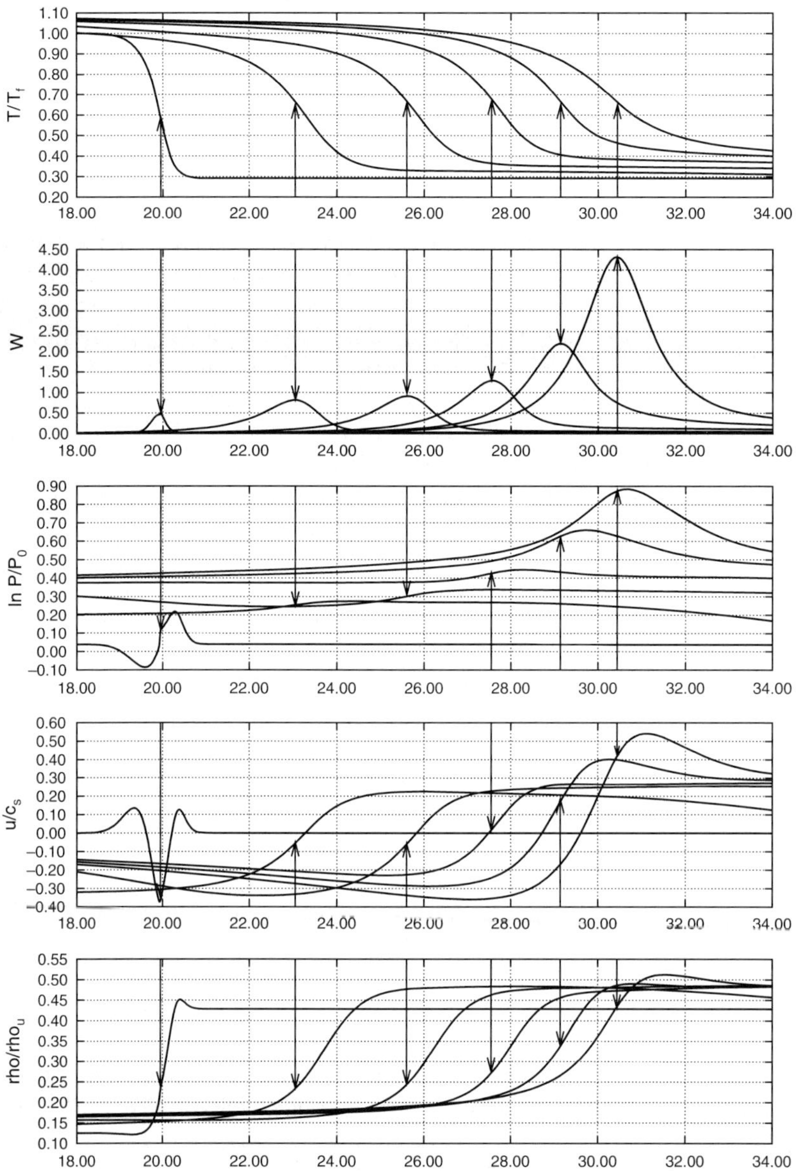

Fig. 10.11 First stage of temporal evolution of temperature, reaction rate, pressure, velocity and density profiles due to preheating zone ahead of the propagating flame before triggering detonation

of formation a shallow temperature gradient from the initial deflagration with adjusted preheating zone in the unburned gas. Evolution of the deflagration wave shown in Fig. 10.11 corresponds to the stage before transition to detonation starts. The figure shows the pressure pulse straddling the reaction region, which is gradually modified. The flame accelerates due to higher temperature upstream and emits pressure waves. Upstream the leading edge of the pressure pulse the temperature is low and the reaction is chemically frozen due to high activation energy, while behind the pressure pulse the reaction is essentially in progress. It is worth to observe that while at the beginning the reaction and the heat release had been localized in a thin reaction zone near the maximum temperature $T = T_b$, soon after emitting the pressure pulse the reaction is in progress within the preheating zone rising temperature behind the edge of the pressure pulse. Thus, in effect, the leading pressure edge overcomes the temperature deficit and causes the reaction, which was chemically frozen before, to be noticeably switched on upstream. After some time the reaction rate and the heat release have grown noticeably upstream and resulted in formation acoustically commensurate zone of shallow temperature gradient. In the main body of the pressure pulse temperature and pressure continue to rise, and so does the reaction rate.

While at the beginning of the evolution the pressure pulses pass ahead of the temperature front, their leading edges coincide after some time, when detonation starts developing. In this way the combustion wave is transformed in a diffusionless deflagration. This later stage of transition to detonation is shown in Fig. 10.12. As a result, now energy release of the thermal explosion is not localized at the narrow reaction zone near $T = T_b$ as it was for deflagration but extent over the preheating zone, so that explosion occurs at different locations at different time instants.

The reaction wave now propagates with the velocity corresponding to the spontaneous regime of combustion, which is defined by the gradient of the induction time

$$u_{sp}(x) = \left(\frac{d\tau_i}{dx}\right)^{-1}. \qquad (10.6.5)$$

The initial steep temperature profile of the deflagration is finally transformed in a shallow temperature gradient such that the local explosion becomes close to the constant-volume one, producing sufficiently strong pressure wave. At the same time the reaction wave accelerates to a velocity close to the local acoustic velocity, so that the shock wave produced by the explosion will overtake the reaction front compressing and heating the fresh mixture upstream.

The shocks generated by the local explosions of different parcels within the formed shallow temperature gradient amplify as they propagate down the temperature gradient. This requires a synchrony between the progress of the shock and sequential release of chemical energy by successive parcels of the mixture. The required synchrony means that the reaction heat release time τ_R becomes smaller

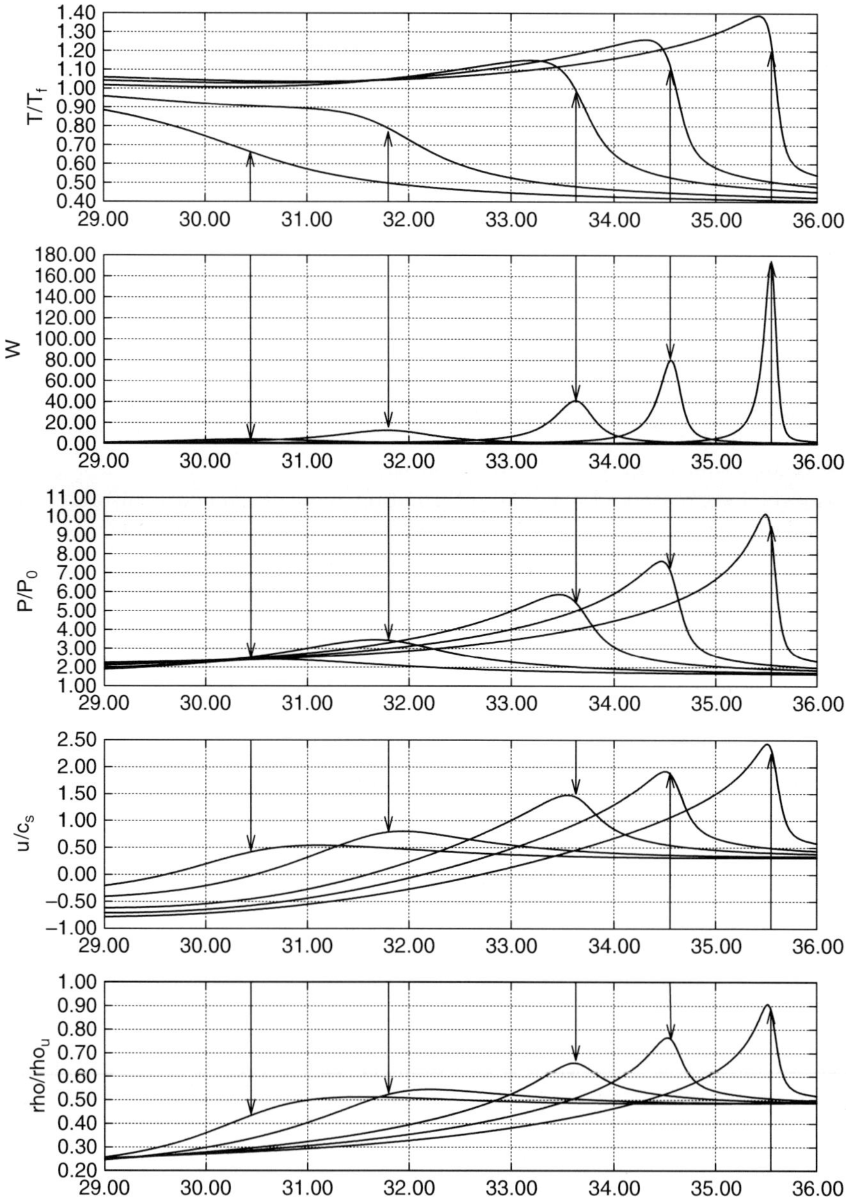

Fig. 10.12 Temporal evolution of temperature, reaction rate, pressure, velocity and density profiles and detonation initiation

than the characteristic acoustic time $\tau_s \simeq T/(dT/dx)a_s$, and the velocity of reaction wave corresponding the formed shallow gradient $u_{sp}(x) \simeq a_s$. The detonation is settled when the traveling shock wave produced by the volume explosion at higher temperature is synchronized with the energy source associated with

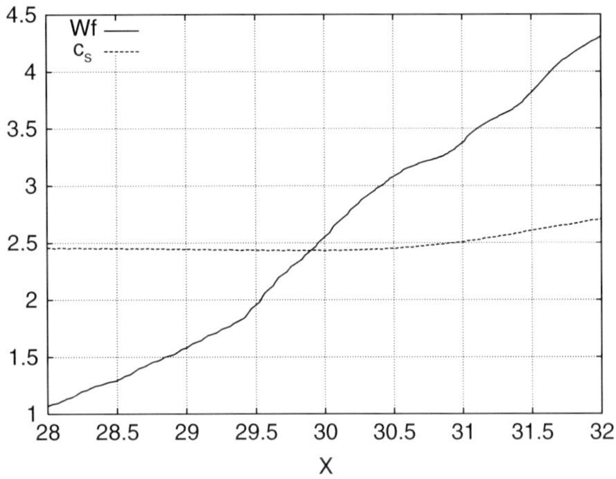

Fig. 10.13 Calculated evolution of the reaction wave velocity and the local sound velocity

self-ignition at lower temperature. In this way the velocity of the reaction wave becomes coupled with the shock, finally forming a detonation wave.

Indeed, the calculations indicate that amplification of the pressure pulses and triggering of the detonation starts form the instant when the velocity of the reaction wave becomes equal to the local sound velocity as it is shown in Fig. 10.13.

Obviously that as faster a flame is, other parameters are identical to those of slower flame, as shorter can be the initial preheating zone or lower temperature in the preheating zone required for transition to detonation since in this case τ_R, which defines the flame velocity, is smaller, and therefore the temperature in the preheating zone rises faster. Since the induction time depends on the reaction order, there is a strong positive feedback of the reaction rate for the higher order reactions, while for the first order reaction the reaction rate decreases with the pressure rise. Consequently, the length of the preheating zone and maximum temperature in the preheated mixture required for transition are smaller for a flame of the second or third order reactions compared to the first order reaction, other parameters being identical to those of the first order reaction flame.

If length of the preheating zone is less than some critical value, or the temperature at the preheated mixture is too low, than the deflagration wave accelerates generating pressure waves, and later will return to the deflagration regime. Figure 10.14a, b show representative results of simulations for two different length of the preheating zone: $L = 9.5L_f$ and $L = 9.0L_f$, with the same maximum temperature at the preheated mixture $T_m = 600K$; other parameters are identical to those of the 2D simulations in Figs. 10.18 and 10.19. The emerging pressure peaks are about $(50/60)P_u$ at Fig. 10.14a and located almost inside the reaction zone when transition settles. As a result the shock-reaction complex is formed, which is strong enough to initiate a self-sustained detonation. As a counter example Fig. 10.14b shows the temperature, pressure and density profiles for a shorter preheating zone, $L = 9.0L_f$. In this case the

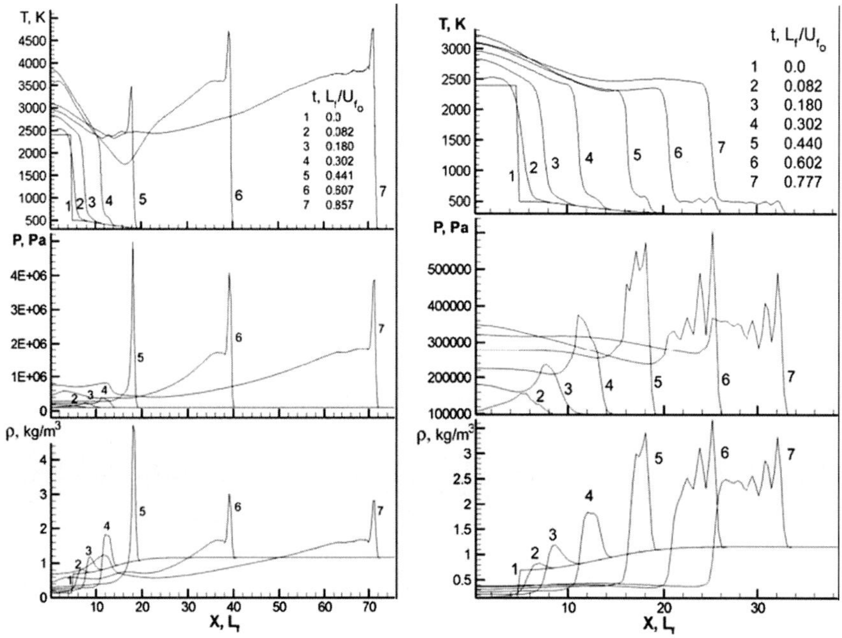

(a) (b)

Fig. 10.14 (a, b) Temporal evolution of the temperature, pressure and density profiles for: (**a**) $T_m = 600K, L = 9.5L_f$; (**b**) $T_m = 600K, L = 9.0L_f$. The temperature profile in the preheating zone is taken similar to the profile formed along the fold axis in 2D modeling shown in Figs. 10.18 and 10.19

emerging pressure peaks appear to be too weak to trigger detonation. Here the pressure waves steepen into a weak shock far ahead of the preheating zone and are not amplified by the accelerating reaction wave. For a successful transition, the pressure wave induced by the gas expansion must steepen into a shock within the time interval shorter than the characteristic time of heat release and while the shock is still positioned within the preheating zone.

The minimum length of the preheating zone required for the successive transition is sensitive to the maximum temperature T_m and the order of the reaction, but only slightly sensitive to the particular temperature profile in it. All the results are quite similar for the case of stepwise, $T(x_f<x<L) = T_m$, temperature distribution in the preheating zone, compared to more realistic temperature distribution observed in 2D modeling. An increase in T_m facilitates the transition for smaller L. The dependence of the critical length of preheating zone on the incipient flame velocity is obvious and was discussed earlier: as faster a flame is, other parameters are identical to those of slower flame, as faster the temperature in the preheating zone rises. Much stronger is sensitivity of the critical preheating length L to the order of the reaction: because of the dependence on the pressure of the reaction rate there is strong feed back of

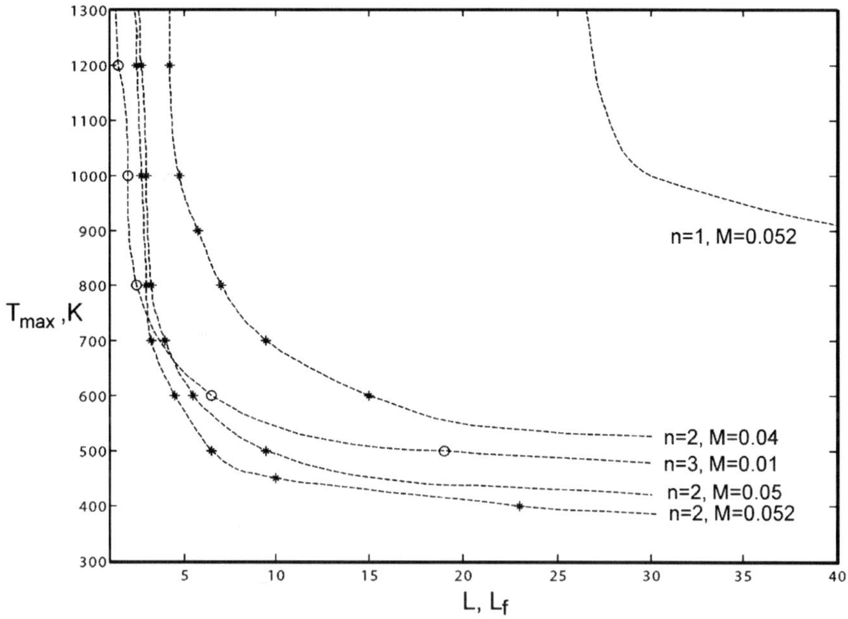

Fig. 10.15 Critical length of the preheating zone: transition boundary evaluated for several reaction orders (n) and incipient flame velocities (M_{f0})

the reaction rate and temperature rise in the preheating zone. The results of the pertinent simulations conducted for flames with different normal flame speed expressed in the Mach number ($M_{f0} = U_{f0}/a_s$) and for different reaction order (n) are plotted on Fig. 10.15.

For the sake of simplicity in calculations shown in Fig. 10.15 a stepwise temperature profile in the preheating zone is taken. The curves shown in this figure separate the upper domain associated with the successful transition to detonation form the lower domain where the transition fails. One readily observes that for a given T_m the critical length L drops drastically for higher values of M_{f0}, but the effect is more profound for the higher order of reaction, which means that the transition is much more feasible for faster flames and for reactions of higher order. Note, that in the above one-dimensional description the preheating zone is prescribed, while for realistic multi-dimensional systems it is a product of the intrinsic pattern forming dynamics sustained either by the flame accelerating under confinement in the presence of adhesive or rough walls or by the hydrodynamic flame instability. In particular, this explains why transition to detonation was never observed in 2D modeling for the flames of the first order reaction in a channel with slip walls, but the transition was feasible in case of adhesive wall. For a flame of the first order reaction the required for successful transition length of the preheating zone is apparently too large for the potential of the DL instability.

10.7 Formation a Preheat Zone and Deflagration to Detonation Transition

Detailed description of the experimentally observed extremely intricate process of flow, consisting the build-up of turbulence its couplings and evolution of a convoluted flame interacting with turbulence and nonlinear acoustics, represents a formidable task, which is beyond the capabilities even the most power computational tools. Clearly, numerical simulation of the complete flame acceleration and detonation onset is extremely daunting problem, greedy of computing resources and not really practical at present, particularly since the turbulence is generated by the previous motion of the flame and a very large range in space and temporal scales exists in both the chemical and fluid dynamic aspects of the problem. Qualitatively, however, the mechanisms of transition can be reasonably well understood using 2D modeling of laminar flames, which can elucidate the main features of the basic physical mechanisms of flame propagating in a tube and triggering the transition from deflagration to detonation.

To single out the impact of preheating zone formation, different sets of modeling were performed exhibiting either impact of preheating zone induced by the large-scale folding of the fast flame, or by the strong flame acceleration due to the adhesive walls or the wall roughness. There are different physical mechanisms capable to form preheating zone ahead of the propagating flame needed to provide condition of synchronization of the accelerating reaction wave and triggering transition to detonation. As an example, it was shown that the large-scale folding of the fast flame produced by the Darrieus-Landau instability alone is quite capable of triggering the transition. Under normal conditions this effect should be feasible in fast burning systems such as acetylene and hydrogen-fluorine mixtures where the normal flame-speed may raise as high as 15 m/s. Such an outcome is clearly a product of a very high flame velocity. An analogous effect may apparently be induced also by the flow turbulence developing in wide channels. Another basic mechanism of the preheating zone formation initiating transition to detonation is the flame acceleration due to the flame stretching in a channel with adhesive walls, which is much stronger than in the case of DL instability. This mechanism may compete with the flame folding, facilitating the transition by shortening the pre-detonation time and distance. For both no-slip and rough walls the transition is triggered after an extended preheating zone is formed, usually along the walls ahead of the advancing flame, and thereby creating conditions for triggering detonation.

To single out the impact of the DL instability, a numerical simulation of premixed gas combustion spreading from the closed end of a semi-infinite, smooth-walled channel was performed with the gas flow subjected to the free-slip and adiabatic boundary conditions, thereby eliminating the possible influence of the flame stretching and acceleration due to the wall friction. The most penetrating inside into the wave phenomena accompanying the onset of detonation is obtained by numerical simulation of a flame with the second order reaction shown in Fig. 10.16, where sequences of the calculated images show temporal evolution of

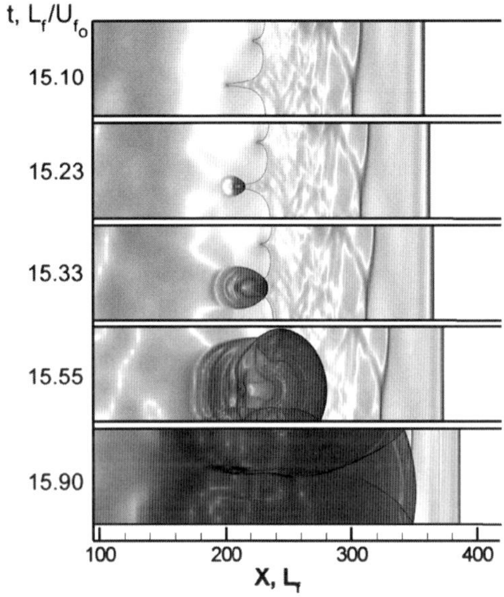

Fig. 10.16 Time sequence of images for the flame/shock dynamics near the transition point. Stronger shading corresponds to higher pressure gradient. The time and distance are referred to L_f/U_f and L_f respectively. The incipient dimensionless flame velocity is $M_{f0} = 0.05$, activation energy $\varepsilon = 8$, $\Theta = 10$, $D = 70L_f$, $\Theta = 10$

the flame shape for the flame with 2 nd order chemical reaction, spreading from a closed to an open end of a rectangular channel of width $D = 70L_f$. The earliest frames with the slightly perturbed flame as well as the left part (far behind the flame) in rest of the computational domain are not shown. The parameters taken in calculations, the incipient dimensionless flame velocity, activation energy, and expansion ratio are: $M_{f0} = U_{f0}/a_{s0} = 0.05$, $\varepsilon = E/RT_b = 8$, $\Theta = 10$. A sequence of events involves development of the flame instability, the flame acceleration and generation of pressure and shock waves.

Temporal evolution of the scaled flame velocity of M_f (solid line) and shock, M_{sh} (dashed line) corresponding to Fig. 10.16, is shown in Fig. 10.17. One observes a characteristic overshoot in the propagation velocity, upon which the process settles into a conventional Chapman-Jouguet (CJ) regime.

A zoomed view of the flame fold dynamics near the transition point is shown in the series of images presented in Fig. 10.18. The profiles of temperature, density, flow velocity, and pressure along the fold axis corresponding to the images of Fig. 10.18, are plotted on Fig. 10.19.

Here one readily observes (i) formation of the large-scale preheating zone (preconditioning) in the unreacted gas trapped within the fold interior, (ii) fast acceleration of the fold-tip reaction zone, and (iii) formation of a high-pressure peak. The preheating zone is formed by the influx of heat from the folded reaction zone increasing temperature of the unburned mixture trapped inside the fold. This requires the fold to be narrow and deep; so that the thickness of the layer of unreacted gas inside the fold to be comparable to the flame thickness, otherwise one ends up with a moderately strong pressure wave emitting by the flame insufficient for triggering the transition. This

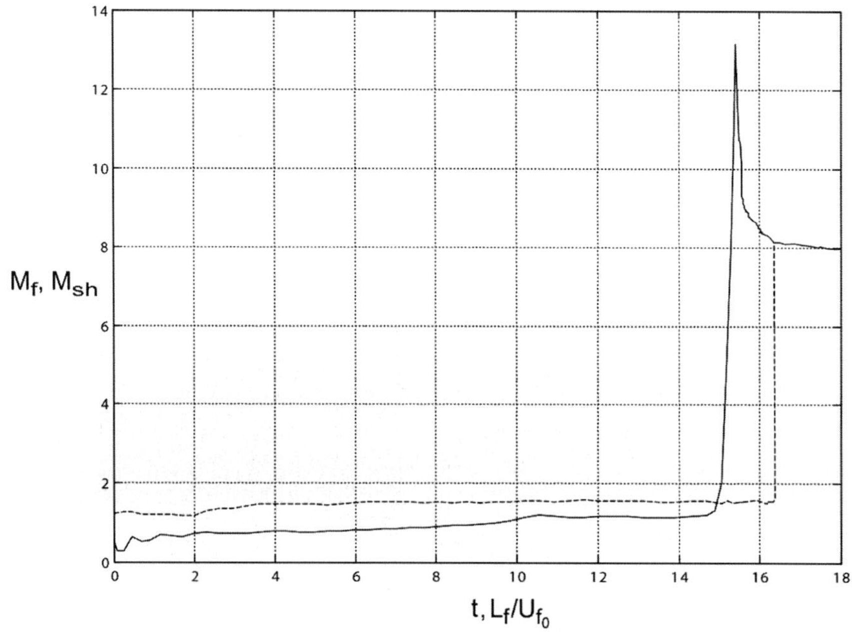

Fig. 10.17 Temporal evolution of the reaction wave (*solid line*) and shock (*dashed line*) velocities for conditions of Fig. 10.16

means that the length of the preheating zone formed by the DL instability inspired folding is limited by about several times of the flame thickness, which in turn limits possibility of transition caused by the DL instability.

Another basic mechanism of the preheating zone formation initiating transition to detonation is the reaction wave acceleration due to flame front stretching by the nonuniform flow formed due to hydraulic resistance, which may compete with the flame folding, facilitating the transition by shortening the predetonation time and distance. In this case the flame dynamics is different from that observed for no-slip walls. Soon after the ignition the flame front develops bulges near the channel's walls, which grow and merge forming the familiar tulip shape. The bulging is a product of a peculiar interaction of the flame and its self-induced boundary layer. In the course of its evolution the tulip may become shallower, and the flame gradually assumes the tongue-like shape stretched along the channel's axis. The transition occurs as soon as the fresh near-wall mixture adjacent to the flame becomes appropriately preconditioned. Quite in line with the results described above, the transition occurs faster for higher initial flame velocities and higher reaction order. As an example, the transition to detonation for the flame of the second order reaction is shown in Figs. 10.20 and 10.21, other parameters are as in Fig. 10.18.

The merging of the bulges, formed near the wall from the very beginning, results in a localized explosion at the tulip's tip. This explosion however does

Fig. 10.18 Time sequence of zoomed images for the flame dynamics near the transition point in Fig. 10.16. Lighter shading corresponds to higher temperature

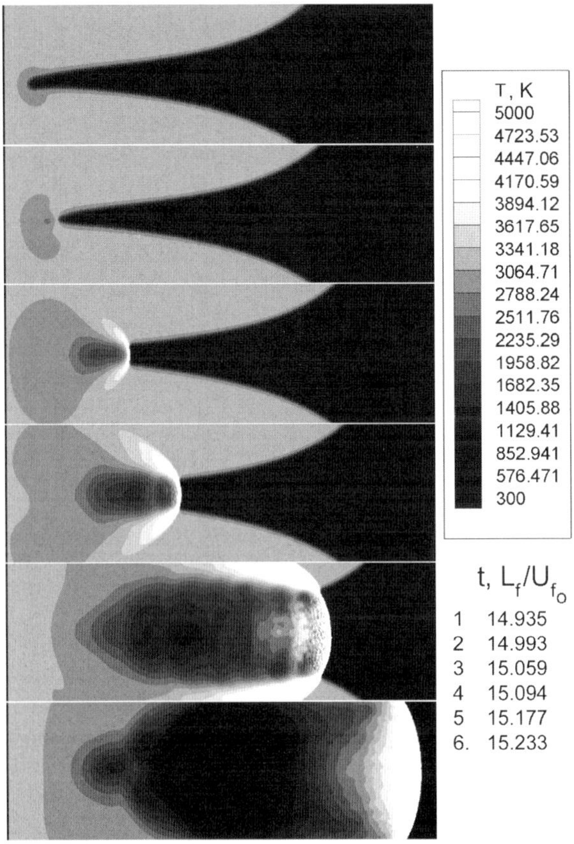

not evolve into a detonation, but only emitting additional compression waves. The transition to detonation occurs somewhat later in the near-wall mixture adjacent to the edge of the tulip's petals where the fresh mixture develops an appropriately extended preheating zone (shown in calculated temperature profiles in Fig. 10.22), where the transition to detonation starts.

It is worthwhile to note that as it is seen form Fig. 10.22, the length of the preheating zone formed in this case is noticeably larger (about $30L_f$) than it can be formed by the folding flame. That is why transition is feasible for the first order reaction flames propagating in tubes with no-slip walls, while it was not observed for the same conditions but for a flame propagating in a tube with slip walls.

Since the classical experiments of Schelkin and coworkers, it was known that the wall roughness may significantly facilitate the transition to detonation. The dynamical picture of the transition in the channels with rough walls shown in Fig. 10.23 is basically similar to that of smooth channels with adhesive walls: the transition is conceived near the wall where the largest acceleration and pre-compression leads to formation of preheating ahead of the advancing flame.

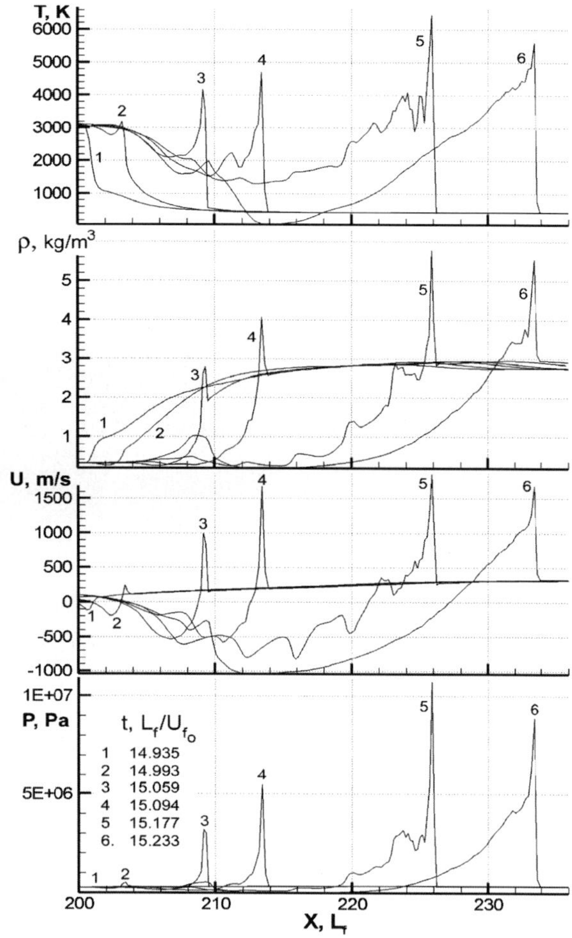

Fig. 10.19 Temperature, density, flow velocity, and pressure profiles for the fold dynamics near the transition point shown in Fig. 10.18

Interesting is that in the case of a rough wall, where the free-slip and adiabatic boundary conditions were taken, it was just a hydraulic resistance producing the wakes of the pressure waves-flow interaction, which is responsible for formation of a preheating zone in the unburned gas. The transition to detonation occurs in the near-wall mixture adjacent to the edge of the tulip's petals where the fresh mixture develops an appropriately extended preheat zone shown in calculated temperature profiles near the channel wall and along the channel axis in Fig. 10.24.

The formation of preheating of any kind and heat losses to the walls exert opposite effects on the transition. Typically the resistance raises the near-wall temperature (through adiabatic compression) and thereby promotes autoignition. The heat loss tends to reverse this trend by reducing the temperature. In smooth channels both mechanisms are of comparable influence.

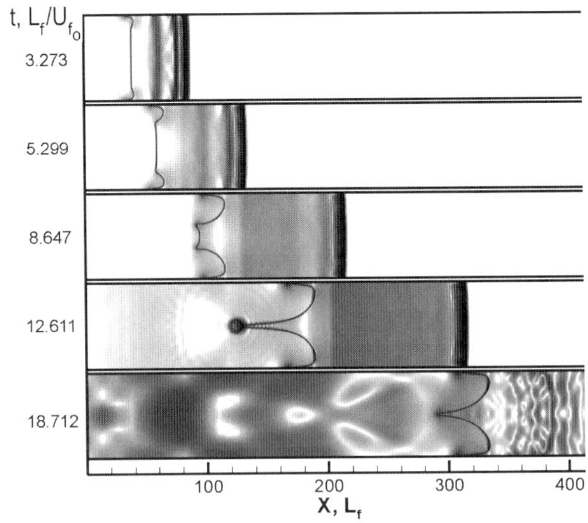

Fig. 10.20 The incipient flame dynamics: formation of the flame bulges and emergence of the tulip flame

Therefore, one cannot be certain about the final outcome of the above competition. The transition may be triggered predominantly by the local formation of an extended preheating zone ahead of the advancing flame, induced either in the fresh mixture trapped within the large-scale fold, or it can be induced somewhere near the wall being induced by the flame acceleration due to adhesive or rough walls.

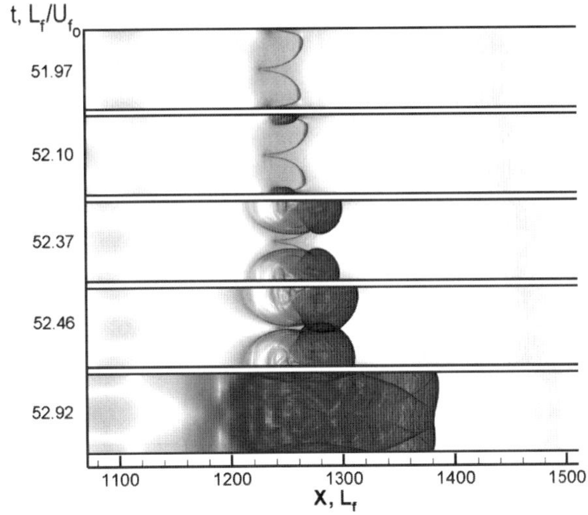

Fig. 10.21 Time sequence of images for the flame transition to detonation in case of adhesive wall boundary conditions; n = 2, $D = 70L_f$, $M_{f0} = 0.05$

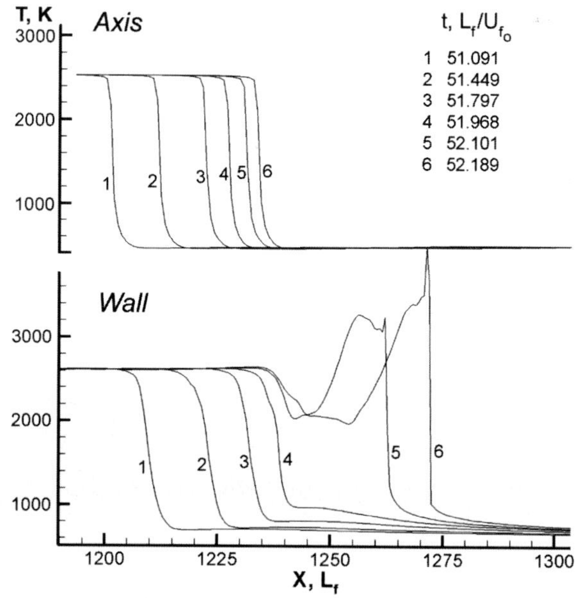

Fig. 10.22 Temperature profiles along the channel wall and axis at several consecutive instants of time for conditions of Fig. 10.21

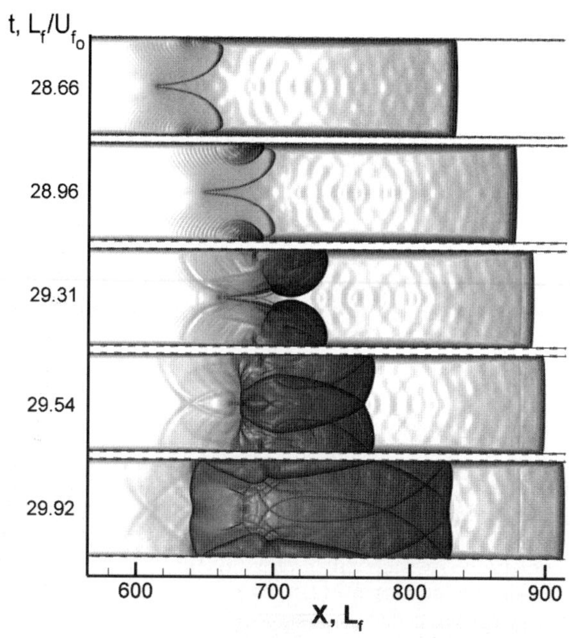

Fig. 10.23 Time sequence of images for the flame-flow evolution and transition to detonation for the flame propagating in a channel with the rough walls. $n = 2$, $D = 70L_f$, $M_{f0} = 0.05$, $\varepsilon = 4$, $\Theta = 8$, wall roughness $\Delta \simeq L_f$

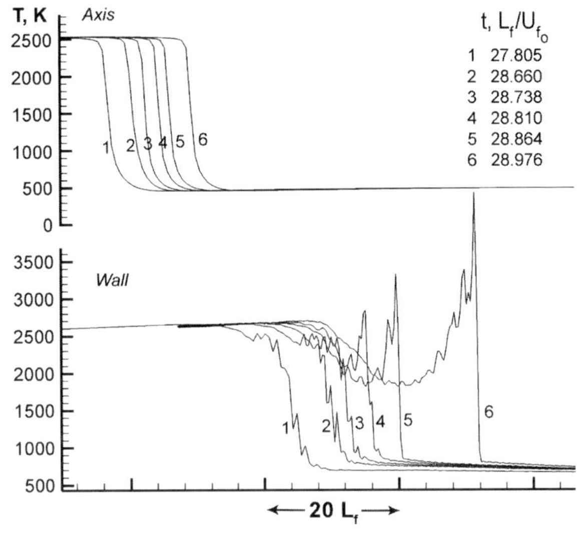

Fig. 10.24 Evolution of temperature profiles along the channel wall and axis at several consecutive instants for Fig. 10.23

10.8 Thermonuclear Burning of the Stars: White Dwarfs

An interesting example of deflagration to detonation transition is the thermonuclear explosion of the white dwarf in Supernovae events. Compact objects – white dwarfs, neutron stars, and black holes are "born" when normal stars "die, " that is, when most of their nuclear fuel has been consumed. Compact objects are the end products of stellar evolution. The primary factor determining whether a star ends up as a white dwarf, neutron star, or black hole is the star's mass. White dwarfs are believed to originate from light stars with masses $M \sim M_{\otimes}$. All three species of compact object differ from normal stars in two fundamental ways. First, since they do not burn nuclear fuel they cannot support themselves against gravitational collapse by generating thermal pressure. White dwarfs support themselves against gravity by the pressure of degenerate electrons, while neutron stars support by the pressure of degenerate neutrons. The predominance of neutrons in their interior following the mutual elimination of electrons and protons by inverse beta-decay, which held together by self-gravity at densities, is comparable to nuclear values. Black holes are completely collapsed stars, that is, stars that could not find any means to hold back the inward pull of gravity and therefore collapsed to singularities. The second characteristic distinguishing compact objects from normal stars is their exceedingly small size. Relative to normal stars of comparable mass, compact objects have much smaller radii and hence much stronger surface gravitational fields. No light or anything else can escape from a black hole. Thus, isolated black holes will appear "black" to any observer. Contrary, neutron stars can be observed as pulsating radio sources ("pulsars") and as gas-accreting, periodic X-ray sources ("X-ray pulsars"). White dwarfs can be observed directly in optical telescopes during their long cooling epoch. Because of their small radii, luminous white dwarfs are characterized by much higher effective temperatures radiating away their thermal energy.

Maximum allowed mass for white dwarfs is around $1.5M_{\otimes}$. The existence of a mass limit for a degenerate stars can be understood from a simple arguments. Let a star of radius R consists of N fermions (baryons and/or electrons), so that the number density of fermions is $n \sim N/R^3$. Then the average momentum of a fermion according to uncertainty principle is $p_F \sim \hbar n^{1/3}$, and the Fermi energy of the gas of relativistic fermion is

$$E_F \sim pc \sim \hbar n^{1/3} c \sim \frac{\hbar c N^{1/3}}{R} \tag{10.8.1}$$

The gravitational energy per fermion (proton) is

$$E_G \sim -\frac{GMm_p}{R} = -\frac{GNm_p^2}{R}, \tag{10.8.2}$$

so that the equilibrium corresponds to a minimum of total energy

$$E = E_F + E_G = \frac{\hbar c N^{1/3}}{R} \frac{GNm_p^2}{R}. \tag{10.8.3}$$

It is obvious from (10.8.3) that the total energy is negative for large N, so that the minimum of the energy is determined by the condition $E = 0$, which corresponds to the maximum baryon number at equilibrium

$$N_{max} \sim \left(\frac{\hbar c}{Gm_p^2}\right)^{3/2} \sim 2 \cdot 10^{57}. \tag{10.8.4}$$

Correspondingly, maximum mass of white dwarfs known as the Chandrasekhar limit is around $M_{max} \sim N_{max} m_p = 1.67 \cdot 10^{-24} g \times 2 \cdot 10^{57} = 3.3 \cdot 10^{33} g \sim 1.5M_{\otimes}$.

The equilibrium radius associated with maximum mass is determined by the onset of relativistic degeneracy, $E_F \geq mc^2$, which gives with account of (10.8.1) and (10.8.4)

$$R \leq \frac{\hbar}{mc}\left(\frac{\hbar c}{Gm_p^2}\right)^{1/2} \sim \begin{cases} 5 \cdot 10^8\,\text{cm} = 5000\text{km} & \text{for } m = m_e \\ 3 \cdot 10^5\,\text{cm} = 3\text{km} & \text{for } m = m_p \end{cases} \tag{10.8.5}$$

White dwarfs no longer burn nuclear fuel. The light thermonuclear fuels such as hydrogen, helium, lithium etc. were already burned during the previous stages of stellar evolution. Instead, they are slowly cooling as they radiate away their residual thermal energy, supporting themselves against gravity by the pressure of degenerate electrons. If the white dwarf accretes mass, then gravity increases compression of the core; the temperature at the white dwarf center increases and the carbon thermonuclear reaction may be ignited. The ignition may occur if the reaction time is shorter than the time of cooling processes, which take the energy away from the star (neutrino losses, for example) or distribute the energy from the hot reacting center of the white dwarf to the cold outer layers (thermal conduction, convection). Among them convection is the fastest of the cooling processes for the conditions of CO white dwarfs. The time of the carbon thermonuclear reaction at the white dwarf center becomes shorter than the typical convection time when the temperature at the star center exceeds $56.89\,\text{keV} \approx 6.6 \cdot 10^8\,\text{K}$.

At the moment of the thermal run-away the typical central density of the white dwarf is about $\rho_c = 3 \cdot 10^9\,\text{g/cm}^3$. The thermonuclear explosion of a carbon-oxygen (CO) white dwarf accreting mass – the Type Ia Supernovae (SNIa) is one of the most fascinating phenomena, which is of greatest interest due to their regularity and brightness. These properties of the SNIa allow using them as distance indicators – standard candle in the Universe, which shed recently new light on our understanding of the geometry of the Universe.

The observations show that during thermonuclear burning most of the white dwarf is converted into Fe-peak elements such as Ni, Co, Fe. At the same time, an important feature of the SNIa spectrum is the presence of lines corresponding to intermediate mass elements such as Si in the outer layers of the ejected matter. The observed production of the intermediate mass elements imposes strict limitations on the possible regimes of white dwarf burning. As in any exothermal reaction, the thermonuclear reaction may spread according to two possible regimes, which are a deflagration or a detonation. When the spherical flame propagates from the white dwarf center, it causes pre-expansion of the outer unburned layers of the star. On the contrary, since the detonation propagates with supersonic velocity, it causes no motion ahead of the detonation front. Thus, the white dwarf consumed by the detonation is burned at high densities equal to the initial densities at the moment when the detonation was ignited.

The choice of the burning regime is important in the theory of thermonuclear Supernovae, because it affects the chemical composition of the "ashes" and the total energy of the explosion. There are observations, which must be explained by the theory. These are the chemical composition of the "ashes" and observed during the explosion mass ejection from the star surface. The characteristic scale of the problem is of the order of $10^3\,\text{km}$, which is much larger than the thickness of detonation and deflagration fronts, which are of the order of $10^{-3}\,\text{cm}$ and $10^{-1}\,\text{cm}$, respectively. Therefore, in any numerical simulation of a thermonuclear Supernova the grid is too rough to determine the hydrodynamic regime of burning, and the choice of the burning regime becomes inevitably an artifact of a particular simulation.

It was found that neither the detonation nor the deflagration models have reproduced the observed features of SNIa events properly. The purely detonation model did not explain the production of intermediate mass elements in the Supernova spectrum properly. The deflagration model reproduced the observed spectrum rather well, but cannot explain the mass ejection. Indeed, the gravitational binding energy per unit mass is comparable to the internal energy per unit mass $GM_{WD} \sim P/\rho \sim a_s^2$. Thus, to overcome gravity the burning must cause flow with velocities comparable to the sound speed, a_s. However, the velocity of the flame in white dwarfs $U_f \sim 0.01a_s$ is much less than sound speed.

The detonation in white dwarfs may also produce a sufficient amount of intermediate mass elements if a considerable part of the star is consumed at low densities. Therefore, the observed SNIa spectrum may be explained if the white dwarf is pre-expanded by some reason prior to the detonation ignition. The desirable pre-expansion may occur if part of the star has been consumed by the flame prior to the detonation was triggered. Bearing this in mind, a good agreement with observations was found in the numerical simulations if the detonation regime was artificially "turned on" after the flame has consumed about 10% of the white dwarf mass. At present the commonly adopted phenomenological model is so-called the delayed detonation model, according to which the explosion starts as deflagration that later undergo transition to detonation, which may reproduce the majority of the observed features in Type Ia supernova. However the major unresolved problem remains the physics of the transition from the slow combustion regime to the detonation regime.

10.9 Burning Ignition and Transition to Detonation in White Dwarfs

Consider the initiation of the self-sustained thermonuclear reaction near the white dwarf center. Since flame is the slow regime of reaction propagation and detonation is the supersonic fast regime, then detonation triggering requires more energy. Flame may be ignited in white dwarfs by hot pockets of a size about $L_f \sim \sqrt{\chi\tau_R} \sim 0.3\text{m} \ll R_{WD}$. On the contrary, initiation of a detonation requires a hot pocket about $L_{Det} \sim a_s\bar{\tau}_R \sim 10^3\text{km}$, which is comparable to the size of the hot center core of the star, therefore flame always starts first in the Supernova explosions. The rate of the thermonuclear reaction depends strongly upon the initial temperature. The kinetic equation for the fuel mass is

$$\frac{da}{dt} = -\frac{a^2}{\tau_0}\exp(\sqrt[-3]{E_a/T}), \tag{10.9.1}$$

where τ_0 is the time constant, which depends slightly on temperature, $E_a = 5.9 \cdot 10^{14}\text{K}$ is the activation energy for carbon reactions.

If Q is the energy release per unit mass, then the energy balance equation is

$$\frac{dH(T)}{dt} = -Q\frac{da}{dt}, \tag{10.9.2}$$

where the enthalpy per unit mass for the degenerate matter of white dwarfs can be expressed as a function of temperature

$$H = \frac{4P}{\rho} = 3.1\frac{\hbar c}{\rho}(\rho Z/m)^{3/4}\left[1 + 2.06\left(\frac{m}{\rho Z}\right)^{2/3}\left(\frac{T}{\hbar c}\right)^{2}\right], \tag{10.9.3}$$

where m is the average mass, eZ is the average electrical charge of ions and temperature is measured in energy units.

Taking into account that at the beginning the process is isobaric and neglecting by the energy losses, the Eqs. (10.9.1), (10.9.2) and (10.9.3) can be combined in

$$\left(\frac{\partial H}{\partial T}\right)_P \frac{dT}{dt} = \frac{Q}{\tau_0}\left[1 + \left(\frac{H_c - H}{Q}\right)^2\right] \exp(-\sqrt[3]{E_a/T}), \tag{10.9.4}$$

where subscript "c" denotes initial values at the white dwarf center.

The characteristic induction time of the reaction is

$$\tau_i \sim 28\tau_0 \sqrt[3]{T_c/E_a}\,\frac{T_c^2}{\rho Q\hbar c}(\rho Z/m)^{2/3} \exp(\sqrt[3]{E_a/T}). \tag{10.9.5}$$

Obviously, due to strong exponential sensitivity of the reaction rate for the large = activation energy, even slight increase of the temperature, $\Delta T = (T - T_c) \ll 3T_c\sqrt[3]{T_c/E_a}) \ll T_c$ leads to considerable increase of the reaction rate. Therefore the characteristic reaction time is determined by the time of the first slow induction stage of the reaction, when temperature exceeds the initial value by $\Delta T \sim T_c\sqrt[3]{T_c/E_a})$. The slow induction stage of burning is followed by the fast stage of the reaction (explosion), which is much shorter than the induction time

$$\tau_{R\,\exp} \ll \tau_i \sqrt[3]{T_c/E_a}. \tag{10.9.6}$$

The temperature increase $\Delta T \sim T_c\sqrt[3]{Tc/E_a})$ on the induction stage of the reaction near the white dwarf center, which may initiate the combustion, also causes the density decrease

$$\Delta\rho = -\left(\frac{\partial\rho}{\partial T}\right)_P 3T_c\sqrt[3]{T_c/E_a}), \tag{10.9.7}$$

which leads to the convection that can be roughly described as rising of the bubbles of light heated matter out of the center of the star.

The convection distributes the energy released in the reaction from the hot white dwarf center to the cold outer layers and may prevent the self-accelerating reaction (thermal run-away) at the star center. The thermal run-away starts when the induction time of the reaction is shorter than the convection time $\tau_i \leq \tau_{conv} \simeq 3/4s$.

The convection establishes the iso-entropy profile of the average temperature a body of large mass in the equilibrium

$$\bar{T} = \bar{T}_c\left(1 - \frac{1}{2}\frac{R^2}{R_{WD}^2}\right), \tag{10.9.8}$$

where $\bar{T}$ and $\bar{T}_c$ are averaged over the surface $R = const$, and the characteristic length scale of the white dwarf core

$$R_{WD} \simeq 0.38\sqrt{\frac{\hbar c}{G\rho_c^{2/3} m_p^{4/3}}} \simeq 600km \tag{10.9.9}$$

is steamed from the equilibrium condition of a body of large mass for $\rho_c = 3 \cdot 10^9 g/cm^3$.

Consider the ignition of the nuclear flame that can propagate in highly degenerate matter of white dwarfs, which as it was shown is the most probable initial stage of the burning. The ignition of the flame means formation of the self-sustained reaction front propagating due to thermal conduction in the star matter. If ignition occurs, the burning becomes localized with a sharp boundary between burned and unburned material. As it was shown in Sect. 9.15, the propagation of a flame in a strong gravitational field can be fully controlled by the bubbles of burned matter. Let us consider a small volume near the center of the star where the flame was ignited. The reaction is completed inside the volume and propagates outside in the deflagration regime. The bubble of the burnt matter is separated from the cold fuel by the flame front, which is the surface of the bubble. The burning is localized at the surface of the bubble, which is almost discontinuous boundary between burned and unburned matter. The density of the burned matter inside the bubble is smaller than the density of the surrounding fuel. The expansion coefficient is determined by the chemical composition of the CO white dwarfs, $\Theta \simeq 1.2$. Therefore, the bubbles near the white dwarf center tend to run away to the surface of the star.

There are competing processes, which determine the destiny of the bubble: the bubbles grow in volume due to the flame propagation, and at the same time the bubbles as a whole run away from the star center. Let us compare the effects of the bubble rising from the star center and growth of the bubble's radius due to the flame propagation. The velocity of a steady planar flame at the center of the carbon-oxygen white dwarf is about $U_f = 100km/s$, and this is the velocity, which define growth of the bubble radius. At the same time since the Rayleigh Taylor instability in a strong gravitational field is the determinative process (for an estimate we can take the gravity acceleration $g \sim G(4\pi/3)\rho_c R_b$), the characteristic time of the bubble formation is of the order of

$$\tau_b \simeq \frac{1}{\sqrt{(4\pi/3)(\Theta - 1)G\rho_c}}. \tag{10.9.10}$$

During this time, the radius of a bubble becomes

$$R_b \simeq U_f \tau_b, \tag{10.9.11}$$

In the case of white dwarf with the central density $\rho_c = 3 \cdot 10^9 g/cm^3$ this means that the flame may incinerate only the bubble of a small radius $R_b \simeq 10km$, which is much smaller than the radius of the hot central part of the white dwarf. The bubbles are running away from the center as faster as larger their radius

$$U_b \simeq 0.67\sqrt{gR_b} = 0.67R_b\sqrt{\frac{4\pi}{3}\frac{\Theta - 1}{\Theta}G\rho_c} \simeq 0.67R_b/\tau_b. \tag{10.9.12}$$

The mass of the burned matter in such bubbles is negligible (less than 0.1%) in comparison with the total mass of in the center of the white dwarf. As a result, almost all matter in the central part of the white dwarf remains unburned.

Thus we come to the conclusion that the deflagration stage of white dwarf burning is very asymmetrical and the hot central part of the white dwarf remains unburned despite of the flame ignition and propagation. The main result of the ignited flame during this stage is the initiation of convection, which forms the nonuniform profile of the average temperature (10.9.8). The unburned thermonuclear carbon-oxygen fuel at the center of the white dwarf may explode later leading to the detonation regime of burning. This occurs since another regime of spontaneous burning is involved.

For the nonuniform profile of the temperature (10.9.8) the average velocity of the spontaneous explosion in the white dwarf is

$$u_{sp} = 3\sqrt[3]{T_c/E_a}\,\frac{R_{WD}^2}{R\tau_R}. \tag{10.9.13}$$

Triggering of detonation may occur either if the velocity of the spontaneous wave becomes equal to the local sound speed, or if the velocity of the spontaneous wave becomes equal to the velocity of the Chapman-Jouguet detonation. Numerical data indicate that the detonation can be triggered near the edge of the hot core of the star. More detailed studies (Kriminskii, Bychkov, Liberman, 1998) have shown that the detonation wave is absolutely unstable at the high densities and it becomes stable only at the densities less than $2 \cdot 10^7 g/cm^3$, near the surface of the star.

Thus we come to the following scenario of the white dwarfs burning shown schematically in Fig. 10.25. At the beginning, the local thermal run-away ignites the flame at several points at the center of the star. The bubbles of burnt matter together with the flame front are pushed out from the center of the star faster than the flame propagates, so that the center of the star remains unburned.

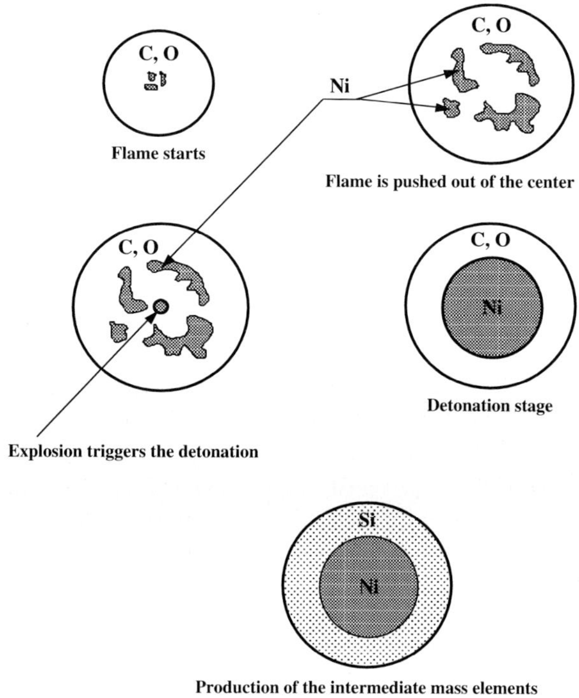

Fig. 10.25 Schematic scenario of white dwarf explosion

At the same time, the growth of bubbles due to the flame propagation ensures the pre-expansion of the star. After a period of time equal to the average induction time at the white dwarf center, spontaneous explosion occurs on a length scale comparable to the size of the white dwarf. The detonation triggered by the spontaneous explosion incinerates the rest of the star. Supersonic velocity of the particles in the detonation overcomes the gravitational binding and leads to mass ejection. When the detonation consumes the outer layers of the white dwarf, pre-expanded due to preceding flame stage of the burning, it produces the intermediate mass elements.

The amount of the intermediate mass elements produced by the detonation in the pre-expanded layers depends upon the fraction of white dwarf mass consumed by the flame in the deflagration stage of white dwarf burning. For the most part of the fuel, the spontaneous explosion occurs after a period of time equal to the average induction time at the white dwarf center. The presented physical scenario shows when and how transition from deflagration to detonation regime in white dwarfs occurs. The scenario includes the consequent stages of the flame, the spontaneous explosion and the detonation. Though, as it was said the full-scale numerical simulation of white dwarfs burning is not available yet, the existing numerical models confirm the tendency of bubble formation and the pushing out of the bubbles from the center of the star similar to those described above.

Problems

10.1. Normal velocities of the planar flames at normal conditions in the mixtures $6\%CH_4 + Air$, $5\%C_3H_8 + Air$, $H_2 + Air$, $H_2 + 0.5O_2{+}0.5N_2$, $C_2H_2{+}25O_2$. are: $U_f = 5.0$ cm/s, 50, 220, 900, and 1330 cm/s, correspondingly. What must be possible oncoming flow velocities for these mixtures to obtain a steady flame in the Bunsen burner? What is the angle of the Bunsen flame cone? Assume a two-dimensional version of the axisymmetric Bunsen burner. How result will change for cylindrically symmetric burner? How would you design the experiment for flame speed measurement using the Bunsen burner?

10.2. A stoichiometric propane-air flame (4% C_3H_8-Air) has normal velocity 40 cm/s. $T_b = 2240C$, $\rho_b/\rho_u = 7.8$ at normal conditions $P = 1atm$, $T = 300K$. Find the velocity of the combustion products relative to the flame front and pressure change across the front.

10.3. Find the velocity of the combustion products relative to the flame front and pressure change across the front for the flame in: 6% CH_4+94%Air, 25%CH_4-O_2, 30% H_2-Air, $H_2{+}0.5O_2$, $C_2H_2{+}2.5O_2$. Use data of Table 5.1 and compare the pressure change across the front for different flames.

10.4. According to Table 5.1 the maximum normal velocity of methane-air flame is 43 cm/s, for the propane-oxygen it is 320 m/s, etc. Also temperature of combustion products is different for different mixtures. Explain what cause this difference in the flame velocities and temperatures.

10.5. The laminar flame velocity in a stoichiometric methane-air flame at normal conditions of 1 atm and 300 K is 43 cm/s. Calculate the reaction rate and the thickness of the flame front and chemical reaction zone. Thermal conduction and density of methane and air at $T = 300K$ are $0.74 \cdot 10^{-4}$cal $cm^{-1}s^{-1}$, $0.56 \cdot 10^{-4}$cal $cm^{-1}s^{-1}$; and 0.717 g/L and 1.29 g/L. Assume that thermal conduction increases with temperature as $\sqrt{T}$.

10.6. A horizontal long tube 3 cm in diameter is filled with a stoichiometric methane-air. The normal laminar flame velocity at normal conditions of 1 atm and 300 K is 43 cm/s and the expansion ratio is 7.5. The planar flame is initiated near the left closed end and starts propagating to the right open end of the tube with ideally slip adiabatic walls. What is the mass burning rate of the mixture and the flame velocity under a laminar flow conditions if the unburned gas mixture density is 0.0063 g/cm^3.

10.7. What is the mass burning rate of the mixture and the flame velocity under a laminar flow conditions of previous problem just after initiating of the flame if the flame propagates in the tube with adiabatic and no-slip walls? How the flame velocity will change in the established laminar flow? Sketch a velocity profile that would exist in a laminar flow ahead and behind the flame front.

10.8. A laminar flame propagates in a horizontal tube 3 cm in diameter filled with a combustible mixture. The normal laminar flame velocity is 43 cm/s (a stoichiometric methane-air mixture). The tube is open at both ends. Due to buoyancy effect the flame tilts at 30° angle to the upper tube wall. What is the mass burning rate of the mixture and the flame velocity under a laminar flow conditions if the unburned gas mixture density is 0.0063 g/cm^3. Assume the walls to be slip and adiabatic.

10.9. The normal laminar flame velocity of a stoichiometric hydrogen-air mixture is 265 cm/s at normal conditions of 1 atm and 300 K. Suppose that the air is replaced with helium, and assume that temperature of the burned products do not change, estimate the burning velocity. Take thermal conductivity of the hydrogen-helium mixture $0.56 \cdot 10^{-4}$cal cm^{-1}s^{-1}, at T = 300K assuming that it increases with temperature as $\sqrt{T}$. For hydrogen and helium use $c_P = \frac{7}{2}R$ and $c_P = \frac{5}{2}R$. Since the specific heat of helium is lower than that of air, explain how will change the flame temperature?

10.10. The normal laminar flame velocity in a stoichiometric methane-air flame at normal conditions of 1 atm and 300 K is 43 cm/s. The activation energy is 18 kcal/mol and the adiabatic flame temperature is 2250 C. An inert diluent gas is added to the mixture so that the flame temperature lowered to 1700 C. Estimate the flame speed after inert diluent gas was added assuming that it has no other effects on the properties of the system.

10.11. A flame propagates in a constant-volume chamber between two parallel plates. Calculate the quenching distance for the flame if distance between plates is 3 cm for the following cases: (a) P = 1atm, T = 300K, U_f = 15cm/s, (b) P = 1atm, T = 500K, U_f = 15cm/s, (c) P = 10atm, T = 300K, U_f = 15cm/s. Assume that the thermal conduction of the mixture is the same as for air $0.56 \cdot 10^{-4}$cal cm^{-1}s^{-1}, at T = 300K.

10.12. Calculate the quenching distance for the flame of Problem 10.10 for cases (a), (b), (c) if the normal velocity of the flame is U_f = 90cm/s. Compare the results and explain the difference.

10.13. A laminar flame propagates through a combustible mixture at P = 1 atm, T = 300K with a velocity U_f = 220 cm/s, a mass burning rate is 0.16 g/cm^{2}s. What will be velocity and mass burning rate of a turbulent flame in the same mixture at P = 10 atm, if reaction is of the second order if due to turbulence the thermal diffusivity increased in 10 times.

10.14. A normal velocity of a laminar flames at P = 1 atm, T = 300K: U_f – 3.2 m/s, U_f = 9 m/s, and U_f = 13.3 m/s. How these velocities will be increased by turbulence if intensity of turbulent pulsation is $u' = 0.8$ m/s?

10.15. Calculate velocity, pressure and temperature of a strong detonation wave propagating in hydrogen-oxygen mixture at normal conditions, P = 1 atm, T = 300K. Take the heat of reaction for the mixture Q = 57.8 kcal/mol, and $\gamma_2 = 1.4$.

10.16. Calculate velocity, pressure and temperature of a strong detonation wave propagating in hydrocarbon-air mixture at normal conditions, P = 1 atm, T = 300K. Take the heat of reaction for the mixture Q = 27.8 kcal/mol, and $\gamma_2 = 1.2$.

10.17. For the conditions of Problems 10.14 and 10.15 calculate ratio of the detonation and adiabatic flame temperatures.

Chapter 11
Internal Combustion Engines

The industrial revolution of the nineteenth century was largely fuelled by coal and, as industrialization .developed, the close relationship between economic growths and increased demand for primary energy sources was established. Since the late nineteenth century petroleum demand has steadily increased, with road and air transport being the biggest users. Few inventions have had as great an impact on society, the economy, and the environment as the reciprocating internal combustion (IC) engine. Yet for decades, IC-engine design and improvement remains largely a cut-and-try experiential process. Engineers develop new combustion systems by making variations in previously successful configurations. However, today the automotive industry has faced numerous challenges. One of the most compelling has been reducing exhaust emissions and reductions in fuel consumption. The current trend is to replace highly expensive experimental investigation by numerical modeling with the main task of raising the thermodynamic efficiency of the power generating cycle with reduced pollutant emission. Analysis and numerical modeling of practical combustion systems together with detailed chemical model require understanding of combustion regimes, in particular, associated with changes in fuel composition for low-emission engines. The theory and design of internal combustion engines: spark ignition (SI) engine, diesel, and gas turbine have been the subject of intense engineering studies and treated exhaustively in many specialized monographs.

The purpose of internal combustion engines is the production of mechanical power from the chemical energy containing in the fuel. This converting of stored chemical energy is released by burning of the fuel-air mixture inside the engine. The fuel-air mixture before combustion and the burned products after combustion are the *working fluids*. The work transfers providing the desired power output occur between these working fluids and the mechanical components of the engine. There are two main types of internal combustion engines: spark-ignition engine (Otto engine) where burning is initiated by a spark, and compression-ignition engine (diesel engine), where burning is initiated by the heating due to adiabatic compression of the mixture. Both these type of engines are widely used in transportation and for power generation. The early engines for commercial use, which burned coal-gas-air at atmospheric pressure, had

M.A. Liberman, *Introduction to Physics and Chemistry of Combustion*,
DOI: 10.1007/978-3-540-78759-4_11,

been invented more than 150 years ago. Next step was in 1867 due to Nicolaus Otto and Eugen Langen, who used the pressure rise resulting from combustion of the fuel-air in the outward stroke to accelerate a free piston so that its momentum generate a vacuum in the cylinder, while atmospheric pressure then pushed the piston inward. Thermal efficiency of these engines was very small, less than 10%. Next step in increasing thermal efficiency was introduction of an engine cycle with four strokes and the increase of the pressure of the compression stroke before ignition. Maximum efficiency for an internal combustion engine is simple consequence of thermodynamic and can be summarized form conditions of minimum heat losses and maximum work transfer in the thermodynamic Carno cycle. These are: the minimum ratio of the boundary surface and cylinder volume; maximum expansion ratio; maximum pressure at the beginning of expansion. In fact almost two times higher efficiency was achieved in invented in 1892 by Rudolf Diesel engine, where injected liquid-air fuel was heated and ignited solely by compression, thus providing much greater expansion ratio compared to spark-ignition engine, where compression ration has been limited from the very beginning by knock.

Considerable progress has been made in course of 100 years of engine development. Nevertheless, nowadays the main concern is related to reduction of fuel consumption and further more considerable increase of thermal efficiency because of very tough situation, limited sources and high prices of oil, and considerable reduction of emission from engines, which may be at least partly responsible for global warming and other unpleasant phenomena. In this chapter we shall discuss briefly some aspects of combustion processes in engines for the purpose of background information.

11.1 Spark Ignition Engine (Otto-Engine)

In reciprocating engines the piston moves back and forth in a cylinder, transmitting chemical energy released from burning of the fuel in the cylinder through a connecting rod and crank mechanism into the mechanical energy of the rotating shaft. The corresponding schematic picture is shown in Fig. 11.1. The piston comes to rest at the top center (TC) of the crank position and at the bottom center (BC) crank position, where the cylinder volume is a minimum and maximum, respectively. The minimum cylinder volume is called the clearance volume, V_c and the difference between the maximum volume and the minimum volume is called swept volume. The ratio of maximum volume to minimum volume is the compression ratio, R_c. Typical values of the engine parameters are: the compression ratio are: $R_c = (8/12)$ for SI engines and $R_c = (15/24)$ for diesel engines; $B/L = 0.8/1.2$ for small and medium size engines; $\ell/R \simeq 3/4$. Important characteristics are the instantaneous piston speed S_p and the mean piston speed $\bar{S}_P$:

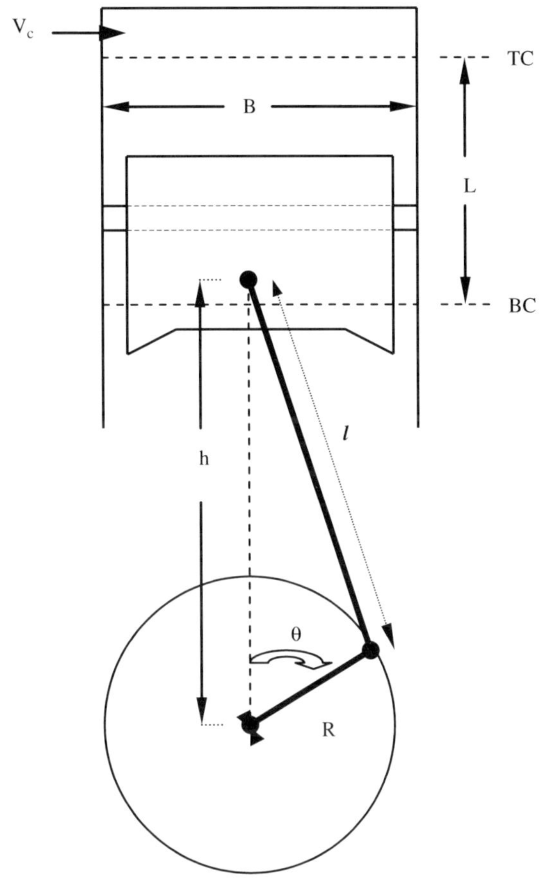

Fig. 11.1 Schematic picture of a connecting rod, crank mechanism, and the rotating shaft

$$S_P = \frac{dh}{dt}, \bar{S}_P = 2LN, \qquad (11.1.1)$$

where N is the rotational speed of the crank shaft in units revolutions per second. From geometrical consideration of Fig. 11.1 follows that

$$S_P = \bar{S}_P \frac{\pi}{2} \sin\theta \left[1 + \frac{\cos\theta}{\sqrt{(\ell/R)^2 - \sin^2\theta}} \right] \qquad (11.1.2)$$

The piston velocity is zero at the beginning of the stroke, reaches a maximum near the middle of the stroke and decreases to zero at the end of the stroke. Majority of the time it changes almost linearly. The maximum piston velocity is limited by stress due to inertia and typically it is within the range 8/15 m/s. The power $\dot{W}$ delivered by the engine is the product of torque and angular velocity in units rad/s: $\dot{W} = \omega T = 2\pi NT$ (Watt)

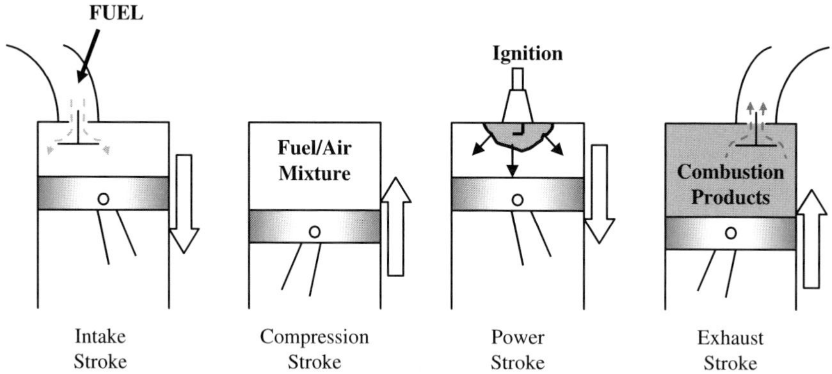

Fig. 11.2 Four-stroke cycle of a reciprocating engine

Majority of reciprocating engines operate on four-stroke cycle, shown schematically in Fig. 11.2. Each cylinder requires four strokes of its position to complete the sequence of events, which produces one power stroke, thus – two revolutions of the crankshaft. An *intake stroke* starts with the position at TC and ends at BC: fuel-air mixture introduced into cylinder through intake valve. A *compression stroke*: fuel-air mixture compressed, valves are closed. A *power stroke* starts with the position at TC and ends at the position BC: combustion occurs and high temperature gases expand doing work. An *exhaust stroke*: burned gases pushed out of the cylinder through the exhaust valve. The four-stroke cycle requires for each engine cylinder two crankshaft revolutions for each power stroke. To obtain a high power output from a given engine size the two-stroke cycle was developed for both SI and diesel engines.

Schematic pressure-volume engine diagram for a four-stroke SI engine is shown in Fig. 11.3, and more realistic diagram is shown in Fig. 11.4. At the beginning of intake stroke the piston start to move at the top center with a

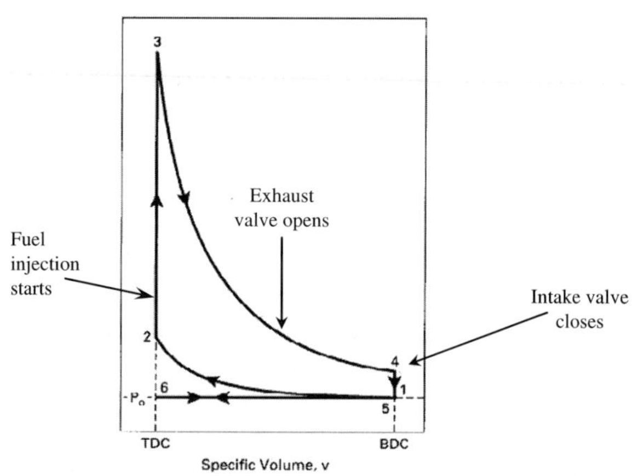

Fig. 11.3 Pressure-volume engine diagram for a four-stroke SI engine

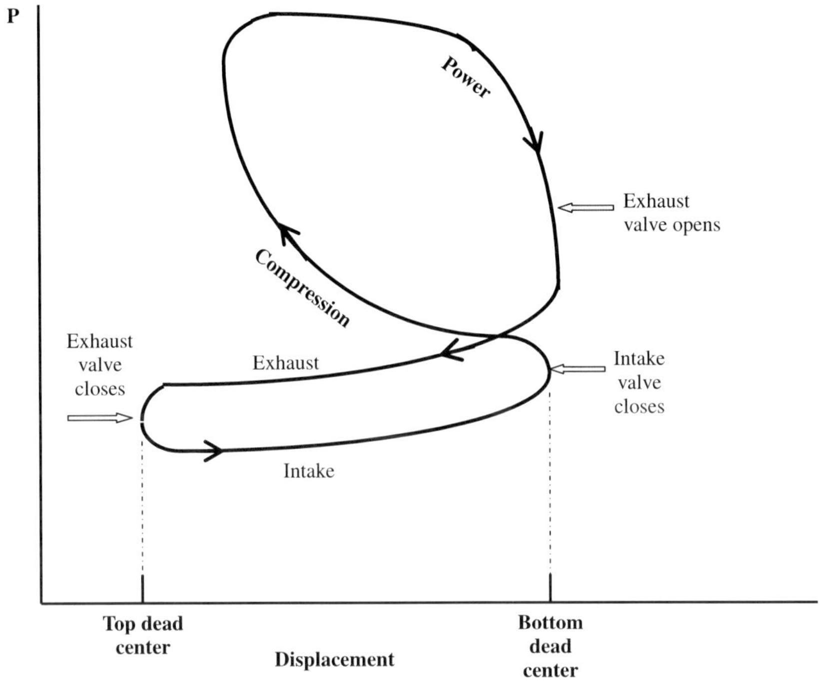

Fig. 11.4 Pressure-volume engine diagram for a four-stroke SI engine

clearance above it, which is filled with burned gas from the previous burning. As the piston moves downward, the exhaust valve closes and the intake value opens through which a charge of gasoline and air in approximately stoichiometric proportions is drawn into the cylinder. The intake valve closes and mixture is compressed by the upward-moving piston in compression stroke. A little before top dead center the spark passes, and the ensuing combustion causes a rapid rise in pressure. During the combustion time the piston moves through top dead center and is subsequently forced downward by the expanding hot gas. When the piston reaches bottom dead center the exhaust valve opens and the burned gas escapes in the subsequent exhaust stroke the piston moves upward. The corresponding sequence of strokes is shown in Fig. 11.2 by arrows.

11.2 Engine Operating Cycles

Let us consider simple models, which provide useful insight into performance and efficiency of engines. For the sake of simplicity we consider ideal engine cycles, assuming that the working mixture is an ideal gas. Typical engine operating cycle represents a consecutive sequence of processes, examples of which are shown in the Figs. 11.3 and 11.4. The cycle consists of: adiabatic compression (1–2), which is isentropic; combustion (2–3), which can occur either at constant volume, or it can be combined – partly at constant volume and partly at constant pressure; adiabatic and isentropic expansion (3–4), adiabatic exhaust (4–5–6); intake (6–1). Reliability of such ideal model for evaluation of an engine performance depends essentially on how realistic is accepted description of these processes. As a matter of fact we consider two limiting cases: the constant volume cycle corresponding to the limit of infinitely fast combustion at TC, and the constant pressure cycle corresponding to slow and late combustion. This means that Fig. 11.3 illustrates the first cycle of constant volume combustion, and Fig. 11.4 is a schematic sketch of the intermediate case.

Let us apply the first and the second laws of thermodynamic to evaluate the engine performance and efficiency assuming the working fluid (fuel-air and combustion products) to be an ideal gas with constant specific heats C_V and C_P. We consider an ideal cycle 1–2–3–4 shown in Fig. 11.3. In position 1 the intake valve closes and the upward moving piston compresses the mixture. The curve of compression stroke, 1–2, is isentropic compression. The spark initiates combustion at point 2, slightly before TC, causing a rapid rise in pressure, which is depicted at Fig. 11.3 as constant volume pass 2–3, which is the constant volume heat addition. During the combustion time the piston moves from the top dead center downward forced by the pressure of expanding hot gas. This is isentropic expansion-work stroke corresponding to curve 3–4. The stage 4–1 is the constant volume heat rejection. The exhaust valve opens when the piston reaches bottom dead center, piston moves upward and pushes away the burned products during exhaust stroke 5–6, and then returns back to 1 during intake stroke. The work gained in the cycle 1–2–3–4 is $R_i = \oint PdV$ given by the area inside 1–2–3–4 in Fig. 11.3. For more realistic cycle it should be difference between areas of upper and lower loops in Fig. 11.4. The area of the lower loop represents the work expended to overcome the flow resistance of the gas during intake and exhaust strokes, which is pumping losses. If the heat of combustion for a given fuel is known, the diagram provides a measure of the transformation of chemical energy into mechanical work, i.e., thermal efficiency or fuel economy. The product of the work done in each cycle and the number of cycles per unit time power determines the engine power.

For the idealized cycle shown in Fig. 11.3, when no heat or pumping losses occurred and assuming the combustion is fast and its energy contents is known, so that fuel is burned at constant volume at top dead center to thermodynamic equilibrium, the thermal efficiency and power can be calculated from the changes of state of the gas during the cycle. A simple estimate of the factors that influence thermal efficiency and power can be obtained by analyzing the standard air cycle. In such idealized cycle we assume that chemical reaction involves only addition of heat to the gas without changing its composition. The actual working fluid in an engine consists largely of nitrogen whose specific heat is a mild function of temperature; furthermore, the change in number of moles due to chemical reaction is small. Therefore, if the specific heat of the gas in the standard air cycle is identified with an average constant specific heat of the actual mixture during the cycle, the quantitative differences between this cycle and the idealized cycle for the actual gas will not be too large. Thus, the following simplifications compared to the real cycle include: (1) fixed amount of ideal gas for working fluid; (2) combustion process not considered; (3) intake and exhaust processes not considered; (4) engine friction and heat losses not considered. This is called Air-Standard Otto cycle, which consists of four stages: (1–2) isentropic compression, (2–3) constant volume heat addition, (3–4) isentropic expansion, (4–1) constant volume heat rejection. The compression ratio of the cycle $R_c = V_1/V_2$ is the ratio of the volumes in the state $BC - V_1$ and in $TC - V_2$.

Let masses of the fresh intake fuel and residual gas are denoted by M_f and M_R, and T_1, T_2 and T_3, T_4 are the temperatures at the beginning and end of compression stroke and at the beginning and end of the work stroke, respectively. The thermal efficiency is defined as

$$\eta_{th} = \frac{\text{heat input} - \text{heat rejected}}{\text{heat input}} = \frac{W_{cycle}}{Q_{in}} = \frac{W_{out} - W_{in}}{Q_{in}}, \tag{11.2.1}$$

where work per cycle, W_{cycle} is the sum of the compression stroke work and the expansion stroke work.

The work during isentropic compression (1–2) is

$$W_{in} = (M_f + M_R)C_V(T_2 - T_1). \tag{11.2.2}$$

The constant volume heat input is chemical energy released in combustion QM_f corresponding to the increase of energy and temperature change between 2–3

$$Q_{in} = QM_f = (M_f + M_R)C_V(T_3 - T_2), \tag{11.2.3}$$

where Q is energy release in combustion per unit mass.

The isentropic expansion work (3–4) is

$$W_{out} = (M_f + M_R)C_V(T_3 - T_4). \tag{11.2.4}$$

We obtain for the net cycle work

$$W_{cycle} = W_{out} - W_{in} = (M_f + M_R)C_V(T_3 - T_4) - m(T_2 - T_1). \tag{11.2.5}$$

Substituting (11.2.5) and (11.2.3) in (11.2.1) we obtain

$$\eta_{th} = \frac{W_{cycle}}{Q_{in}} = \frac{W_{out} - W_{in}}{Q_{in}} = \frac{(T_3 - T_4) - (T_2 - T_1)}{(T_3 - T_2)} = 1 - \frac{T_4 - T_1}{T_3 - T_2}. \tag{11.2.6}$$

It follows from (11.2.6) that the work done in the cycle is $W_{cycle} = \eta_{th}Q_{in}$. To determine the power one needs to know the amount of fresh fuel mixture M_f

drawn into the cylinder each cycle. The fresh fuel comes with temperature T_f and the residual gas left in the cylinder at the end of the exhaust stroke at the temperature T_6, therefore

$$T_1 = (M_f T_f + M_R T_6)/(M_f + M_R) \tag{11.2.7}$$

For the two isentropic processes (1–2) and (3–4) in the cycle, assuming ideal gas with constant specific heat and using equations for ideal gas ($PV^{\gamma} = \text{const.}$, $PV = RT$), or

$$TV^{\gamma-1} = \text{const}, \tag{11.2.8}$$

we can find relations between temperatures for (1–2)

$$\frac{T_2}{T_1} = \left(\frac{V_1}{V_2}\right)^{\gamma-1} = {R_c}^{\gamma-1}, \tag{11.2.9}$$

and

$$\frac{T_4}{T_3} = \left(\frac{V_3}{V_4}\right)^{\gamma-1} = \left(\frac{1}{R_c}\right)^{\gamma-1} \tag{11.2.10}$$

From the last two equations we have $T_1/T_2 = T_4/T_3$, and (11.2.6) can be rewritten using (11.2.9) as

$$\eta_{th} = 1 - \frac{T_4 - T_1}{T_3 - T_2} = 1 - \frac{(T_1/T_2)T_3 - T_1}{T_3 - T_2} = 1 - \frac{T_1}{T_2} = 1 - \frac{1}{{R_c}^{\gamma-1}}. \tag{11.2.11}$$

It is seen that the thermal efficiency increases with the increase of compression ratio and value of adiabatic constant γ. For air, which can be considered as two-atomic gas $\gamma = 1.4$. Therefore γ is smaller for reach mixtures and larger for leaner mixtures. For typical compression ratio $R_c = 8$ the ideal model gives for the thermal efficiency the value of 56% which is about twice of the actual value. It is seen that in the ideal cycle where no heat and pumping losses occur, the thermal efficiency is not dependent on engine factors other than the compression ratio. As far as mixture composition is concerned, it is not possible to effect more than a compromise between power and economy. The power can be increased further by supercharging, for example, by increasing initial pressure, however both thermal efficiency and power are increased by increasing the compression ratio. As far as design of the Otto engine is concerned, the compression ratio is limited not by engineering factors but by the increase of knocking tendency of fuels with increasing the compression ratio. Knock is the term used to describe a pinging noise emitted from a SI engine undergoing abnormal combustion. The practical limit of compression ratio seems to be reached in the modern high-speed Diesel engine, which actually operates very nearly on an Otto cycle.

11.3 Diesel Cycle

The compression-ignition and diesel engines are superior to that of the port-fuel-injected SI engines due to the use a higher compression ratio. The diesel engines, however, generally exhibits higher particulate and NO_x emissions than the SI engines. The combustion process in the compression-ignition engines proceeds by the following stages shown schematically in Fig. 11.5 and in the corresponding diagram of ideal Diesel cycle in Fig. 11.6. Compression stroke – isentropic compression: (a–b); fuel injection and combustion – constant pressure heat addition stroke: (b–c); power stroke – isentropic expansion: (c–d); exhaust stroke-constant volume heat rejection (d–a).

In *compression stroke* – the fuel is compressed adiabatically along (a–b). The liquid fuel atomizes into small drops and penetrates into the combustion chamber. The fuel vaporizes and mixes with the high-temperature high-pressure air. Combustion of the fuel is going at constant pressure along (b–c). The fuel mixed with the air during the ignition delay period, which occurs rapidly in a few crank angles. The rate of burning is controlled in this phase primarily by the fuel-air mixing process.

The heat added from b to c is chemical energy released in combustion QM_f, so that according to the first law of thermodynamic temperature change between (b–c) is

$$\begin{aligned} Q_{in} &= QM_f = M_f(h_c - h_d) \\ &= M_f[C_P(T_c - T_b) + P_2(V_3 - V_2)] \end{aligned} \tag{11.3.1}$$

From the equation of state of ideal gas we can write also

$$P_b = P_c = \frac{R_B T_b}{V_2} = \frac{R_B T_c}{V_3} \tag{11.3.2}$$

or,

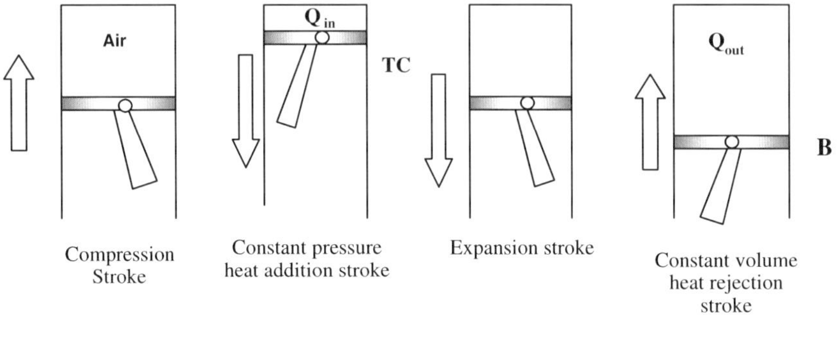

Fig. 11.5 Scheme of Diesel cycles

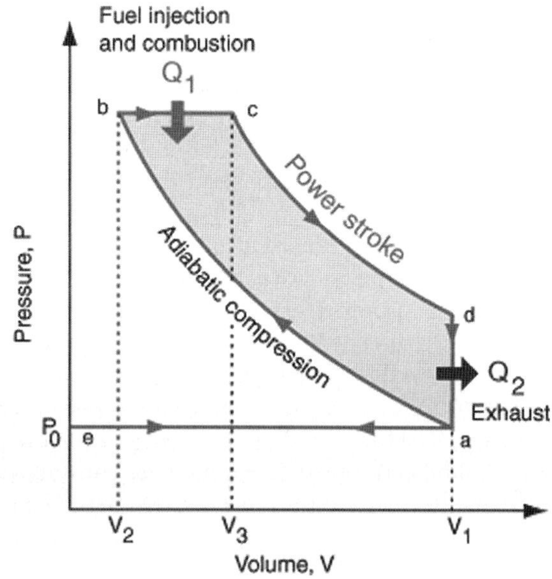

Fig. 11.6 PV-diagram of ideal Diesel cycle

$$\frac{V_3}{V_2} = R_{bc} = \frac{T_c}{T_b}. \tag{11.3.3}$$

Since the compression and power strokes of this idealized cycle are adiabatic, the efficiency can be calculated from the constant pressure and constant volume processes. The input and output energies and the efficiency can be calculated from the temperatures and specific heats in a way similar to how we did for ideal Otto engine cycle. Similar to (11.2.1) thermal efficiency for Diesel cycle is defined as

$$\eta_{\text{Diesel}} = 1 - \frac{Q_{out}}{Q_{in}} \tag{11.3.4}$$

Equations for processes (a–b) and (c–d) are the same as those (1–2) and (3–4) for the Otto cycle Eqs. (11.2.8), (11.2.9) and (11.2.10). Repeating calculations for the work per cycle delivered to the piston over the compression and expansion strokes similar to Eqs. (11.2.8), (11.2.9) and (11.2.10) we can write down for isentropic expansion (c–d)

$$W_{out} = M_f(T_c - T_d). \tag{11.3.5}$$

The relations between compression ratios can be written as

$$\frac{V_d}{V_c} = \frac{V_d}{V_2} \cdot \frac{V_2}{V_c} = \frac{V_1}{V_2} \cdot \frac{V_2}{V_3} = \frac{R}{R_{bc}}, \tag{11.3.6}$$

where $R \equiv R_{ab} \equiv R_{12} = V_1/V_2$.

Since $V_a = V_d = V_1$, we can write using equation of state

$$\frac{P_d V_d}{T_d} = \frac{P_c V_c}{T_c} \Rightarrow \frac{P_d}{P_c} = \frac{T_d}{T_c} \cdot \frac{R}{R_{bc}}. \tag{11.3.7}$$

For thermal efficiency we have

$$\eta_{Diesel} = 1 - \frac{Q_{out}}{Q_{in}} = 1 - \frac{T_d - T_a}{h_3 - h_2}. \tag{11.3.8}$$

Taking into account that according to the thermodynamic gas law

$$T_c/T_b = V_b/V_c, \tag{11.3.9}$$

and for the adiabatic compression and expansion

$$\frac{T_a}{T_b} = \left(\frac{V_b}{V_a}\right)^{\gamma-1} = \left(\frac{V_2}{V_1}\right)^{\gamma-1} = R^{1-\gamma}, \tag{11.3.10}$$

and for adiabatic expansion

$$\frac{T_d}{T_c} = \left(\frac{V_c}{V_a}\right)^{\gamma-1} = \left(\frac{V_3}{V_1}\right)^{\gamma-1}, \tag{11.3.11}$$

and combining the Eqs. (11.3.6), (11.3.7), (11.3.8), (11.3.9), (11.3.10) and (11.3.11) we obtain for the thermal efficiency

$$\eta_{Diesel} = 1 - \frac{1}{\gamma R^{\gamma-1}} \cdot \frac{(R_{bc}^{\gamma} - 1)}{(R_{bc}^{\gamma} - 1)} \tag{11.3.12}$$

Note, that for a given compression ratio and mixture composition the efficiency of the Diesel cycle is less than the efficiency of the Otto cycle, since the term in the square bracket in (11.3.12) is always larger than one, however compression ratio used in a diesel engines is always much larger than that in SI-engines. The difference depends on the magnitude of the pass (b–c) in the PV diagram Fig. 11.6, which itself is a function of compression ratio. With increasing compression ratio the efficiencies of the two cycles approach each other. The comparison also shows that in order to obtain the highest efficiency in the Otto cycle, combustion should take place at as nearly constant volume at top dead center. For equal compression ratios the peak temperature and pressure obtained in the Otto cycle are much higher than in the Diesel cycle. Because of the poorer mixing of the fuel and air, Diesel engines are always operated with an excess of air. However, in general the efficiency of Diesel engines is higher than the Otto engines since in Diesel engines much higher compression ratio can be used.

11.4 Knock in SI-Engines

A thermodynamics of the engine cycle shows that overall efficiency can be increased with the increase of the compression ratio. Yet, as the compression ratio is raised the onset of a phenomenon called "knock" occurs, which can be destructive to the engine and should be avoided. The knock phenomenon in SI engines has constituted a major and the most serious limitation upon increasing efficiency of SI engines by increasing the compression ratio from the very beginning of car technology. Even though knock in SI engines was considered for decades as one of the most challenging problems, our ability to extend the knock limits of a spark-ignition engine is limited so far by the lack of fundamental knowledge of the processes, which cause knock in an engine. One of the main difficulties in analyzing the problem implies strong coupling of multidimensional hydrodynamics of a gas fuel and turbulence in an engine cylinder with chemical kinetics, making it hard to reveal the key mechanisms governing the knock occurrence. A number of chemical kinetic models of autoignition at high pressure and temperature fuel-air mixtures are currently available with the complete models including up to several hundred reactions. As a result, even given the remarkable development of computational facilities, numerical simulations of comparatively realistic models often meet with formidable difficulties.

Knock is currently believed is the result of the spontaneous thermal ignition of a certain amount of fuel-air mixture in the combustion chamber before it can be consumed by a primary flame propagating through the cylinder charge. At moderate compression ratios, the temperature of the end gas may be within the temperature range of (600–800) K, and knock may be related to two-stage ignition and cool flame phenomena, being caused by autoignition of the end-gas. The heat release accompanying the autoignition may be so rapid that it produces strong pressure rise, which is sometimes followed by exciting of shock waves. This abnormal combustion, known as knock, which got this nickname from the noise that is transmitted from the colliding of the multiple flame fronts and the increased cylinder pressure that causes the piston, connecting rod and bearings to resonate, has been the limiting factor in internal combustion engine power generation since the discovery of the Otto cycle itself. The noise is generated by shock waves produced in the cylinder when unburned gas ahead of the flame auto-ignites. Since the engine thermal efficiency is directly related to the compression ratio, but engine knock occurs more easily if the compression ratio is increased, it is important to find possible ways of how to avoid the knocking combustion. It was found that autoignition depends on the sensitivity of the induction time on the range of temperature change, and that the main factors that influence knock are the combustion chamber size, the location of the spark plug, and the ratio of specific heats.

As the flame propagates away from the spark plug, the pressure and temperature of the unburned gas increase. The unburned gases are compressed by the piston and additionally compressed by the burned gases that expand behind the spark-ignited flame front. The last remaining unburned gas is called the *end gas*. With higher compression ratios, the end-gas temperature increases until spontaneous ignition occurs. This sudden ignition leads to the formation of pressure peaks in the cylinder that cause the audible knocking noise. Knock must be avoided since these pressure peaks damage the piston and engine. The end-gas autoignites after a certain *induction time* which is dictated by the chemical kinetics of the fuel-air mixture. If the flame burns all the fresh gas before autoignition in the end-gas can occur then knock is avoided. Therefore knock is a potential problem when the burning time is long enough.

Fuels differ in their tendency to produce knock. Isooctane (with a low knock tendency) has an octane number of 100 while n-heptane, which has a high knock tendency, has an octane number 0. Thus, a fuel with (octane number) ON = 80 has the same knocking tendency as a mixture of 80% isooctane and 20% n-heptane. Engine parameters that affect occurrence of knock are as following. (1) *Compression ratio*: at high compression ratios, the fuel-air mixture is compressed to a high pressure and temperature, which promotes autoignition. (2) *Engine speed*: at low engine speeds the burn time is long and this results in more time for autoignition. However, as higher engine speed as less the heat losses, so the unburned gas temperature is higher, which promotes autoignition. Thus, just simple rise of the engine speed does not lead to knock mitigation.

There is no complete explanation of the knock phenomenon yet, however it is generally agreed that knocking in SI-engines is caused by the end-gas autoignition, which results in the extremely rapid release of much energy contained in the end gas ahead of the flame propagating from the spark and results in a very fast rise of local pressure. In reacting mixture as hydrocarbon-air used in engines, the reaction is not a single or even few-step process, but actual chemical mechanism consists of many hundreds reactions between a large amount of species. In such chain reactions there are *initiating* reactions, where highly reactive intermediate radicals are produced from stable molecules of fuel and oxygen. If due to chain branching, the number of radicals increases sufficiently rapidly, the reaction rate becomes extremely fast and results in chain-branching explosion.

Among various low-temperature kinetics models the Shell model developed by Halstead et al. (1975) should be distinguished. Despite of its mathematical simplicity the model catches the essence of the mechanism of low-temperature oxidation and gives a good agreement for hydrocarbon fuels autoignition for the experiments in a rapid compression machines. It was first developed phenomenologically as generalization of experimental data obtained on rapid compression machines, and later on Cox and Cole (1985) have shown that generalized species used in the Shell model can be related to particular chemical species and that the Shell model can be reformulated in terms of the elementary reactions of certain species and radicals, which makes this mathematical model well-grounded from the point of view of real chemical kinetics. The Shell model is based on skeletal mechanisms, which represent the most important generic species, though since some of the reaction rates are not known, they must be adjusted to the experimental data according to the fuels used and the engine operation conditions. With such an interpretation and a good fit to experimental observation the Shell model can be treated as a well reliable for modeling low-temperature kinetics.

The Shell model is formulated in terms of generalized species, their variation being described by the equations based on detailed analysis of the experimental results on hydrocarbon autoignition.

$$\frac{1}{V}\frac{dn_R}{dt} = 2\left\{k_q[RH][O_2] + k_B[B] - k_t[R]^2\right\} - f_3 k_p[R], \tag{11.4.1}$$

$$\frac{1}{V}\frac{dn_B}{dt} = f_1 k_p[R] + f_2 k_p[Q][R] - k_B[B], \tag{11.4.2}$$

$$\frac{1}{V}\frac{dn_Q}{dt} = f_4 k_p[R] - f_2 k_p[Q][R], \tag{11.4.3}$$

$$\frac{1}{V}\frac{dn_{O_2}}{dt} = -pk_p[R], \tag{11.4.4}$$

$$n_{RH} = \frac{n_{O_2} - n_{O_2}(t=0)}{pm} + n_{RH}(t=0), \tag{11.4.5}$$

where n_s and [...] denote the concentration and number of moles of various species, [RH] is the concentration of the hydrocarbon moles, [R] is the total concentration of radicals participating in the reactions, B denotes the intermediate agents of the branching chains, Q is the autocatalysis product which can be identified with the aldehyde radical RCHO; m is the number of the hydrocarbon moles; p is the corresponding number of oxygen moles needed to form 1 mol of water by the cool flame. The dependence of the reaction rates k_i is taken in the form of the Arrhenius law $k_i = A_i \exp(-E_i/R_B T)$, and the coefficients f_i can be written in the form $f_i = A_{f_i} \exp(-E_{f_i}/R_B T)[O_2]^{\alpha}[RH]^{\beta}$. In the model they are considered as the fitting parameters, with the constants being chosen for the best fitting of the first and second induction times, which are measured from the experiments with the adiabatic compression of the Primary Reference Fuel. Fuel consumption is assumed to occur at a rate of single entity for each propagation cycle. The number of O_2 moles consumed per propagation cycle in cool flame is defined from the assumed overall reaction stoichiometry

$$\frac{1}{n+1} C_n H_{2n+2} + pO_2 \rightarrow \frac{zn}{n+1} CO + \frac{(1-z)n}{n+1} CO_2 + H_2O, \tag{11.4.6}$$

where $z = 0.67$, and $p = [(3-z)n + 1]/2(n+1)$.

The heat release per one cycle of the chemical reaction (per 1 mol of water) is $Q_f = qk_p[R]$ with the value of the specific heat release for the standard fuel $q = 9.4 \cdot 10^4 \text{cal/cycle}$.

The chain branching reaction, which dominates combustion at high temperatures, is too slow to explain autoignition at temperatures below 1200 K. The initiating steps proceed primarily through hydrogen peroxide (H_2O_2) to form the hydroxyl radicals (OH) at low temperatures. The process by which hydrocarbon is oxidized can exhibit different types of behavior, typically two-stage ignition of cool flames with slightly exothermic reactions, followed by a hot flame. Sensitivity analyses and analyses of reaction paths indicate that the chain branching responsible for the autoignition after an initiation reaction

$$RH + O_2 \rightarrow R + HO_2 \tag{11.4.7}$$

are the following

$$RH + HO_2 \rightarrow R + H_2O_2, \tag{11.4.8}$$

$$H_2O_2 + M \rightarrow 2OH + M. \tag{11.4.9}$$

The OH-radicals can reproduce the HO_2 radicals in reactions

$$OH + H_2 \rightarrow H_2O + H, \tag{11.4.10}$$

$$H + O_2 + M \rightarrow HO_2 + M. \tag{11.4.11}$$

This branching via the HO_2-radical can explain the knock process in an engine at temperatures of about 1100 K.

Though these mechanisms can explain the observations of so-called *two-stage ignition* and a *negative temperature coefficient of the* ignition delay time, the process involves too complex chemistry and also depends on hydrodynamic of the propagating flame, in particular non-uniformities formed by the pressure waves generated in the end gas by the propagating flame. Therefore, numerical modeling of knock in engines is usually based on temperature and pressure histories taken from experimental studies. The pressure is measured directly, whereas the temperature is calculated assuming nearly adiabatic compression in the cylinder and a certain heat loss to the cylinder walls.

Thorough analysis of the flame dynamics in engines (Liberman et al., 2006) has shown that development of the autoignition is tightly connected to the formation of hot spots that evolved from the nonuniformities caused by pressure waves emitted by the propagating flame. It was shown that there is a considerable positive feedback: a propagating flame is accelerated by the temperature increase due to development of the cool flames in the end gas, and the development of the autoignition is enhanced by the flame acceleration. The numerical modeling has shown that the calculated dependence of the temperature and pressure in the end gas on crank angle and predicted time of the autoignition onset for different engine operation conditions, in particular, for different percentage of the exhaust gas recycled (EGR) were in a good agreement with the experimental data.

To increase the engine efficiency and mitigate knock some chemical additives can be used to increase the octane number of gasoline. Among these alcohols, ethanol and methanol have high knock resistance. Besides the compression ratio, the occurrence of knock is influenced the engine speed. At high engine speeds there is less heat loss so the unburned gas temperature is higher which promotes autoignition. These effects are competing; some engines show an increase in propensity to knock at high speeds while others don't. Maximum compression from the piston advance occurs at TC, so that the increasing the spark advance makes the end of combustion crank angle approach TC and thus get higher pressure and temperature in the unburned gas just before burnout. If the fuel-air mixture is leaned out with excess air or is diluted with increasing amounts of residual gas or exhaust gas recycle, then the burn time increases and the cycle-by-cycle fluctuations in the combustion process increase. As dilution increases, the burning slows and combustion is only just completed prior to the exhaust valve opening. As dilution increases further, in some cycles combustion is not complete prior to the exhaust valve opening and flame extinguished before all the fuel is burned. As the dilution is further increased, the proportion of partial burns and misfires increase to a point where the engine no longer runs.

Chapter 12
Combustion and Environmental Concerns

Recent climate catastrophes indicate serious warning of the global warming underway. It is generally believed that burning of hydrocarbon fuels, required by industry and transport, is the main source of the pollutants, which may influence or aggravate global warming. Of the total emissions of NO_x, CO, CO_2, aerosols, and other chemical species almost all are attributable to the fossil-fuel combustion. At the same time the energy requirement is growing and it is expected to triple over the next 20 years. Advance in computational power together with theoretical combustion models offers an opportunity to revolutionize the design and performance of combustion systems, which will considerably lower emissions and increase the thermodynamic efficiency of new combustion technologies.

All internal combustion engines and combustion processes used in the industry produce harmful emissions. Those include unburned hydrocarbons, carbon monoxide (CO), oxides of nitrogen NO and NO_2 collectively called NO_x, sulfur dioxide, and solid carbon particulates (soot). They pollute the environment and contribute to the acid rain, smog, and respiratory and other health problems. Hydrocarbons' emissions from gasoline-powered vehicles include a number of toxic substances such as benzene, polycyclic aromatic hydrocarbons (PAHs), butadiene and aldehydes (formaldehyde, acetaldehyde, acrolein). Carbon dioxide is an emission, which while not being regulated, is the primary greenhouse gas. The NO_x-s participate in a chain reaction removing ozone from the stratosphere allowing more ultraviolet radiation to reach the Earth's surface. The NO_x is a major contributor of photochemical smog and ozone in the urban air. Consequently, reduction of the NO_x production has become one of the most important topics in combustion research. With the steady increase in combustion of hydrocarbon fuels, the products of combustion are distinctly identified as a severe source of environmental damage.

12.1 Formation of Hydrocarbons and Soot

The carbon particles formed from the gas phase are generally referred as soot. Hydrocarbon (HC) emissions result from the presence of unburned fuel in the burner or in engine exhaust. Some of the exhaust hydrocarbons are not

M.A. Liberman, *Introduction to Physics and Chemistry of Combustion*,
DOI: 10.1007/978-3-540-78759-4_12, © Springer-Verlag Berlin Heidelberg 2008

found in the fuel, but are hydrocarbons derived from the fuel whose structure was altered due to a chemical reaction, which did not proceed to a completion. The examples are acetaldehyde, formaldehyde, butadiene, and benzene, which are all classified as toxic emissions. About 9% of fuel supplied to the engine is not burned during the normal combustion phase of the expansion stroke. Only 2% ends up in the exhaust, the rest is consumed during the other three strokes. Thus, hydrocarbon emissions ending up as an air pollutant also decrease the thermal efficiency.

In general, unburned hydrocarbons are a consequence of local flame extinction at walls and in gaps. High strain caused by turbulence or other events may also lead to local extinction of the flame. When flame interacts with turbulent flow turbulence is modified by combustion. This happens because of the strong flow acceleration through the flame front induced by heat release, and because of the large changes in kinematic viscosity associated with temperature changes. On the other hand, turbulence alters the flame structure, and it may enhance chemical reaction but also, in certain cases, completely inhibit it, leading to flame quenching. The study of the effect of turbulence on premixed combustion is of great practical importance because turbulence may increase the propagation rate of the turbulent flame to a value well above the laminar burning velocity in the same mixture. This in turn increases the heat release rate and thus the power available from combustor or internal combustion engine of a given size. Yet, it has long been observed that for each gaseous mixture there is a certain level of turbulence at which the flame speed reaches its maximum. Further increase in the flow intensity leads to a drop of the flame speed, followed by the flame disintegration and quenching. This can be one of the reasons for the high emission of hydrocarbons from lean combustion engines.

Flame extinctions at walls and in gaps are caused by cooling of the reaction zone due to the heat losses to the walls as well as by the removal of reactive radicals by surface reactions with the quenching distance being of the order of the flame thickness. The issue is controversial since the experimental studies have shown that the flame extinction at walls contributes much less to the emission of unburned hydrocarbons from SI engines than it was expected. Nevertheless, it is generally believed that the principal mechanisms responsible for hydrocarbon emissions in engines are: crevices, representing a narrow regions in the combustion chamber located around the piston, head gasket, spark plug and valve seats, inside which the flame cannot propagate because it is smaller than the quenching distance, oil layers, liquid fuel, flame quenching, and exhaust valve leakage. Since the piston ring is not 100% effective in preventing oil penetration into the cylinder above the piston, oil layers may appear within the combustion chamber. This oil layer traps fuel and releases it later during expansion. With time carbon deposits build up on the valves; cylinder and piston head, and these deposits are porous. The pores' sizes are smaller than the quenching distance so that trapped fuel cannot burn. The fuel is released later during expansion. For some fuel injection systems there is a possibility that liquid fuel is introduced into the cylinder past an open intake valve. The less volatile fuel constituents may not vaporize during engine warm-up and be absorbed by the crevices or by carbon deposits.

A high concentration of particulate matter is manifested as visible smoke in the exhaust gases. Particulates are substance other than water that can be collected by filtering the exhaust: solid carbon material or soot, condensed hydrocarbons and their partial oxidation products. Diesel particulates consist of solid carbon at exhaust gas at temperatures below 500 C. Diesel engine designers experience the following problem: an attempt to reduce NO_x increases the amount of soot in the exhaust. Particulate can arise if leaded fuel or overly rich fuel-air mixture are used. Also the highest CO emission occurs during engine warm-up when the engine is run fuel-rich to compensate for poor fuel evaporation. For fuel-rich mixtures there is insufficient oxygen to convert all the carbon of the fuel to carbon dioxide. Any carbon not oxidized in the cylinder ends up as soot in the exhaust. In a properly adjusted SI engines soot is usually not present in the exhaust. Most particulate material results from incomplete combustion of fuel, which occurs in fuel-rich mixtures when the C/O ratio exceeds 1. Experimentally it is found that the critical C/O ratio for onset of soot formation is between 0.5 and 0.8.

Methods used to control engine emissions include: advances in fuel injectors, oxygen sensors and on-board computers, spark retard and exhaust gas recirculation (EGR), and the catalytic converter. The catalytic converters are built in a honeycomb or pellet geometry to expose the exhaust gases to a large surface made of one or more noble metals: platinum, palladium and rhodium. Rhodium is used to remove NO and platinum is used to remove HC and CO. A catalyst forces a reaction at lower temperature. As the exhaust gases flow through the catalyst, the NO reacts with the CO, HC and H_2 via a reduction reaction on the catalyst surface, e.g.,

$$NO + CO \rightarrow \frac{1}{2}N_2 + CO_2 \quad NO + H_2 \rightarrow \frac{1}{2}N + H_2O$$

The remaining CO and HC are removed through an oxidation reaction forming CO_2 and H_2O products with air added to exhaust after exhaust valve.

12.2 Processes of Particulate and Soot Formation

The various phenomena involved in carbon particulate formation have been under extensive study. The carbon particulates formation may be as harmful as well as beneficial. The presence of particulates in gas turbines can severely affect the lifetime of the blades, soot particulates in diesel engines are carcinogenic and otherwise harmful to the environment. Higher temperatures and pressures, such as in diesel engines, may lead to increased amounts of soot, which may be itself carcinogenic or absorb other carcinogenic polycyclic aromatic hydrocarbons. The extent of soot formation is related to the type of flame in a given process.

However, the presence of carbon particulates in industrial furnaces increases the radiative power of the flame and therefore increases the heat transfer rates. Notice that hydrocarbon diffusion flames are luminous. The luminosity arises due to carbon particulates that strongly radiate at high combustion temperature; that is why candle is lighting. The solid phase particulate cloud has a very high emissivity approaching the black body conditions compared to a pure gaseous mixture. That's why diffusion flames in the presence of particulates are yellow. Note a yellow color of the burning in a strove and fireplaces. Thus, for some industrial furnaces it is beneficial to operate in a particular diffusion flame burned-off in the later stage with additional air.

The appearance of soot and smoke is the result of unburned hydrocarbons in combustion, and they depend on time and temperature. By allowing for more time at high temperatures with good mixing, one is usually assured of oxidizing soot and other hydrocarbons. However, at high temperatures a larger amount of NO is produced.

The words "unburned hydrocarbon, carbon and soot" are used to describe a wide variety of solid materials, which contain appreciable amounts of hydrogen and compounds that may be present in original hydrogen fuel. The unburned hydrocarbon can be directly formed from the fuel or the hydrocarbon can be generated during the combustion process. The problem can be even more important if combustion is used to destroy toxic compounds or medical wastes.

A remarkable feature of the soot formation is the rapidity with which soot particles form in flames. They appear in less than 1 ms and rise up to size about 0.1 μm in less than 10 ms. There is still great controversy about chemical reaction steps leading to soot formation. It is believed that the main element in the overall process is the formation of the aromatic ring. Aromatic rings are formed by reaction of CH or CH_2 with C_2H_2 to C_3H_3, which can form the first ring – benzene C_6H_6 after recombination. An important class of higher hydrocarbons is the *polycyclic aromatic hydrocarbons* (PAH). These compounds are usually formed under fuel-rich conditions and can be carcinogenic. Acetylene formed in high amounts under rich conditions is the most important precursor of PAH's. PAH formation is started by C_3H_4 decomposition or reaction of CH or CH_2 mentioned above. It is believed that further growth of the PAH leads to soot. Subsequently the particles grow by surface growth by addition of mainly acetylene and by coagulation. The first step here is formation of particle-like structures by conglomeration of molecules. The process of coagulation may be considerably affected by turbulence, leading either to enhance or to weakening of the clustering depending on the intensity of turbulence.

The tendency of fuels to soot formation may be different for premixed and diffusion flames. Usually the tendency to soot in premixed flames is correlated with equivalence ratio at which luminosity just begins, so that the luminosity is attributed to the formation of soot. The smaller is the equivalence ratio at the point where luminosity begins, the greater is the tendency to soot for the fuel.

The major physical effect on soot formation is the temperature. Increasing temperature for premixed flame conditions decreases the soot production. The opposite effect is true for diffusion flames. In the latter case fuels are vaporized to form a gaseous jet. The mass flow of the fuel jet increases until soot breaks through the top of the flame. In non-premixed combustion systems soot oxidation is taking place after mixing with oxygen. If too much radiation is allowed, the soot is too cool for rapid oxidation and thus appears in the exhaust. This overcooling is responsible for the smoking of kerosene lamps if the wick is turned up too high. The CO, H_2, and C (s) are oxidized in the diffusion flame to CO_2 and H_2O via the reactions

$$CO + \frac{1}{2}O_2 \rightarrow CO_2 \quad C(s) + O_2 \rightarrow CO_2 \quad H_2 + \frac{1}{2}O_2 \rightarrow H_2O$$

12.3 NO_x Formation and Reduction

The remedy to the appearance of soot is more time, higher temperature and better mixing in combustion, which assure the soot and other hydrocarbons oxidizing. However, higher temperatures result in a larger production of NO. The major producer of NO_x is automobile emission. Nitrogen oxides can be formed from the atmospheric nitrogen and also from fuel bound nitrogen, when fuel compounds contain nitrogen atoms bound to other atoms. The flame structure, combustion mixture ratio, and combustion temperature are the prime parameters in determining the quantities of NO_x formation. It was recognized that nitrogen oxides are the major contributors of photochemical smog in troposphere and urban air, as well as NO_x is participates in a chain reactions removing ozone from the stratosphere and causes the increase of ultraviolet radiation on the earth's surface. Therefore reduction of NO_x production became of one of the most important problem in combustion research.

The kinetic rout of NO formation is not simply the attack of an oxygen molecule on a nitrogen molecule according to reaction $N_2 + O_2 = 2NO$, but these are oxygen atoms formed from the dissociation of O_2 which attack nitrogen molecules to start the chain reaction

$$O + N_2 \overset{(k_1)}{\rightarrow} NO + N, \; k_1 = 1.8 \times 10^{14} \exp(-76500/RT), \quad (12.3.1)$$

$$N + O_2 \overset{(k_2)}{\rightarrow} NO + O, \; k_2 = 9.0 \times 10^{9} \exp(-6300/RT), \quad (12.3.2)$$

$$N + OH \overset{(k_3)}{\rightarrow} NO + H, \; k_3 = 2.8 \times 10^{13} \exp(-2600/RT)\, cm^3/(mol\ s). \quad (12.3.3)$$

This mechanism was first postulated by Ya. B. Zel'dovich (1946) and it is referred to as *Zel'dovich* or *thermal* mechanism (see Chap. 3). It is thought to be the main source of NO in combustion systems. The mechanism name is *thermal* because the reaction (12.3.1) has very high activation energy due to the strong bond in the N_2-molecule. Reaction (12.3.1) is the rate-limiting step of the thermal NO-formation. Therefore NO is only formed at high temperatures and the reaction rate is relatively slow. At temperatures below 2000 K the reaction rate is extremely slow. Thus, the best practical means of controlling NO according to Eq. (12.3.1) is to either reduce the combustion temperature or oxygen concentration.

Experimental measurements indicated that reactions other than the *Zel'dovich* mechanism might play certain role in the flame, especially in the fuel-rich flames. The mechanism of *prompt* NO was proposed by C. P. Fenimore (1979), who measured NO concentration above a hydrocarbon flat flame and noted that the [NO] did not approach zero level as the probe approached the flame from the downstream side as the Zel'dovich mechanism predicts. The additional mechanism that is promptly producing NO at the flame front is more complicated than that of thermal NO, because the prompt NO results from the CH radicals. The CH radicals are formed as an intermediate at the flame front and react with the nitrogen of the air forming hydrocyanic acid (HCN). The nitrogen compound then formed in the following mechanism

$$CH + N_2 = HCN + N, \tag{12.3.4}$$

$$C_2 + N_2 = 2CN. \tag{12.3.5}$$

The N atoms can form further NO by reactions (12.3.2) and (12.3.3), and CN could yield NO by oxygen atom attack. The activation energy of the reaction (12.3.4) is about 75 kJ/mol, compared to 318 kJ/mol for the formation of the thermal NO, therefore, in contrast to thermal NO, prompt NO is produced at relatively low temperatures (about 1000 K). Precise information about the rate-limiting reaction (12.3.4) is rather poor, and predictions of Fenimore NO are less accurate. Because C_2H_2 as a CH-radical precursor is accumulated under fuel-rich conditions due to CH_3-recombination, prompt NO is favored in rich flames. Lean conditions can suppress the formation of CH, hence leading to less prompt NO, and low temperatures can suppress the thermal NO.

The conversion of fuel-nitrogen into NO is mainly observed in coal combustion, because coal contains about 1% of chemically bound nitrogen. The nitrogen-containing compounds evaporate during the gasification process and lead to NO formation in the gas phase.

If combustion modifications are not efficient or not possible, post-combustion measures are necessary to remove NO. The well known of NO removal method is the catalytic converter that is in the exhaust system of many automobiles. The catalyst is a combination of noble metals that oxidize CO to CO_2 and simultaneously reduce some amount of NO to N_2.

In SI engines the dominant component of NO_x is NO, which is formed as a result of dissociation of molecular nitrogen and oxygen. Since the activation energy of the limiting reaction $O + N_2 \rightarrow NO + N$ is high the reaction rate is very sensitive to temperature, and NO is only formed at high temperatures. At temperatures below 2000 K the reaction rate is extremely slow, so NO formation not important. Since the cylinder temperature changes throughout the cycle, the NO reaction rate also changes. The actual NO concentration tends toward its equilibrium value but never achieves the equilibrium value since the chemistry is not fast enough. If NO concentration is lower than equilibrium value then NO forms. If NO concentration is higher than equilibrium value than NO decomposes. Once the temperature rises up to 2000 K the reaction rate becomes so slow that the NO concentration effectively freezes at a value greater than the equilibrium value. The total amount of NO that appears in the exhaust can be calculated by summing the frozen mass fractions for the whole process. Typically peak NO concentrations occur for slightly lean mixtures. Since the formation of NO is highly dependent on cylinder gas temperature, the effective method is to increase residual gas or exhaust gas recirculation.

In attempt to reduce NO generated by industrial burners, engineers use staged combustion technologies. The fuel rich conditions are used in order to reduce amount of NO_x, HCN and NH_3 compounds. Then oxygen rich condition is used to achieve stoichiometric combustion in the overall process. The combustion temperature is steadily reduced due to radiation and convective heat transfer, so that N_2 is not converted to thermal NO during the first stage. A further reduction of NO is achieved by using excess of air in the second stage. On the third stage an additional reburned can be used similar to EGR used in SI engines.

One should remember that the reduction of NO production and the thermodynamic efficiency are competing processes. To reduce NO production we are interested in systems, which burn at lower temperature and pressure, while thermodynamic efficiency is higher at higher temperature and pressure. The optimum can be obtained for some intermediate systems, which require thorough theoretical and computational analysis.

Appendix A
Conversion Formulas and Constants

Prefixes:
10^{-3}– milli [m], 10^{-6}– micro [μ], 10^{-9}– nano [n], 10^{-12}– pico [p], 10^{-15}– femto [f]
10^{3}– kilo [k], 10^{6}– mega [M], 10^{9}– giga [G], 10^{12}– tera [T], 10^{15}– peta [P]

Length units: $1\,km = 10^3 m = 10^5 cm$; $1\,cm = 10\,mm = 10^4 \mu m = 10^7 nm = 10^8 Å$

Volume units: $1\,m^3 = 10^3 L = 10^6 cm^3 = 10^9 mm^3$

Mass units: $1\,ton = 10^3 kg$, $1\,kg = 10^3 g = 10^6 mg = 10^9 \mu g$

Energy units:
$1\,cal = 4.1868\,J = 4.1868 \cdot 10^7 erg = 2.6126 \cdot 10^{19} eV$
$1\,J = 1\,N \cdot m = 1\,W \cdot sec = 10^7\,erg = 6.2419 \cdot 10^{11} eV$
$1\,eV = 11600\,K = 6.021 \cdot 10^{-12}\,erg = 6.021 \cdot 10^{-19}\,J = 3.827 \cdot 10^{-20}\,cal$

Temperature units: $T[K] = T[°C] + 273.15$

Density units: $1\,kg/m^3 = 10^{-3}\,g/cm^3$

Pressure units:
$1\,atm = 1.0133\,bar = 1.0133 \cdot 10^5\,N/m^2 = 1.0133 \cdot 10^6\,dyn/cm^2 = 760\,Torr = 760\,mmHg$
$1\,bar = 10^5\,N/m^2 = 10^5 Pa$

Force units: $1\,N = 10\,kg \cdot m/sec^2 = 10^5 dyn$

A.1 Fundamental Constants

Universal gas constant:
$R_B = 8.3144\,J/g\text{-}mol \cdot K = 8.3144\,kJ/kmol \cdot K = 1.987\,cal/g\text{-}mole \cdot K$
$= 8.3144 \cdot 10^7\,erg/mol \cdot K$

Boltzmann constant:
$k = R_B/N_A = 1.38 \cdot 10^{-23}\,J/K$

Avagadro number:
$N_A = 6.022 \cdot 10^{23}$ molecules/g-mol. Number of molecules in $1\,cm^3$:
$n = 2.686 \cdot 10^{19}\,cm^{-3}$
Volume of 1 mole at normal conditions: $V_0 = 2.2414 \cdot 10^4 cm^3/mol$

Stefan-Boltzmann constant:
$\sigma = \pi^2 k^4/60\hbar^3 c^2 = 5.670 \cdot 10^{-8}\,\mathrm{W/m^2 \cdot K^4} = 5.670 \cdot 10^{-5}\,\mathrm{erg/cm^2 \cdot sec \cdot K^4} =$
$= 1.355 \cdot 10^{-12}\,\mathrm{cal/cm^2 \cdot sec \cdot K^4}$

Atomic mass unit:
$m_a = 1.660 \cdot 10^{-24}\,\mathrm{g} = 1.660 \cdot 10^{-27}\,\mathrm{kg}$
$m_p = 1.67264 \cdot 10^{-24}\,\mathrm{g}$, $m_n = 1.67495 \cdot 10^{-24}\,\mathrm{g}$, $m_e = 9.109534 \cdot 10^{-28}\,\mathrm{g}$

Speed of light: $c = 2.997 \cdot 10^{10}\,\mathrm{cm/sec} = 2.997 \cdot 10^8\,\mathrm{m/sec}$

Electron charge: $e = 1.602189 \cdot 10^{-19}$coulombs $= 10^{-20}$cgsm units

Max Plank constant:
$h = 6.626 \cdot 10^{-27}\,\mathrm{erg\ sec} = 6.626 \cdot 10^{-34}\,\mathrm{J\ sec}$, $\hbar = h/2\pi = 1.054 \cdot 10^{-27}\,\mathrm{erg\ sec}$

A.2 Density, Thermal Conductivity, Viscosity

Density of some gases at $T = 0°C$, $P = 760\,\mathrm{mmHg}$ $\rho\,[\mathrm{kg/m^3}]$

Gas	Air	H_2	O_2	C_2H_2	CH_4	C_2H_4	C_2H_6	C_3H_8	CO	CO_2	NO
$\rho\,[\mathrm{kg/m^3}]$	1.29	0.089	1.429	1.170	0.716	1.260	1.356	2.003	1.250	1.976	1.34

Thermal conduction of some gases at $T = 0°C$, $P = 760\,\mathrm{mmHg}$: $\lambda\left[10^{-6}\dfrac{\mathrm{cal}}{\mathrm{cm\ sec\ K}}\right]$

Gas	Air	H_2	O_2	C_2H_2	CH_4	C_2H_4	C_2H_6	C_3H_8	CO	CO_2	NO
$\lambda\,[10^{-6}\mathrm{cal/cm\,s\,K}]$	57.5	401	57.5	44	69.6	41.5	43.5	35.5	51.1	34.6	56.4

Coefficient viscosity of some gases at $T = 0°C$, $P = 760\,\mathrm{mmHg}$: $\eta\left[10^{-8}\dfrac{\mathrm{kg}}{\mathrm{m\ sec}}\right]$

Gas	Air	H_2	O_2	C_2H_2	CH_4	C_2H_4	C_2H_6	C_3H_8	CO	CO_2	NO
$\eta\,[10^{-8}\mathrm{kg/m\,s}]$	1720	866	1993	955	1028	855	1223	1390	1660	1463	1800

Temperature dependence is approximated by Sutherland's formula $\eta_T = \eta_0\sqrt{\dfrac{T}{T_0}}\dfrac{1 + a/T_0}{1 + b/T}$

Appendix B
Useful Formulas of Vector Analysis

A, **B**, **C** are vectors, f is a scalar function, **r** is the radius vector $\mathbf{r} = x\hat{\mathbf{e}}_x + y\hat{\mathbf{e}}_y + z\hat{\mathbf{e}}_z$

$$\mathbf{A} \cdot (\mathbf{B} \times \mathbf{C}) = (\mathbf{A} \times \mathbf{B}) \cdot \mathbf{C} = \mathbf{B} \cdot (\mathbf{C} \times \mathbf{A})$$

$$\mathbf{A} \times (\mathbf{B} \times \mathbf{C}) = (\mathbf{C} \times \mathbf{B}) \times \mathbf{A} = \mathbf{B}(\mathbf{A} \cdot \mathbf{C}) - \mathbf{C}(\mathbf{A} \cdot \mathbf{B})$$

$$\nabla \cdot (\mathrm{f}\mathbf{A}) = \mathrm{f}\nabla \cdot \mathbf{A} + \mathbf{A} \cdot \nabla \mathrm{f}$$

$$\nabla \times (\mathrm{f}\mathbf{A}) = \mathrm{f}\nabla \times \mathbf{A} + \nabla \mathrm{f} \times \mathbf{A}$$

$$\nabla \cdot (\mathbf{A} \times \mathbf{B}) = \mathbf{B} \cdot (\nabla \times \mathbf{A}) - \mathbf{A} \cdot (\nabla \times \mathbf{B})$$

$$\nabla \times (\mathbf{A} \times \mathbf{B}) = \mathbf{A}(\nabla \cdot \mathbf{B}) - \mathbf{B}(\nabla \cdot \mathbf{A}) + (\mathbf{B} \cdot \nabla)\mathbf{A} - (\mathbf{A} \cdot \nabla)\mathbf{B}$$

$$(\mathbf{A} \cdot \nabla)\mathbf{A} = \nabla \frac{1}{2} A^2 - \mathbf{A} \times (\nabla \times \mathbf{A})$$

$$\mathbf{A} \times (\nabla \times \mathbf{B}) = \nabla(\mathbf{A} \cdot \mathbf{B}) - (\mathbf{A} \cdot \nabla)\mathbf{B} - (\mathbf{B} \cdot \nabla)\mathbf{A} - \mathbf{B} \times (\nabla \times \mathbf{A})$$

$$\nabla \times (\nabla \times \mathbf{B}) = \nabla(\nabla \cdot \mathbf{B}) - \nabla^2 \mathbf{A}$$

$$\nabla \times (\nabla f) = 0, \; \nabla \cdot (\nabla \times \mathbf{A}) = 0$$

$$\nabla \cdot \mathbf{r} = 3, \nabla \times \mathbf{r} = 0, \; \nabla \mathrm{f}(\mathrm{r}) = \frac{\mathrm{df}}{\mathrm{dr}} \frac{\mathbf{r}}{\mathrm{r}}$$

$$\int_V \nabla \mathrm{f} \cdot \mathrm{dV} = \int_S \mathrm{f} \cdot \mathrm{d}\mathbf{S}$$

$$\int_V \nabla \cdot \mathbf{A}\, \mathrm{dV} = \int_S \mathbf{A} \cdot \mathrm{d}\mathbf{S}$$

$$\oint_C \mathbf{A} \cdot \mathrm{d}\mathbf{r} = \int_S \nabla \times \mathbf{A} \cdot \mathrm{d}\mathbf{S}$$

B.1 Cylindrical Coordinates

r is the distance from the axis; θ is the azimuthal angle about the axis; z is the coordinate along the axis

$$\nabla\cdot\mathbf{A}=\frac{1}{r}\frac{\partial}{\partial r}(rA_r)+\frac{1}{r}\frac{\partial A_\theta}{\partial\theta}+\frac{\partial A_z}{\partial z}$$

$$(\nabla f)_r=\frac{\partial f}{\partial r},\ (\nabla f)_\theta=\frac{1}{r}\frac{\partial f}{\partial\theta},\ (\nabla f)_z=\frac{\partial f}{\partial z}$$

$$(\nabla\times\mathbf{A})_r=\frac{1}{r}\frac{\partial A_z}{\partial\theta}-\frac{\partial A_\theta}{\partial z}$$

$$(\nabla\times\mathbf{A})_\theta=\frac{\partial A_r}{\partial z}-\frac{\partial A_z}{\partial r}$$

$$(\nabla\times\mathbf{A})_z=\frac{1}{r}\frac{\partial}{\partial r}(rA_\theta)-\frac{1}{r}\frac{\partial A_r}{\partial\theta}$$

$$\nabla^2 f=\frac{1}{r}\frac{\partial}{\partial r}\left(r\frac{\partial f}{\partial r}\right)+\frac{1}{r^2}\frac{\partial^2 f}{\partial\theta^2}+\frac{\partial^2 f}{\partial z^2}$$

$$(\nabla^2\mathbf{A})_r=\nabla^2A_r-\frac{2}{r^2}\frac{\partial A_\theta}{\partial\theta}-\frac{A_r}{r^2}$$

$$(\nabla^2\mathbf{A})_\theta=\nabla^2A_\theta+\frac{2}{r^2}\frac{\partial A_r}{\partial\theta}-\frac{A_\theta}{r^2}$$

$$(\nabla^2\mathbf{A})_z=\nabla^2A_z$$

$$[(\mathbf{A}\cdot\nabla)\mathbf{B}]_r=A_r\frac{\partial B_r}{\partial r}+\frac{A_\theta}{r}\frac{\partial B_r}{\partial\theta}+A_z\frac{\partial B_r}{\partial z}-\frac{A_\theta B_\theta}{r}$$

$$[(\mathbf{A}\cdot\nabla)\mathbf{B}]_\theta=A_r\frac{\partial B_\theta}{\partial r}+\frac{A_\theta}{r}\frac{\partial B_\theta}{\partial\theta}+A_z\frac{\partial B_\theta}{\partial z}+\frac{A_\theta B_r}{r}$$

$$[(\mathbf{A}\cdot\nabla)\mathbf{B}]_z=A_r\frac{\partial B_z}{\partial r}+\frac{A_\theta}{r}\frac{\partial B_z}{\partial\theta}+A_z\frac{\partial B_z}{\partial z}.$$

B.2 Spherical Coordinates

$$\nabla\cdot\mathbf{A}=\frac{1}{r^2}\frac{\partial}{\partial r}(r^2A_r)+\frac{1}{r\sin\theta}\frac{\partial}{\partial\theta}(\sin\theta A_\theta)+\frac{1}{r\sin\theta}\frac{\partial A_\alpha}{\partial\alpha}$$

$$(\nabla f)_r=\frac{\partial f}{\partial r},(\nabla f)_\theta=\frac{1}{r}\frac{\partial f}{\partial\theta},(\nabla f)_\alpha=\frac{1}{r\sin\theta}\frac{\partial f}{\partial\alpha}$$

$$(\nabla\times\mathbf{A})_r=\frac{1}{r\sin\theta}\frac{\partial}{\partial\theta}(\sin\theta A_\alpha)-\frac{1}{r\sin\theta}\frac{\partial A_\theta}{\partial\alpha}$$

$$(\nabla\times\mathbf{A})_\theta=\frac{1}{r\sin\theta}\frac{\partial A_r}{\partial\alpha}-\frac{1}{r}\frac{\partial}{\partial r}(rA_\alpha)$$

$$(\nabla \times \mathbf{A})_\alpha = \frac{1}{r}\frac{\partial}{\partial r}(rA_\theta) - \frac{1}{r}\frac{\partial A_r}{\partial \theta}$$

$$\nabla^2 f = \frac{1}{r^2}\frac{\partial}{\partial r}\left(r^2\frac{\partial f}{\partial r}\right) + \frac{1}{r^2 \sin\theta}\frac{\partial}{\partial \theta}\left(\sin\theta\frac{\partial f}{\partial \theta}\right) + \frac{1}{r^2 \sin^2\theta}\frac{\partial^2 f}{\partial \alpha^2}$$

$$(\nabla^2 \mathbf{A})_r = \nabla^2 A_r - \frac{2}{r^2}\frac{\partial A_\theta}{\partial \theta} - \frac{2A_r}{r^2} - \frac{2\cot\theta A_\theta}{r^2} - \frac{2}{r^2 \sin\theta}\frac{\partial A_\alpha}{\partial \alpha}$$

$$(\nabla^2 \mathbf{A})_\theta = \nabla^2 A_\theta + \frac{2}{r^2}\frac{\partial A_r}{\partial \theta} - \frac{A_\theta}{r^2 \sin^2\theta} - \frac{2\cot\theta}{r^2 \sin\theta}\frac{\partial A_\alpha}{\partial \alpha}$$

$$(\nabla^2 \mathbf{A})_\alpha = \nabla^2 A_\alpha + \frac{2}{r^2 \sin\theta}\frac{\partial A_r}{\partial \alpha} - \frac{A_\alpha}{r^2 \sin^2\theta} + \frac{2\cot\theta}{r^2 \sin\theta}\frac{\partial A_\theta}{\partial \alpha}$$

$$[(\mathbf{A}\cdot\nabla)\mathbf{B}]_r = A_r\frac{\partial B_r}{\partial r} + \frac{A_\theta}{r}\frac{\partial B_r}{\partial \theta} + \frac{A_\alpha}{r\sin\theta}\frac{\partial B_r}{\partial \alpha} - \frac{A_\theta B_\theta + A_\alpha B_\alpha}{r}$$

$$[(\mathbf{A}\cdot\nabla)\mathbf{B}]_\theta = A_r\frac{\partial B_\theta}{\partial r} + \frac{A_\theta}{r}\frac{\partial B_\theta}{\partial \theta} + \frac{A_\alpha}{r\sin\theta}\frac{\partial B_\theta}{\partial \alpha} + \frac{A_\theta B_r}{r} - \cot\theta\frac{A_\alpha B_\alpha}{r}$$

$$[(\mathbf{A}\cdot\nabla)\mathbf{B}]_\alpha = A_r\frac{\partial B_\alpha}{\partial r} + \frac{A_\theta}{r}\frac{\partial B_\alpha}{\partial \theta} + \frac{A_\alpha}{r\sin\theta}\frac{\partial B_\alpha}{\partial \alpha} + \frac{A_\alpha B_r}{r} + \cot\theta\frac{A_\alpha B_\theta}{r}$$

Appendix C
Equations of Fluid Mechanics in Curvilinear Coordinate Systems

C.1 Cylindrical Coordinates

The continuity equation

$$\frac{\partial u_r}{\partial r} + \frac{u_r}{r} + \frac{1}{r}\frac{\partial u_\theta}{\partial \theta} + \frac{\partial u_z}{\partial z} = 0.$$

The Navier-Stokes equation

$$\frac{\partial u_r}{\partial t} + (\mathbf{u}\cdot\nabla)u_r - \frac{u_\theta^2}{r} = -\frac{1}{\rho}\frac{\partial P}{\partial r} + \frac{\mu}{\rho}\left(\nabla^2 u_r - \frac{u_r}{r^2} - \frac{2}{r^2}\frac{\partial u_\theta}{\partial \theta}\right) + \frac{1}{\rho}F_r$$

$$\frac{\partial u_\theta}{\partial t} + (\mathbf{u}\cdot\nabla)u_\theta + \frac{u_r u_\theta}{r} = -\frac{1}{\rho r}\frac{\partial P}{\partial \theta} + \frac{\mu}{\rho}\left(\nabla^2 u_\theta - \frac{u_\theta}{r^2} + \frac{2}{r^2}\frac{\partial u_r}{\partial \theta}\right) + \frac{1}{\rho}F_\theta,$$

$$\frac{\partial u_z}{\partial t} + (\mathbf{u}\cdot\nabla)u_z = -\frac{1}{\rho}\frac{\partial P}{\partial z} + \frac{\mu}{\rho}\nabla^2 u_z + \frac{1}{\rho}F_z.$$

The operator $(\mathbf{u}\cdot\nabla)$ and the operator ∇^2 acting on a scalar

$$(\mathbf{u}\cdot\nabla)f = u_r\frac{\partial f}{\partial r} + \frac{u_\theta}{r}\frac{\partial f}{\partial \theta} + u_z\frac{\partial f}{\partial z},$$

$$\nabla^2 f = \frac{1}{r}\frac{\partial}{\partial r}\left(r\frac{\partial f}{\partial r}\right) + \frac{1}{r^2}\frac{\partial^2 f}{\partial \theta^2} + \frac{\partial^2 f}{\partial z^2}.$$

The components of the viscous stress tensor in cylindrical coordinates are:

$$\zeta_{rr} = 2\mu\frac{\partial u_r}{\partial r}, \qquad \zeta_{r\theta} = \zeta_{\theta r} = \mu\left(\frac{1}{r}\frac{\partial u_r}{\partial \theta} + \frac{\partial u_\theta}{\partial r} - \frac{u_\theta}{r}\right),$$

$$\zeta_{\theta\theta} = 2\mu\left(\frac{1}{r}\frac{\partial u_\theta}{\partial \theta} + \frac{u_r}{r}\right), \qquad \zeta_{z\theta} = \zeta_{\theta z} = \mu\left(\frac{1}{r}\frac{\partial u_z}{\partial \theta} + \frac{\partial u_\theta}{\partial z}\right),$$

$$\zeta_{zz} = 2\mu\frac{\partial u_z}{\partial z}, \qquad \zeta_{zr} = \zeta_{rz} = \mu\left(\frac{\partial u_z}{\partial r} + \frac{\partial u_r}{\partial z}\right).$$

C.2 Spherical Coordinates

The continuity equation

$$\frac{1}{r^2}\frac{\partial}{\partial r}\left(r^2 u_r\right)+\frac{1}{r\sin\theta}\frac{\partial}{\partial\theta}\left(u_\theta\sin\theta\right)+\frac{1}{r\sin\theta}\frac{\partial u_\alpha}{\partial\alpha}=0.$$

The Navier-Stokes equation

$$\frac{\partial u_r}{\partial t}+(\mathbf{u}\cdot\nabla)u_r-\frac{u_\theta^2+u_\alpha^2}{r}=-\frac{1}{\rho}\frac{\partial P}{\partial r}$$

$$+\frac{\mu}{\rho}\left(\nabla^2 u_r-\frac{2u_r}{r^2}-\frac{2}{r^2\sin^2\theta}\frac{\partial}{\partial\theta}\left[u_\theta\sin\theta\right]-\frac{2}{r^2\sin\theta}\frac{\partial u_\alpha}{\partial\alpha}\right)+\frac{1}{\rho}F_r,$$

$$\frac{\partial u_\theta}{\partial t}+(\mathbf{u}\cdot\nabla)u_\theta+\frac{u_r u_\theta}{r}-\frac{u_\alpha^2}{r}\cot\theta=\frac{1}{\rho r}\frac{\partial P}{\partial\theta}+$$

$$\frac{\mu}{\rho}\left(\nabla^2 u_\theta\frac{u_\theta}{r^2\sin^2\theta}+\frac{2}{r^2}\frac{\partial u_r}{\partial\theta}\frac{2\cos\theta}{r^2\sin^2\theta}\frac{\partial u_\alpha}{\partial\alpha}\right)+\frac{1}{\rho}F_\theta,$$

$$\frac{\partial u_\alpha}{\partial t}+(\mathbf{u}\cdot\nabla)u_\alpha+\frac{u_r u_\alpha}{r}+\frac{u_\theta u_\alpha}{r}\cot\theta=-\frac{1}{\rho r}\frac{\partial P}{\partial\alpha}+$$

$$\frac{\mu}{\rho}\left(\nabla^2 u_\alpha-\frac{u_\alpha}{r^2\sin^2\theta}+\frac{2}{r^2\sin\theta}\frac{\partial u_r}{\partial\alpha}+\frac{2\cos\theta}{r^2\sin^2\theta}\frac{\partial u_\theta}{\partial\alpha}\right)+\frac{1}{\rho}F_\alpha.$$

The operator $(\mathbf{u}\cdot\nabla)$ and the operator ∇^2 acting on a scalar.

$$(\mathbf{u}\cdot\nabla)f=u_r\frac{\partial f}{\partial r}+\frac{u_\theta}{r}\frac{\partial f}{\partial\theta}+\frac{u_\alpha}{r\sin\theta}\frac{\partial f}{\partial\alpha},$$

$$\nabla^2 f=\frac{1}{r^2}\frac{\partial}{\partial r}\left(r^2\frac{\partial f}{\partial r}\right)+\frac{1}{r^2\sin\theta}\frac{\partial}{\partial\theta}\left(\sin\theta\frac{\partial f}{\partial\theta}\right)+\frac{1}{r^2\sin^2\theta}\frac{\partial^2 f}{\partial\alpha^2}.$$

The components of the viscous stress tensor in cylindrical coordinates are:

$$\zeta_{rr}=2\mu\frac{\partial u_r}{\partial r},\quad \zeta_{r\theta}=\zeta_{\theta r}=\mu\left(\frac{1}{r}\frac{\partial u_r}{\partial\theta}+\frac{\partial u_\theta}{\partial r}-\frac{u_\theta}{r}\right),$$

$$\zeta_{\theta\theta}=2\mu\left(\frac{1}{r}\frac{\partial u_\theta}{\partial\theta}+\frac{u_r}{r}\right),\qquad \zeta_{\alpha\theta}=\zeta_{\theta\alpha}=\mu\left(\frac{1}{r\sin\theta}\frac{\partial u_\theta}{\partial\alpha}+\frac{1}{r}\frac{\partial u_\alpha}{\partial\theta}-\cot\theta\frac{u_\alpha}{r}\right),$$

$$\zeta_{\alpha\alpha}=2\mu\left(\frac{1}{r\sin\theta}\frac{\partial u_\alpha}{\partial\alpha}+\frac{u_r}{r}+\cot\theta\frac{u_\theta}{r}\right),$$

$$\zeta_{\alpha r}=\zeta_{r\alpha}=\mu\left(\frac{1}{r\sin\theta}\frac{\partial u_r}{\partial\alpha}+\frac{\partial u_\alpha}{\partial r}-\frac{u_\alpha}{r}\right).$$

Index

Acetylene, 36, 39, 40, 93, 302, 344
Acetylene-oxygen flame, 42
Activation energy, 57, 61, 65, 67, 69, 74, 77, 78, 92, 93, 95, 103, 104, 216, 220, 221, 222, 241, 242, 279, 280, 282, 296, 302, 303, 312, 346, 347
Activation reaction, 59
Active centers, 60, 61, 63, 67, 68, 69, 70, 73, 74, 75, 89, 99
Adapted grid, 311
Adiabatic compressibility, 17, 23, 146, 265, 308, 319, 323, 329, 332, 333
Adiabatic expansion, 24, 25, 329
Adiabatic exponent, 120, 124, 199, 216
Adiabatic flame temperature, 40–42, 43, 46–50, 80, 91, 242, 267, 281, 293
Adiabatic flow, 121, 126
Adiabatic motion, 114, 115, 118
Adiabatic process, 9, 23, 41, 76, 149
Adsorption rate, 70
Airfoil, 129
Alcohols, 39
Aldehydes, 332
Aliphatic hydrocarbon combustion reaction, 343, 344
Alkynes, 39
Ammonia, 347
Aromatic ring, 344
  formation, 344
Arrhenius equation (law), 57
Arrhenius law, 54, 57, 77, 79, 80, 81, 92, 103, 215, 216, 272, 332
Ash, 311
Asymmetric airfoil, 129
Atmospheric aerosols, 341
Atomic mass unit, 350
Autoignition, 272, 308, 330, 331, 332, 333, 334
  temperature, 330, 333
Automobile engines, 319
Average molecular velocity, 67, 93, 147
Avogadro's number, 8, 27

Benzene, 36, 93, 100, 272, 341, 342, 344
Bernoulli
  equation, 126, 129, 134, 138, 192, 246, 260, 261
  integral, 128, 129
Beta function, 309
Bimolecular collision (binary collision), 59
Bimolecular reaction (binary reaction), 55, 56, 57, 59, 62, 65, 68, 76, 106
Boltzmann
  constant, 8, 58, 349
  factor, 8, 57, 58
Boltzmann L., 4
Boltzmann's exponential factor, 58
Bond energy, 38, 39, 40, 60, 71
Borghi diagram, 276, 277
Boundary conditions, 86, 94, 95, 102, 104, 116, 138, 141, 161, 165, 170, 172, 173, 190, 192, 194, 199, 209, 219, 221, 223, 230, 234, 235, 237, 243, 248, 250, 259, 278, 285, 302, 306
Boundary layer, 190–196, 249, 259, 286, 287, 288, 289, 305
Branched-chain explosion of hydrogen, 272
Bubble radius, 143, 314, 315
Bunsen
  burner, 272, 273, 274, 278
  flame, 269, 317
Buoyancy effect, 270, 317
Burning velocity, 273, 278, 285, 286, 342
  mixtures, 342
Butane, 36, 42, 93

C_2H_5OH (ethanol) reactions, 39
Candle, 311, 344

Carbon black, 40
Carbon monoxide, 30, 36, 42, 93, 106
Carnot's
cycle, 16–18
theorem, 16–18
Catalytic combustion, 346
Catalytic converter, 343, 346
Catalytic surface reaction, 343
Cellular flame structure, 245, 246, 254
Cellular structure, 241, 245, 284, 285, 287, 288
CH_2CO reactions, 29
CH_2O reactions, 345
CH_2 reactions, 58, 344
CH_3 reactions, 39
CH_4 oxidation, 39
CH_4 reactions, 94
Chain-branching reaction, 65–66
Chain-initiating reactions, 60
Chain reaction, 59–62, 68, 69, 71, 73, 76, 331, 341, 345
Chain-terminating reaction (chain-breaking, reaction), 61, 65, 68, 70, 72, 73, 74, 83
Chaotic process, 180, 271
Chapman-Jouguet, 304
detonations, 205, 207, 208, 209, 210, 293, 316
point, 205, 206, 207, 210
Characteristic length, 122, 124, 167, 230, 231
Characteristic time of combustion, 92, 275
Characteristic time of reaction, 84, 92, 275, 299, 314
Chemical equilibrium, 44, 45, 46–50, 53, 102
constant, 45
Chemical kinetics, 54, 65, 71, 215, 217, 218, 256, 280, 330, 331
Chemical potential, 19, 20, 44
Chemical rate of production, 63, 70
Chemical reaction, 29, 53–59, 63, 66, 67, 73, 79, 80, 83, 89, 90, 92, 95, 97, 98, 103, 164, 202, 241, 242, 271, 272, 275, 276, 281, 283, 324, 333, 342, 344
Chemical reaction rate, 54, 98
Chemical species, 56, 341
Clausius – Clapeyron equation, 21
Clausius R., 4
Closed system, 2, 3, 4, 7, 8, 10, 25
Coagulation, 344
Coal, 319, 346
combustion, 346
Collision
frequency, 67
integral, 106
number, 55, 106
Combustion bomb, 326
Compressibility, 125, 146, 177, 201, 217
Compressible flow, 121, 122
Compression
ratio, 320, 325, 326, 327, 328, 329, 330, 331, 334
work, 325
Conservation equation, 116
Conservation equation of velocity circulation, 129–131
Constant pressure specific heat, 6, 11, 13, 24, 31, 41, 92, 95, 97, 120, 163, 216, 324, 328
Constant volume specific heat, 6, 11, 13, 23, 24, 31, 76, 81, 120, 216, 324, 328
Continuity equation, 115, 116, 117, 118, 123, 138, 162, 169, 170, 173, 187, 188, 191, 192, 194, 227, 272, 355, 356
Control emissions of pollutants, 343
Convection, 24, 161, 164, 259, 310, 313, 315
Cool flame, 272, 330, 332, 333
Correlation function, 187–190
Coward, 245, 264, 265
Critical heat production, 82
Critical Mach number, 300
Critical pipe diameter, 268
Critical temperature, 66, 74
Critical tube width, 249, 250, 254
Cylindrical coordinates, 173, 243, 352, 355, 356

Dalton law of partial pressure, 31
Damkohler number, 276, 280
Darrieus-Landau (DL) theory of flame instability, 246
Deflagration, 89, 96, 97, 202, 271, 281, 283
Deflagration-to-detonation transition, 284, 285
Deflagration wave, 285, 294, 295, 299
Density, 1, 91, 93, 95, 106, 109, 110, 111, 112, 113, 115, 118, 119, 120, 121, 122, 123, 124, 141, 142, 143, 145, 146, 148, 149, 150, 151, 152, 162, 163, 164, 165, 176, 179, 185, 186, 197, 198, 199, 201, 203, 208, 210, 211, 212, 216, 217, 220, 221, 229, 230, 232, 259, 261, 263, 266, 275, 284, 295, 296, 297, 299, 300, 304, 305, 309, 310, 313
Density variation in sound wave, 146
Detonating mixture, 66

Detonation
adiabatic, 202–206, 207
velocity, 285
wave, 89, 202, 203, 204, 205, 206, 208, 209–210, 281, 283, 292, 297, 316
speed, 202, 281, 292, 316
velocity, 287, 292, 299
Diatomic gas, 200
Diesel cycle, 327–329
Diesel engine, 319, 320, 322, 326, 327, 329, 343
Diffusion
coefficient, 98, 100, 101
equation, 97, 98
flame, 91, 95, 96, 344, 345
flux, 202, 272
time, 94
Direct numerical simulation (DNS), 216, 256
Discontinuity, 83, 132, 151, 152, 153, 197–198, 199, 202, 205, 209, 210, 220, 224, 226, 243, 249, 255
Dispersion relation, 141, 142, 178, 224, 225, 230, 234, 235, 237, 238, 239, 240, 241, 250
Dissipation, 114, 131, 143, 152, 157–196
rate, 185
Dissociation reactions, 43, 50
Divergence, 125
DL theory (Darrieus-Landau theory), 215, 230, 231, 246
Droplet burning, ix
Dynamic viscosity, 159

EGR – exhaust-gas recirculation, 334, 343, 347
Eigenvalue, 101, 220, 221, 222, 234, 235
Einstein's equation, 94
Electron, 27, 39, 57, 65, 66, 309, 310
Elementary reaction, 67, 331
End-gas, 331
pressure, 331
temperature, 331
Endothermic reaction, 33, 65
Energy
equation, 101, 243
flux density, 119, 160, 162
transfer, 162, 164, 178, 185, 217
Engine knock, 330
Enthalpy, 13–14, 16, 19, 22, 31, 32, 34, 36, 37, 76, 97, 99, 100, 101, 115, 118, 119, 199, 203, 216, 217, 218, 280, 312
Entropy, 1–4, 7, 8, 9, 10, 11, 12, 15, 16, 17, 18, 19, 20, 23, 31, 44, 46, 114, 115, 148, 161, 162, 198, 200, 201, 205, 206, 313
flux density, 115
Entry length, *see* Poiseuille flow
Equation
of continuity, 98, 111–113, 120, 121, 122, 143, 157, 161, 173, 179
of motion, 114, 115, 118, 132, 137, 157, 179, 191, 211, 212
of state, 14, 109, 111, 120–122, 146, 200, 216, 217
of state for ideal gas, 21, 22, 24, 120, 146, 327
Equilibrium, 2, 4, 5, 7, 8, 9, 10, 11, 12, 14, 16, 17, 18, 25, 37, 43–45, 46–50, 53, 59, 62, 64, 65, 69, 95, 102, 110, 123, 124, 125, 134, 137, 138, 143, 151, 185, 310, 313, 314, 324, 347
conditions, 314
composition, 47, 50
constant, 43–45, 47, 48, 59, 62, 65
equivalence ratio, 43
for isentropic flow, 116, 130
mechanical, 5, 10, 123, 125
Equivalence ratio, 42, 43, 344
Ethene, 36
Ethyl, 39
Ethylene (C_2H_4), 290
Euler equation, 113–116, 118, 120, 121, 122, 123, 126, 130, 132, 133, 137, 144, 148, 161, 226
Exhaust-gas recirculation, 343, 347
Exothermic reaction, 33, 64, 66, 67, 92, 333
Expanding spherical flames, 255
Explosion
limits, 66–70, 85
thermal, 66, 70, 73, 81–83, 84–85, 272, 281, 282, 297
Extinction, 271, 280, 342

Fenimore NO -prompt NO, 346
First-order reaction, 59
Flame
extinction, 342
front, 90, 92, 96, 97, 101, 131, 215, 219, 222–231, 234, 241, 242, 243, 244, 245, 246, 247, 248, 249, 250, 252, 254, 255, 257, 259, 261, 262, 263, 265, 267, 272, 275, 276, 279, 280, 285, 288, 291, 292, 293, 294, 305, 314, 316, 330, 331, 342, 346
laminar, 86, 89–107, 275, 276, 277, 279, 280, 281, 284, 286, 294, 302
laminar premixed, 271–318, 344
length, 251

Flame (*cont.*)
pressure dependence, 56
propagation, 89, 90, 94, 96, 101–105, 109, 235, 243, 245, 246, 247, 250, 255, 258, 261, 265, 267, 268, 271, 276, 287, 293, 314, 316
quenching, 267–268, 342
radius, 255, 257, 258
stability, 231, 234
stabilization, 241, 244
stretch, 259, 280, 289, 302
structure, 245, 252, 263, 266, 275, 276, 342, 345
temperature, 40–43, 46–50, 80, 91, 242, 267, 280, 281, 293
temperature dependence, 57, 80
thickness, laminar, 281
turbulent, 273, 279–280, 288, 342
velocity, 90, 92, 93, 94, 95, 101, 105, 106, 220, 222, 223, 228, 229, 233, 244, 248, 249, 250, 251, 252, 254, 255, 256, 258, 259, 261, 263, 264, 265, 266, 267, 280, 289, 291, 298, 300, 302, 303
laminar, 280
see Normal flame velocity
turbulent, 280
wrinkled, 254, 255, 256, 257, 272, 276, 277, 279, 280
Flamelet model, 280
Flamelet regime, 276, 277, 279
Flammability
of laminar flames, 267–269
limits, 267–268
Flow
in a pipe, 152–154
reactor, 276
velocity, 126, 128, 169, 170, 171, 191, 225, 273, 274, 275, 278, 279, 289, 295, 304, 305
Fluctuations, 1, 2, 334
Fluid
dynamics, 28, 109–111, 116, 120, 121, 122, 126, 134, 135, 161, 176, 179, 187, 302
element, 110, 113, 114, 123, 127, 128, 131, 137, 151, 158, 161, 169, 186, 190
particle, 109, 114, 115, 117, 121, 127, 129, 131, 137, 139, 141, 143, 144, 157, 161, 187
Flux density, 113, 115, 119, 160, 162, 199, 203
Fourier
equation, 164, 165
integral, 165
series, 147
Fractal flame structure, 254, 255, 257
Frank-Kamenetsky
theory, 220
transformation, 80, 84
Free energy, 12, 13, 14, 19, 21, 22, 23, 44, 46, 125
Froude number, 177, 238
Fuel
-bound nitrogen, 345
evaporation, 343
jet, 262, 345
Fuellean, 59, 275
Fuel-rich, 343, 344, 346
Fundamental, 9, 57, 69, 83, 111, 184, 215, 228, 244, 254, 309, 330

Gas
constant, 6, 30, 57, 146
equation, 200
ideal, 21–22, 23, 24, 30, 31, 33, 37, 44, 46, 56, 120, 124, 146, 149, 153, 157, 163, 199, 206, 210, 218, 229, 323, 324, 325, 326, 327
-turbine engines, ix
Gauss theorem, 113, 117, 119
Gibbs free energy, 14, 22, 44, 125
Global reaction rates, 272
Global warming, 320, 341
Gravitational potential, 124
Gravity, 114, 115, 117, 118, 123, 125, 128, 137, 141, 145, 157, 177, 216, 237, 238, 239, 240, 259, 262, 263, 265, 309, 310, 311, 314
acceleration, 123, 124, 142, 143, 177, 216, 237, 239, 240, 262, 314
waves, 137, 141, 142

Harmful effects, 63
Hartwell, 245, 264, 265
Heat
of combustion, 324
conduction, 102
conduction equation, 102
conductivity, 85
of formation, *see* Standard heat of formation
flux, 81, 102, 104, 105, 162, 165
loss, 41, 81, 82, 83, 84, 250, 267, 268, 275, 308, 320, 325, 331, 333, 334, 342

production, 81, 82, 83, 84, 103
of reaction, 33, 35, 38, 76, 98, 106
release, 69, 77, 81, 82, 83, 89, 91, 92, 94, 97, 101, 102, 103, 105, 218, 267, 275, 297, 299, 330, 333, 342
thermal conductivity, 89, 102
transfer, convective, 347
transfer, radiative, 344, 347
transport, 92
Helmholtz free energy, 12
Heptane-air flame, 331
HO_2 formation reaction, 70
Homogeneous reactions, 89, 271, 283
Homogenous ignition, 89, 271, 283
Hot spot, 283, 285, 333
Hugoniot adiabatic, 199, 201
Hydrocarbon
-air mixtures, 90
-air reactions, 90, 275, 284, 331
combustion, xiii
oxidation, 343
kinetics, 331
Hydrocyanic acid (HCN), 346
Hydrodynamic instability, 224, 227, 242, 244, 247, 250, 254, 255, 258, 267, 276, 287
Hydrogen/air premixed flames, 344
Hydrogen fluoride, 32, 42
Hydrogen-oxygen explosive mixture, 284
Hydrogen-oxygen reactions, 66–70
Hydrogen peroxide (H_2O_2) formation reactions, 333
Hydrostatic, 123–125
Hydroxyl radical, 36, 65, 333

IC engines, 319
Ideal equation of state, 111, 120–122, 146, 218, 327
Ideal fluid, 109–155, 157, 161, 201, 237
Ideal gas, 21–22, 23, 24, 30, 31, 33, 37, 44, 46, 56, 120, 124, 146, 149, 153, 157, 163, 199, 206, 210, 218, 229, 323, 324, 325, 326, 327
Ignition, 37, 66, 67, 68, 69, 73, 76–79, 82, 83, 85–86, 89, 243, 254, 255, 271, 272, 281, 282, 283, 284, 285, 287, 294, 297, 305, 308, 310, 312–316, 319, 320–323, 327, 330, 331, 332, 333, 334
delay time, 333
energy, 85–86
temperature, 66, 83
Implicit solution, 149
Incompressible flow, 121, 122, 175, 177
Incompressible fluids, 121–122, 135, 159, 160, 163–165, 168, 169, 178, 179, 191
Independent variable, 5, 6, 7, 9, 10, 14, 18, 19, 20, 102, 211, 212
Induced ignition, 294, 305
Induction time, 75, 79, 84, 85, 96, 281, 282, 297, 299, 313, 316, 330, 331, 332
Industrial furnace, 344
Inertia force, 137, 177
Initial conditions, 74, 79, 134, 258, 284, 295
Instabilities, 91, 134–136, 225, 250, 252, 255, 266, 284, 285
Instabilities hydrodynamic instability, 250, 255
Intensity of turbulence, 276, 277, 344
Internal combustion engine, 319–334, 341, 342
Internal energy, 4, 5, 21, 22, 31, 116, 119, 120, 121, 128, 199, 213, 216, 311
Irreversible process, 5, 161
Irrotational flow, 131
Isentropic, 115, 116, 130, 131, 133, 197, 323, 324, 325, 326, 327, 328
Isobaric specific heat, 216
Isolated system, 4, 16
Isothermal, 12, 17, 57, 59, 121, 122, 123, 124, 146
Isothermal flow, 121, 122
Isothermal gas, 124

JANNAF Tables, 35, 36, 41, 46, 62
Joule, 8, 30

Karlovitz number, 276
Kelvin theorem, 227
Kinematic viscosity, 159, 163, 169, 175, 178, 185, 276, 342
Kinetic energy, 4, 5
Kinetic energy turbulent, 184
Kinetic theory of gases, 6
Knock, 272, 320, 326, 330–334
in Otto engines, 320, 326
tendency, 331
Koch fractal structure, 254
Kolmogorov length, 186
Kolmogorov spectrum (turbulence), 185
Kolmogorov theory of turbulence, 184–187

Laminar burning velocities, 342
Laminar flame (premixed laminar flame), 89, 91, 271, 273, 275, 279, 302

Laminar flame speed
effect of initial temperature, 90
effect of pressure, 284
flame temperature, 90
measurement, 90
mixture composition, 90, 284
thermal diffusivity, 90
Landau (DL) flame instability, 224, 241, 244, 245, 302
Laplace equation, 260
Laplacian, 173
Law of thermodynamics, 4–7, 32, 35, 76
first, 4–7, 32, 76
second, 4
third, 14–16
Lean mixtures, 43, 100, 437
Lean-premixed combustion, 271, 342
Le Chatelier principle, 18
Length scale, 89, 110, 122, 124, 163, 180, 184, 186, 187, 189, 190, 191, 222, 230, 231, 255, 256, 282, 314, 316
of the flow, 110, 163
Taylor, 314
turbulent, 184, 186, 187, 189, 190
Lewis number, 98, 105, 214, 218, 222, 231, 235, 236, 238, 239, 240, 241
Lewis and von Elbe theory, 90
Lifetime, 106, 343
Lifting work, 129
Linear waves, 134–136
Liouville theorem, 3
Liquid fuels, combustion, 327, 342

Mach number, 91, 177, 200, 216, 217, 281, 291, 293, 300
Mallard and LeChatelier theory, 220
Markstein
length, 242, 243
numbers, 258
Mass
-average velocity, 174
conservation, 112, 113, 138, 219
density, 28, 29, 91, 106, 111, 112, 113, 115, 118, 119, 123, 124, 125, 160, 162, 185, 203, 212, 229, 232, 260, 263, 309, 314, 315
diffusion coefficients, 98, 101
flux density, 113, 199, 203
fraction, 28, 29, 216, 347
Material derivative, *see* Substantial derivative
Maximum flame
speed, 280, 342
temperature, 42, 280
Max Plank quantum constant, 3
Maxwell-Boltzmann distribution, 57
Mean free path, 94, 106, 110, 168
Mean mass velocity, 101
Mechanical energy, 16, 17, 160, 178, 320
Mechanical equilibrium, 5, 10, 123, 125
Methane, 36, 39, 42, 93, 95, 245, 264
air flame, 93
oxidation, 39
Minimum ignition
energy, 85–86
temperature, 66, 83
Mixture fraction, 93, 216, 217, 286, 347
Molar-average velocity, 174
Molar concentration, 29
Molar concentration, 55, 64
Molar fraction, 28, 29, 47
Molar standard free energy of formation, 46
Momentum
equation, 159
flux, 97, 160, 198, 218, 223, 229, 232, 237, 238
Monatomic gas, 200
Multicomponent
diffusion coefficients, 98
thermal diffusion, 92, 164

N_2O mechanism, 40
NASA Air Force Thermochemical Tables, 35
Natural gas-air flame, 275
Navier-Stokes equations, 161, 175, 191, 256
Negative temperature coefficient, 333
Nernst's theorem, 14–16
Newton law, 113, 139, 212
Nitric oxides, NO, NO, 36, 38, 40, 43, 47, 48, 63, 64, 66
Nitroether, 40
Nitrogen dioxide, 36, 63
Nitrous oxide formation, influence of temperature, 345–347
NO_2 mechanism, 66
NO emissions, 327, 341, 345
NO, formation, 345, 346, 347
NO formation reaction mechanism, 345–349
Nonequilibrium thermodynamics, 1–25
Non-linearity, 105, 133, 135, 139, 142, 143, 154, 180, 188, 218, 243–244, 245–249
Nonpremixed flame, 345
NO production mechanism, 347
NO reduction, 341, 345–347

Normal flame velocity, 90, 101, 105, 106, 244, 250, 265, 266, 267, 279
Normalized velocity, 90
NO_x formation mechanism, 341, 345–347
Number density, 67, 309

Octane number, 331, 334
Olefins, 39
One-dimensional planar propagating premixed flames, 148–152, 294
One-dimensional travelling wave, 148–152
One-step (single-step) chemical reaction, 53, 54, 215, 256
Onset of retonation in stoichiometric hydrogen-oxygen mixture, 42, 43, 50, 53, 54, 66, 68, 95
Order of chemical reaction, 66, 103, 105, 106, 135
Ordinary differential equations, 59, 101, 120, 126, 197
Orthogonal curvilinear coordinate system, 355–356
Oscillations, 132, 133, 137, 140, 142, 148
Otto engine, 319, 320–323, 326, 328, 329
Overall reaction
  global, 65, 272, 332
  order, 58, 59, 79
Oxidation, 39, 40, 63–65, 343, 345
  high-temperature, 327
  of hydrocarbons, 41, 341, 344, 345
  low temperature, 102, 162, 331, 332
Ozone, 341, 345

PAH – polycyclic aromatic hydrocarbons, 341, 344
Partial pressure, 30, 31, 45, 50
Peclet number, 178
Photochemical initiation of detonation, 294
Pipe flow, 152–154
Piston, 32, 85, 96, 152, 153, 259, 285, 286, 291, 294, 320, 321, 322, 323, 324, 328, 330, 331, 334, 342
Planck constant, 3
Poiseuille flow, 170–172, 173, 174, 184, 196, 259, 274
Poisson adiabatic, 22–25, 200
Pollutant formation, 66, 341
Polycyclic aromatic hydrocarbone (PAH), 341, 344
Potential energy, 12, 32, 57, 128
Potential flow, 131–134, 137, 201
Prandtl
  equation, 192, 193
  number, 164, 216
Pre-exponential factor, 57, 58
Premixed flame, 89, 91, 95, 244, 250, 271–318, 344
  laminar, 89, 91, 275, 277, 279, 302
  turbulent, 273, 279, 342
Pressure, 1, 4, 5, 6, 9–12, 13, 14, 15, 18, 19, 21, 22, 23, 24, 31, 33, 40–42, 45, 49, 57, 66, 67–68, 70, 76, 82, 86, 89, 91, 95, 97, 99, 106, 109, 110, 113, 114, 116, 117, 119, 120, 121, 123–124, 125, 128, 129, 135, 137–139, 142, 143, 144, 145, 146, 147, 148, 149, 151, 159, 161, 162, 163, 165, 168, 169, 170, 172, 173, 175, 192, 197, 198, 201, 203, 208, 210, 211, 212, 216, 217, 218, 222, 225, 226, 246, 261, 265, 266, 272, 275, 281, 284, 285, 286, 287, 290, 291, 293, 294, 295, 296, 297, 298, 299, 300, 302, 303, 305, 307, 309, 319, 320, 322, 323, 324, 326, 327, 328, 329, 330, 331, 333, 334, 343, 344, 347
  balance, 138, 139, 261
Production rate, 103
Prompt NO mechanism, 346
Propagation of detonations in rough-walled tubes, 209
Propane, 36, 38, 42, 93, 263
Propane-air flame, 317
Proton mass, 27, 309, 310

Quasi-equilibrium, 2
Quenching distance, 268, 269, 342
Quenching distance between parallel plates, 268
Quenching of flame, 267–269, 342

Radiation, 56, 96, 267, 275, 341, 345, 347
Radical recombination reaction, 346
Radicals, 56, 57, 59, 60, 65, 66, 67, 68, 69, 70, 73, 271, 272, 331, 332, 342
Random process, 157, 161
Rarefaction wave, 151, 209, 210, 259
Rate coefficient
  pressure dependence, 56
  temperature dependence of, 58
Rate limiting reaction, 346
Rayleigh, 136, 141–143, 225, 237, 245, 267, 314

Rayleigh relation, 142, 225, 237
Rayleigh-Taylor instability, 136, 141–143, 225, 245, 267
Reacting flow, 275, 276
Reaction, 1, 18, 29, 30, 32–37, 38, 39, 40, 41, 43, 44, 45, 46, 47, 48, 49, 50, 53–59, 60, 61, 62, 63, 64, 65, 66, 67, 68, 70, 71, 73–87, 89–91, 92, 93, 94, 95, 96, 97, 98, 99, 101, 102, 103, 104, 105, 106, 164, 202, 204, 206, 215, 216, 218, 219, 220, 221, 222, 224, 231, 241, 242, 244, 265, 266, 267, 268, 271, 272, 275, 279, 280, 281–283, 284, 285, 286, 287, 289, 292, 293, 294, 295, 296, 297, 298, 299, 300, 301, 302, 303, 304, 305, 306, 307, 310, 311, 312, 313, 314, 324, 331, 332, 333, 341, 342, 343, 344, 345, 346, 347
  bimolecular, 55, 56, 57, 59, 62, 65, 68, 76, 106
  energy, 39, 73, 85
  enthalpy, 32, 34, 35, 36, 37, 76, 97, 99, 100, 216, 218, 280, 312
  global, 272
  mechanism, 54, 67, 272
  mechanism of H_2 oxidation, 67
  mechanisms of H_2/O_2 systems, 67
  molecularity, 56
  order, 58, 216, 299, 300, 306
  rate, 54, 55, 56, 57, 58, 59, 61, 62, 65, 66, 67, 71, 74, 75, 76, 77, 78, 79, 80, 81, 82, 92, 98, 99, 100, 101, 102, 106, 202, 216, 220, 241, 268, 280, 281, 295, 296, 297, 298, 299, 313, 331, 332, 346, 347
  trimolecular, 63, 106
Reactors, 275, 277
Real gas, 157
  equation of state, 200
Recombination reaction, 47, 344, 346
Reduced, 50, 58, 59, 75, 82, 137, 169, 187, 194, 206, 217, 227, 232, 235, 242, 293, 319, 347
Reference: state, 31, 33, 34, 35, 36, 50
Reforming, xiv
Residence time, 95
Resolution, 147
Retonation wave, 293
Reverse reaction, 35, 58, 59, 65
Reversible process, 5, 11, 17, 161
Reynolds number, 172, 174, 175, 176, 178, 179, 180, 181, 184, 185, 186, 195, 196, 258, 259, 276
  turbulent, 179, 180, 181, 184, 185, 186, 187, 190, 195, 276
Reynolds stress, 189
Riemann, 149
Rocket propellants, ix

Scalar, 120, 127, 133, 277
Schlieren photographs, 285
Second explosion limit, 70
Second-order reaction, 58, 105, 222, 302, 306
Self-ignition induction time, 76–79, 281, 282
Self-similarity, 106, 211, 258, 259
Self-sustained detonation, 299
Semenov's theory of thermal explosion, 81–83, 84
Sensitivity analysis, 333
Shear flow, 195
  turbulent, 195, 286, 288
Shear layer, 195, 286, 288
Shock
  adiabatic, 199–201, 202, 203, 204, 205
  front, 200, 201, 202, 203, 212
  strength, 177, 200, 202, 281, 293
  tube, 209, 210, 259, 284, 285, 286, 287, 288, 289, 290, 292, 294, 302
  waves, 86, 89, 96, 153, 168, 177, 198, 199, 200, 201, 202, 203, 204, 205, 206, 209, 210, 211, 212, 281, 282, 284, 285, 288, 291, 292, 293, 294, 297
Shock Wave Amplification by Coherent Energy Release (SWACER), 294
SI engine – spark-ignited engine, 331
Simple wave, 149, 150, 151, 152, 153
Single-step reaction, 56
Solid fuels combustion, 341
Sonic velocity, 205, 209, 311, 316
Soot, 40, 341–345, 344, 345
  formation, 343–345
Sound
  velocity, 41, 91, 144, 146, 149, 151, 168, 200, 205, 206, 297, 299
  wave, 121, 133, 143–147, 148, 149, 154, 281
Spark
  -ignited engine, 271, 319, 320–323, 330, 331
  ignition, 85–86, 271, 319, 320–323, 330
  plug, 330, 342
Specific enthalpy, 13, 22, 32, 76, 97, 99, 115, 199, 203, 216
Specific heat, 6, 13, 15, 22, 23, 24, 31, 41, 46, 76, 81, 92, 95, 97, 99, 105, 120, 163, 199, 210, 216, 324, 326, 328, 330, 333

Specific heat at constant pressure, 13, 46, 95, 120, 163
Specific heat at constant volume, 6, 11, 13, 31, 76, 81
Specific internal energy, 22, 120
Speed
  of light, 350
  of sound, 97, 163
Spherical coordinates, 352, 356
Spherical flame, 254, 255, 258, 272, 311
Spherical flame propagation, 311
Spherically expanding flame propagation, 254–258
Spontaneous ignition, 85, 331
Spontaneous reaction, 96, 265, 281–283
Stabilization, 231–236, 238, 239, 241, 243–244
Stagnation point, 126, 128, 261
  flow-field, 126, 128, 261
Standard
  enthalpy of formation, 34
  heat of formation, 34
  state, 33, 34, 35, 41, 46, 65
    entropy, 46
    free energy of formation, 46
Stationary concentrations method, 59, 61, 63
Stationary flame, 92, 215, 216, 219, 220, 223, 232, 235, 244–249, 250–254, 255, 256, 258, 259
Stationary solution, 172
Steady flow, 32, 122, 126, 127, 131, 134, 169, 170, 175, 178, 179, 180, 181–184, 185, 278
Steam reforming, xiv
Stefan-Boltzmann constant, 350
Stirred reactor, 275, 276, 277
Stoichiometric coefficient, 29, 50, 53, 54, 56, 71
Stoichiometric equation, 54, 97
Stoichiometric mixture, 41, 42, 43, 66, 267, 268
Stoichiometry, 29, 35, 43, 47, 49, 332
Streamlines, 126, 127, 128, 131, 132, 133, 134, 154, 224, 261
Stream-tube, 127
Stress tensor, 158, 159, 189, 216, 355, 356
Substantial derivative, 110, 114, 169
Supernova
  explosion, 96, 312
  Type Ia, 311, 312
Surface reaction, 342

Taylor length scale, 314
Taylor series (expansion), 80
Temperature, 1, 5, 6, 7–9, 10, 12, 14, 16, 17, 18, 19, 21, 22, 24, 30, 32, 33, 34, 40–43, 44, 45, 46–50, 57, 58, 59, 62, 65, 66, 67, 68, 70, 73, 74, 76, 77, 78, 79, 80, 81, 82, 83, 84, 85, 86, 89–107, 110, 120, 121, 124, 125, 142, 146, 159, 162, 163, 164, 165, 168, 202, 203, 206, 208, 209, 216, 217, 219, 220, 221, 232, 233, 234, 240, 242, 243, 265, 266, 267, 268, 269, 271, 272, 275, 280, 281, 282, 284, 292, 293, 294, 295, 296, 297, 298, 299, 300, 304, 305, 307, 308, 309, 310, 312, 315, 324, 325, 327, 328, 329, 330, 331, 333, 334, 342, 343, 344, 345, 346, 347
  fluctuations, 1, 334
Tensor, 158, 159, 160, 189, 215, 355, 356
Theory of Semenov, 81–83, 84
Theory of Zel'dovich, Frank-Kamenetsky, 90, 101–105, 220
Thermal conductivity, 83, 89, 90, 95, 97, 99, 100, 101, 102, 103, 162, 163, 168, 178, 201, 210, 241, 276
Thermal diffusion, 92, 164
  coefficient, 164
  instability, 92
Thermal diffusivity, 92, 93, 95, 98, 241, 247
Thermal expansion coefficient, 15
Thermal explosion (thermal ignition), 66
Thermal flame propagation, 91
Thermal NO mechanism, 66, 345
Thermal self-ignition, 76–79
Thermodynamic data, 54
Thermodynamic equilibrium, 7, 95, 110, 125, 324
Thermodynamic functions, 2, 110, 120
Thermodynamic potentials, 12, 13, 14, 19, 20, 22, 23
Thermodynamic variables, 14, 219
Thickness of a flame, 89, 91–97, 221, 222, 223, 231, 232, 234, 235, 236, 237, 238, 244, 247, 250, 258, 265, 266, 276, 277, 281, 286, 304, 305, 342
Third explosion limit, 71
Third law of thermodynamics, 14–16
Third-order reaction, 216, 222, 266, 299
Thomson circulation theorem, 230
Three-body reaction, 55, 56
Time-dependent problem, 196
Time scales of chemical reactions, 281
Transition from a deflagration to detonation (DDT), 284, 285, 286, 288, 293–301

Transport
  coefficients, 240
  processes, 90, 92, 161, 215, 223
Traveling plane wave, *see* One-dimensional travelling wave
Travelling wave, 146, 267
Trimolecular reaction, 63
  rate, 63, 106
Trinitrotoluene, 40
Turbulence
  dissipation rate, 185
  intensity, 276, 280
  kinetic energy, 184–187
  model, 279, 280, 284, 302, 330
  Reynolds stresses, 189, 190
Turbulent boundary layer, 190–196, 286, 288, 289
Turbulent eddy viscosity, 189, 190
Turbulent energy spectrum, 185, 187, 311
Turbulent flames, 273, 279–280, 288, 342
Turbulent flow, 131, 184, 185, 186, 187, 188, 189, 275, 279, 280, 342
Turbulent kinetic energy, 184
Turbulent length scale, 189, 190, 196
Turbulent reacting flow, 275
Turbulent scale external, *see* Fundamental
Turbulent transport, 215, 277
Two stage ignition, 330
Two-step reaction, 59

Unit normal vector, 223, 277
Universal gas constant, 6, 30, 57, 146
Universe, 311

Velocity
  circulation, 129–131, 130, 133
  measurements, 90, 280
  potential, 133, 137, 140, 144, 145, 147, 260
  of sound, 97, 144, 146, 150, 151, 205, 210
    in ideal gas, 146, 210
Viscosity
  coefficient, 158, 159, 160, 177, 189
  dynamic, 159
  kinematic, 159, 163, 169, 175, 178, 185, 276, 342
Viscous dissipation, 157–161, 185, 187
Viscous fluid, 157–161, 162, 168–174, 179
Volume viscosity, 216
Von Karman coefficient, 190
von Neumann, 202, 203
Vortex, 229, 230, 256, 257, 288
Vortex generation, 256, 257, 288
Vorticity, 126, 131, 133, 138, 228, 230, 251, 260

Wave equation, 145
White dwarfs, *see* Supernova explosion
Wrinkled flame front, 256, 257

Zel'dovich, Frank-Kamenetsky theory, 101–105, 220, 222, 232, 235
Zel'dovich mechanism for NO formation, 346
Zel'dovich NO thermal mechanism, 66, 346
Zel'dovich-von Neumann-Döring (ZND) model, 202, 203
ZND detonation wave structure, 284, 285